Nondestructive Testing

Louis Cartz

Marquette University
College of Engineering
Milwaukee WI USA

Manager, Book Acquisitions
Veronica Flint

Manager, Book Production
Grace M. Davidson

Production Project Coordinator
Cheryl L. Powers

**The Materials
Information Society**

ASM International

ISBN: 0-87170-517-6
SAN: 204-7586

ASM International®
Materials Park, OH 44073-0002

Printed in the United States of America

Nondestructive Testing
Radiography, Ultrasonics, Liquid Penetrant, Magnetic Particle, Eddy Current

III. Ultrasonic Testing

IV. Liquid Penetrant Inspection

Preface

Nondestructive testing (NDT), which produces no alteration of the part being tested, is concerned with a very wide range of practical problems, from troubleshooting to the most sophisticated characterizations of materials and systems behavior. It is useful for the skilled technician from industry, research manager, factory supervisor, engineer from an industrial laboratory, university or research institute, or from a government quality control background, to have available a detailed review covering:

- The basic principles of the many NDT methods, many well-established, others very new and highly specialized
- A wide ranging set of examples to provide ideas on how to proceed in the investigation of a particular NDT problem

Many of the same NDT methods are used in the authentication of art objects, in forensic studies, and in biological and medical fields.

The different NDT methods are dealt with in separate chapters, so that each chapter is complete in itself. The most important and frequently used methods are:

- Radiography in Chapter II
- Ultrasonic testing in Chapter III
- Liquid penetrant inspection in Chapter IV
- Magnetic particle inspection in Chapter V
- Eddy current inspection in Chapter VI

Each of the Chapters II to VI is presented in two or three parts:

- Part I covers the outline of the method and its application.
- Part II is concerned with technical discussions of the derivation and explanation of the science involved, with the necessary theory.
- Part III, in some chapters, covers applications and typical calculations.

In general, the reader will wish to use Parts I and III, and only consider Part II for more basic information. There is a companion book in this series on *Visual Inspection* by R.C. Anderson which covers this technique in a most useful way (Ref 285).

Acknowledgements

The final version of this book is certainly more readable and scientifically correct with the helpful and considerate readings and reviews carried out by several colleagues and friends.

Frank G. Karioris (Professor Emeritus, Dept. Physics, Marquette University) has reviewed the entire text very carefully and with considerable insight and his comments have rendered the whole more reasonable and useful.

Many of the present and past workers at Magnaflux Corporation, IL, USA, have reviewed the manuscript and made useful and helpful suggestions. Dr. Joe Hahn and Adolf Fijalkowski (Magnaflux Corporation) have reviewed the chapter on Liquid Penetrant Inspection (LPI) making many helpful comments. Arthur Lindgreen (L&L Consultants, previously of Magnaflux Corporation) is well known in the field of Magnetic Particle Inspection (MPI), and he has reviewed most carefully the chapter concerned with this method, adding many comments on the practical application of the technique. These two techniques, LPI and MPI, are intimately related to the historical development and exploitation of these methods by Magnaflux Corporation. Ken Strass (Centurion NDT, Inc., Schaumburg, IL, USA) and Douglas A. McLane (Transonic, Inc.), both previously from Magnaflux, Inc., have aided with the discussion and description of ultrasonic testing, again giving many suggestions, practical advice, and helpful comments.

Dr. Srini R. Gowda, of G.E. Aircraft Engines, Cincinnati, OH, USA, has most kindly arranged for several of his highly qualified colleagues, with considerable experience in NDT, to review several chapters and NDT techniques presented in the text. Dr. Derek Sturgess has reviewed the chapter on Ultrasonic Testing with considerable insight, and his detailed comments have been most useful in the development of this technique. Charlie Loux has reviewed the chapters on LPI and MPI in detail and provided ideas and worthwhile corrections to improve these chapters. Bernard G. Nightingale reviewed the chapter on Radiography and has made a considerable effort to help in the presentation of a correct and useful description of this technique. David Copley has read, reviewed, and commented on several chapters, in particular Ultrasonic Testing, and helped improve and clarify the discussions. It is clear that the staff concerned with NDT at G.E. Aircraft Engines is very proficient in these techniques and have shown themselves to be very willing to help develop this NDT work.

Robert C. Anderson (Anderson & Associates, Houston, TX, USA) wrote the first two very successful volumes in this series on *Destructive Testing* and *Visual Examination* (Ref 285), and he has kindly reviewed and advised on the complete work, with a view to making the present text as clear as possible with many useful examples and practical hints.

Several fine radiographs are reproduced with the aid of Kodak which is gratefully acknowledged. In particular:

- Figure 1.3: Liberty Bell
- Figure 1.4 and Figure 1.5: Paintings
- Figure 1.6: Lead coffin from 17th century

Ed Bielecki and staff of MQS Inspection, Milwaukee (Ref 10) have been exceptionally helpful in providing information, and in very kindly reviewing the text. They have prepared several fine radiographs and examples of magnetic particle inspection and liquid penetrant testing carried out by Simon Teale, Theresa Sprades, John Petroske, and Don Cywinski. These are as follows:

- Figure 1.7: Radiograph of a teddy bear showing metal inclusions
- Figure 1.8: Training for NDT inspection
- Figure 2.5: Radiograph of a wheel using multifilms
- Figure 2.6(a, b, c): Photographs of x-ray generators in typical working configurations

The chapter "Magnetic Particle Inspection" has benefitted from considerable detailed and careful advice by Dr. Lyndon J. Swartzendruber (NIST, Office of Nondestructive Testing) which has improved the clarity and usefulness of the description of the magnetic method.

Extensive and very helpful reviews have also been carried out on many chapters by the very capable staff of MQS Inspection, Elk Grove Village, IL, USA. James Nash, manager of their eddy current division has provided most useful comments, as has Gary Nunes of their MQS training division.

John Flaherty, Flare Technology, Illinois, has very carefully reviewed the chapters on ultrasonics and on eddy current, providing many useful comments and additions.

Robert Levy, Intercontrôle and La Cofrend France, has considerable knowledge of the eddy current method, and has revised with great care the chapter devoted to this technique. Christian Gondard, French Atomic Energy Commission, CEA, DEA, France, at the suggestion of Robert Levy, has revised the chapter on the ultrasonic method, providing many useful corrections and additions.

This text has benefitted from the considerable assistance of Dolores E. Marrari and Sandra K. Grebe who have carried out the extensive typing requirements as well as the many revisions. Nick Schroeder has been very indulgent in preparing and revising the many diagrams and figures, and Dan Johnson has prepared the photographs. The staff of ASM International has played a crucial role in pursuing the writing and editing the text to completion.

Chapter I

Quality Control and NDT

NDT in Industry

Flaws and cracks can play havoc with the performance of structures, so that the detection of defects in solids is an essential part of quality control of engineering systems for their safe and successful use in practical situations. This is known variously as nondestructive testing (NDT), nondestructive evaluation (NDE), nondestructive characterization, or nondestructive inspection. Quality control, quality technology, and noncontact measurements are related subjects that include or use NDT techniques. However, applications of NDT go much deeper and are much broader in scope than the detection of gross defects. They concern all aspects of the characterization of solids, their microstructure, texture, morphology, chemical constituents, physical and chemical properties, as well as their methods of preparation. There is concern for the most minute detail that may affect the future performance of the object in service, so that all properties need to be under control and all factors understood that may lead to breakdown. Nor is it appropriate to rely on general statements because each study and each example needs to be treated individually, proceeding by the use of all known properties and information about the component. The abbreviation NDT will be used in this text for "nondestructive testing."

A description of the early evolution of NDT is given by Mullins (Ref 1, 2) and his review still covers most of the methods presently used. The established test methods include radiography, ultrasonics inspection, magnetic particle inspection, liquid penetrant inspection, thermography, electrical and magnetic methods, and visual-optical testing. In the case of radiography, x-ray and γ-ray are well established, but neutron, proton, and Compton scattering are also used and there have been recent important advances in tomography. Many of the NDT methods are highly sophisticated, yet there are many techniques that are relatively simple. One such method is visual examination, with a hand lens. The simple methods are stressed because it is important not to overlook the obvious in examining an engineering component.

Applications of NDT in industry concern metals, nonmetals, very small to very large objects, and stationary as well as moving components. In medicine, NDT includes mammography, nuclear magnetic resonance scans, general x-radiography, and microangiography. Noncontact measurements using sensors are important in a wide range of subjects from geology, forensic studies, aerial temperatures, and weather surveys, to thickness measurements and art authentication. The examinations are concerned with detecting cracks, tears, imperfect welds and junctions, inclusions, tomography, and surface contamination effects without altering the piece in any way.

The operator of the tests is another important factor, and operator fatigue, as well as training, represents both an essential ingredient and a severe problem affecting all aspects of NDT. Because NDT is all-encompassing, it is most useful to have a library of as many examples or case studies as possible, all concerned with practical situations.

A survey has been carried out by the Institute of Metallurgists (Ref 3) of the many different NDT techniques used in engineering industry and found that liquid penetrant and magnetic particle testing accounts for about one-half of all the NDT testing, ultrasonics and x-ray methods about another third, eddy current testing accounted for about 10%, and all other methods accounted for only about 2%. It is not necessarily the most sophisticated method that is most suitable; this is illustrated in Fig. 1.1 where a surface crack in a forging is identified using liquid penetrant inspection, yet x-radiography carried out most carefully using multiple film techniques, surface crack next to film, adjustment of exposure and film type could barely observe the crack. (The radiograph was not good enough to permit reproduction in this text.)

Table 1.1 is a simplified breakdown of the complexity and relative requirements of the five most frequently used NDT techniques. Table 1.2 gives a comparison of common nondestructive evaluation methods as judged by the Office of Nondestructive Evaluation, NIST, USA (Ref 4).

NDT in Everyday Life

NDT is well-known as a part of industrial procedures, but it is also of importance in examinations of a more general interest in everyday life. Art objects such as paintings, sculptures, furniture, pottery, and ceramics need to be authenticated, and the tests of necessity must not damage or destroy the object.

In Table 1.3, references are listed of examinations of oil paintings, wooden statues, ancient coins, archeological pottery, and ceramic fragments. The techniques used include x-radiography, neutron activation analysis, and thermolu-

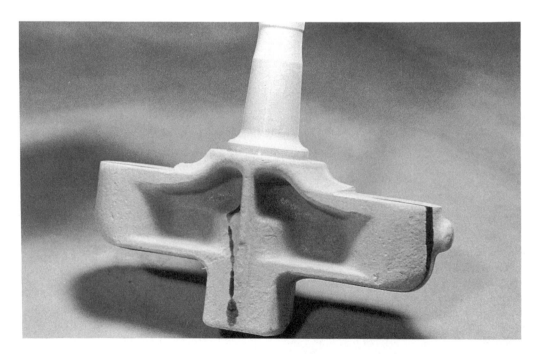

Fig. 1.1 The valve bridge forging, seen at the corner of the table in Fig. 2.6(a), has been examined by several procedures for surface cracks. Liquid penetrant with visible dye reveals the crack very readily. Comparable results are obtained by magnetic particle inspection. However, the same surface crack was barely visible using the most careful x-radiographic procedures.

minescence. Other studies listed in Table 1.3 are of cultural interest. The Statue of Liberty as well as the crack in the Liberty Bell have been examined by γ-radiography; see Fig. 1.2 and 1.3. The bullets used in the assassination of President Kennedy have been characterized using neutron activation analysis, and other forensic studies also make use of this technique (Ref 14). Radiography has been used to examine objects as diverse as colocynth scales, explosive devices, and locks (Ref 20). Radiographic analysis of paintings is illustrated in Fig. 1.4 and 1.5 (Ref 12). In Fig. 1.4, the radiograph reveals only the image of the lady; inorganic pigments, which absorb x-rays, were used in the portrait, and organic pigments for the rest of the painting, probably at a later date. In Fig. 1.5, the radiograph of the *Farmyard Animals* shows a totally different subject, so this is a painting on an older painting. The contrast depends on the x-ray energies used, as illustrated in (b) and (c). X-ray laminography has been used to examine the working of clocks and imperfections in printed circuit boards. Neutron activation analysis has been used in geological studies in situ to determine the chemical composition at depths within the earth's crust by lowering a radioactive source spectrometer system down a drill hole of diameter only 4 in. The internal structure of the earth is itself analyzed using the effects of seismic waves generated by earthquakes (Ref 38). The use of polarized light enables examinations of large objects to be carried out avoiding the interference of scattered and reflected light as well as for stress analysis. The three-dimensional shape of large objects can be portrayed using a shadow-moiré technique.

Ultrasound waves can be used to detect submarines, schools of fish, and as navigational aids (Ref 39). Baggage control at airports as well as metal detectors utilize eddy current examination. Infrared techniques are used in detecting heat losses from buildings, the performance of steam traps, hot spots in electrical equipment, friction in mechanical systems, defects in metals, stresses in metals, as well as aerial photography. Gamma-radiography has been used to examine lead-coated coffins from the seventeenth century; see Fig. 1.6. An x-radiograph is shown in Fig. 1.7 of a Teddy Bear when the filling was found to contain metal scraps. A γ-radiograph, using Iridium-192, of the Freedom Statue of the U.S. Capitol Building has revealed the presence and the condition of the iron skeleton of this bronze structure (Ref 26).

Thus, NDT applies in almost all aspects of everyday life.

History of NDT

The historical development of NDT is outlined for each technique at the beginning of each

Table 1.1   The relative uses and merits of various NDT methods

	Test method				
	Ultrasonics	**X-ray**	**Eddy current**	**Magnetic particle**	**Liquid penetrant**
Capital cost	Medium to high	High	Low to medium	Medium	Low
Consumable cost	Very low	High	Low	Medium	Medium
Time of results	Immediate	Delayed	Immediate	Short delay	Short delay
Effect of geometry	Important	Important	Important	Not too important	Not too important
Access problems	Important	Important	Important	Important	Important
Type of defect	Internal	Most	External	External	Surface breaking
Relative sensitivity	High	Medium	High	Low	Low
Formal record	Expensive	Standard	Expensive	Unusual	Unusual
Operator skill	High	High	Medium	Low	Low
Operator training	Important	Important	Important	Important	
Training needs	High	High	Medium	Low	Low
Portability of equipment	High	Low	High to medium	High to medium	High
Dependent on material composition	Very	Quite	Very	Magnetic only	Little
Ability to automate	Good	Fair	Good	Fair	Fair
Capabilities	Thickness gauging; some composition testing	Thickness gauging	Thickness gauging; grade sorting	Defects only	Defects only

Source: Ref 3

3

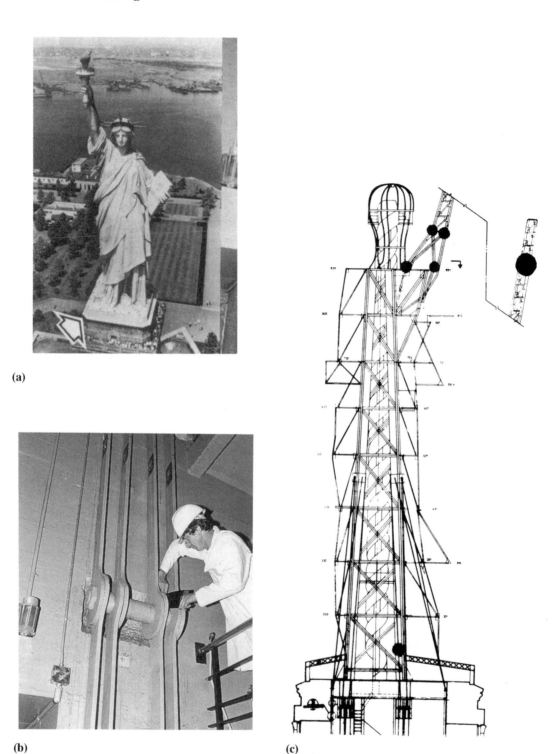

(a)

(b)   (c)

Fig. 1.2   During the renovation of the Statue of Liberty seen in (a), x-radiographs were taken using an Ir-192 radioactive source of activity 100 Curie. A workman is seen locating the source in (b), and the locations examined are indicated in (c). Radiographs were taken at key points of tie-rods, rivets, bolts, and welds to determine the effect of oxidation, inclusions, and holes in the structure. Extensive details of the procedures used are given in Ref 21. Careful visual examination was also carried out. Many of the rivets had failed due to the effects of corrosion. Reprinted with permission of The American Society for Nondestructive Testing

chapter. In almost all cases, the bulk of the techniques have been developed in this century with very little utilization in earlier years. Table 1.4 contains some references covering the history of NDT.

International Organizations of Quality Control and NDT

There are many organizations concerned with NDT and quality control and some of these are listed in Appendix 1.1. The International Organization for Standardization (ISO) is listed first in the table since most countries of the world are members of the NDT committee of ISO. Appendix 1.1 gives the names and addresses of organizations closely related to NDT, including standardization, quality control, and the testing of materials. Other organizations deal with radiation protection and radiological procedures; see Ref 55 to 58. The American Welding Society and the International Institute of Welding are listed, because NDT is such an important factor in the examination of welds.

Appendix 1.2 lists national societies that are members of the International Committee for Nondestructive Testing (ICNDT), which is based in Columbus, OH, USA.

There are many organizations concerned with standards, and indeed all countries of the world are full members or correspondent members of the International Organization for Standardization (see Appendix 1.1). A recent review of standards for NDT is given in Ref 59. An up-to-date listing of International Committee for Nondestructive Testing and related organizations is given in *Materials Evaluation* (Ref 60) on a yearly basis.

Standards for NDT

An important source of practice codes, standards, and recommendations for NDT is given in

Table 1.2 Comparison of some NDT methods

Method	Characteristics detected	Advantages	Limitations	Example of use
Ultrasonics	Changes in acoustic impedance caused by cracks, nonbonds, inclusions, or interfaces	Can penetrate thick materials; excellent for crack detection; can be automated	Normally requires coupling to material either by contact to surface or immersion in a fluid such as water. Surface needs to be smooth.	Adhesive assemblies for bond integrity; laminations; hydrogen cracking
Radiography	Changes in density from voids, inclusions, material variations; placement of internal parts	Can be used to inspect wide range of materials and thicknesses; versatile; film provides record of inspection	Radiation safety requires precautions; expensive; detection of cracks can be difficult unless perpendicular to x-ray film.	Pipeline welds for penetration, inclusions, voids; internal defects in castings
Visual-optical	Surface characteristics such as finish, scratches, cracks, or color; strain in transparent materials; corrosion	Often convenient; can be automated	Can be applied only to surfaces, through surface openings, or to transparent material	Paper, wood, or metal for surface finish and uniformity
Eddy current	Changes in electrical conductivity caused by material variations, cracks, voids, or inclusions	Readily automated; moderate cost	Limited to electrically conducting materials; limited penetration depth	Heat exchanger tubes for wall thinning and cracks
Liquid penetrant	Surface openings due to cracks, porosity, seams, or folds	Inexpensive, easy to use, readily portable, sensitive to small surface flaws	Flaw must be open to surface. Not useful on porous materials or rough surfaces	Turbine blades for surface cracks or porosity; grinding cracks
Magnetic particles	Leakage magnetic flux caused by surface or near-surface cracks, voids, inclusions, material or geometry changes	Inexpensive or moderate cost, sensitive both to surface and near-surface flaws	Limited to ferromagnetic material; surface preparation and post-inspection demagnetization may be required	Railroad wheels for cracks; large castings

Source: Ref 4

the *Annual Book of the American Society of Testing and Materials*, ASTM (Ref 61). The Volume 03.03 *Nondestructive Testing* is revised annually, covering many NDT procedures including acoustic emission, eddy current, leak testing, liquid penetrants, magnetic particle, radiography, thermography, and ultrasonic. For example, there are standard guides on radiographic testing (see Table 2.17). Reference radiographs, essential to NDT, are available for a wide range of alloys; these are listed in Table 2.17(a). Detailed reviews are given in documents:

- ASTM E 94 "Radiographic Testing"
- ASTM E 242 "Radiographic Images"
- ASTM E 748 "Thermal Neutron Radiography"
- ASTM E 1000 "Radioscopy"
- ASTM E 1030 "Radiographic Examination of Castings"

Detailed reviews of ultrasonic testing (see Appendix 3.1) are given in:

- ASTM E 164 "Ultrasonic Examination of Weldments" (p 34-54)
- ASTM E 494 "Ultrasonic Velocity in Materials" (p 177-188)

Reference photographs of liquid penetrant inspection are presented in ASTM E 433, and an extensive review of the process is given in ASTM E 165. A complete description of magnetic particle inspection (see Table 5.14) is given in ASTM E 709 "Standard Guide for Magnetic Particle Inspection" (p 277-308). ASTM standards for eddy current testing are listed in Table 6.3. The many applications are listed in some detail.

The evaluation of NDT procedures and agencies is discussed in ASTM E 543 "Evaluating Agencies Performing NDT." Definitions and terminology of NDT are covered by ASTM E 1316 "Standard Terminology for NDT Examination," prepared by ASTM subcommittee E07.92 (latest revision 1992). All of these documents are in fact very useful reviews of NDT methods and reference data.

Table 1.3　NDT studies of general interest

Technique	Subject examined (Fig.)	Ref
X-radiography	Examination of oil paintings and wooden statues	5, 6, 7, 8, 9
	Teddy bear (Fig. 1.7)	10
	Analysis of paintings (Fig. 1.4, 1.5)	11, 12
	Colocynth scales with denticles and microtubes	13
Activation analysis	Assassination of President Kennedy; forensic studies of silver and antimony contents of bullets	14
Activation analysis, ion beam analysis	Authenticity, age, provenance of oil paintings considering the purity of the white lead paints	15, 16, 17
Neutron activation analysis	Authenticity of ancient coins. Gold contaminant in silver. Silver contaminant in copper	6, 9, 18
Neutron activation using small source	Gas- and oil-bearing geological strata	19
X-ray and neutron radiography compared	Locks, explosive devices. Cooling tubes in turbine blades	20
γ-radiography	Statue of Liberty (Fig. 1.2)	21
	Crack in Liberty Bell (Fig. 1.3)	22
	Ancient Egyptian mummies	23, 24
	Lead-coated coffins of 17th century (Fig. 1.6)	25
	Freedom statue (U.S. Capitol Building)	26
X-ray fluorescence; x-ray absorption	Thickness of metal coatings, glaze on paper, ink on paper, inorganic coated paper	27, 28
X-ray laminography	Clock	29
Underwater sound	Submarine detection, fish detection, navigation, echo depth sounding, object in muddy water	30
Infrared techniques	Heat losses in buildings, steam traps. Hot spots in electrical equipment. Friction in mechanical systems. Aerial photography, defects in metals, stress in solids	31, 32
Thermoluminescence	Dating in archeology, pottery, and ceramics	6, 33, 34
Polarized light	Suppress reflected light interference in inspection of general objects	35
	Recycling of plastics	36
Shadow moiré	Three-dimensional shapes of general objects such as a spoon, stones, tires, human body	37

Table 1.4　Reference list concerning the history of nondestructive testing

Method	Ref
X-rays, x-radiography, and x-ray diffraction	40-47
Nuclear magnetic resonance	48
Liquid penetrant	49
Ultrasonic	50
Magnetic particle	51, 52
General	2, 53, 54

Training NDT Personnel

The skill of the NDT inspector is crucial to the efficiency of the examination, so that training is important. A series of very useful publications is available from the American Society for Nondestructive Testing (ASNT), and the ASNT regulations, recommendations, and proceedings are followed by industry worldwide. In particular, the ASNT Classroom Training Handbooks (Ref 62) are very helpful and details are given in the following ASNT references:

- "Liquid Penetrant NDT" (Ref 63)
- "Magnetic Particle Testing" (Ref 64)
- "Ultrasonic Testing" (Ref 65)
- "Eddy Current Testing" (Ref 66)
- "Radiographic Testing" (Ref 67)

Many of the figures presented in this text are reproduced with permission from these ASNT Handbooks.

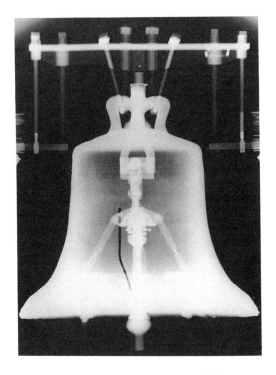

Fig. 1.3 The Liberty Bell was moved into new accommodations in 1976, and a radiograph was taken before the move to determine whether the bell could withstand the stresses involved (Ref 21). Courtesy of Eastman Kodak Co.

Certification of NDT personnel is generally carried out following the recommendation of ASNT document Recommended Practice Number SNT-TC-1A (Ref 68). Three levels of achievement of NDT inspectors are recommended.

- *Level I*: Ability to perform specific calibrations, specific tests, and specific evaluations according to written instructions
- *Level II*: Ability to set up and calibrate equipment, interpret, and evaluate results in accordance with codes, standards, and specifications, and also to report results
- *Level III:* Competent to establish techniques, interpret codes, and designate the test methods and techniques to be used, having a practical background in the technology and be familiar with other commonly used methods of NDT (Ref 68)

The overall ASNT scheme for NDT inspector training is presented in Fig. 1.8. There are several introductory texts on NDT (Ref 82).

A Wealth of NDT Methods

Several of the basic methods of NDT are described extensively in Chapters 2 to 6, covering radiography, ultrasonics, liquid penetration, magnetic particle, and eddy current inspection. It must be stressed that it has not been possible to cover all of the most frequently used techniques in this text.

Alongside the basic NDT methods, there exist a wealth of other methods, and these cover a very diverse range of techniques of examining solids without any adverse effects or damage to the objects involved. Many of these methods are briefly listed with appropriate references in Table 1.5.

Selecting an NDT Process

The various NDT tasks are best studied by a particular testing procedure. A listing is presented in Table 1.6 giving the NDT process selected for a range of metallurgical problems (Ref 65).

Indications: False, Nonrelevant, Relevant

Indications obtained during NDT testing need to be interpreted and evaluated. Any indication that is found is called a discontinuity. Discontinuities are not necessarily defects, but need to be identified and evaluated to decide whether the part is at or below specification; see ASTM standard terminology in Ref 61, p 595-628.

- *False*: Indication not due to the testing procedure. It may be due to improper processing, incorrect procedure, also known as a "ghost," an artifact, "spurious," or "electrical interference"
- *Nonrelevant*: An indication which has no relation to a discontinuity that is considered a defect in the part being tested; a defect within acceptable tolerance levels
- *Relevant*: Indication of a defect that may have an effect on the serviceability of the part
- *Discontinuity*: An interruption, intentional or unintentional in the configuration of the part
- *Indication*: Observation of a discontinuity that requires interpretation, for example, cracks, inclusions, gas pockets
- *Interpretation*: Determination whether an indication is relevant, nonrelevant, or false.
- *Evaluation*: Assessment of a relevant indication to determine whether specifications of the serviceability of the part are met
- *Defect, flaw*: One or several discontinuities that do not meet specifications

The relationship between these terms is illustrated in Fig. 1.9.

(a)

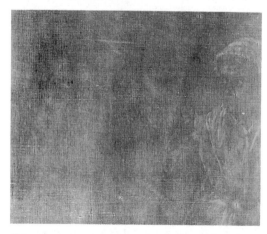

(b)

Fig. 1.4 *Woman in Chair* by J.G. Brown (Ref 12, p 9). The radiograph (b) reveals the portrait of the lady that was painted using inorganic pigments which have a relatively high x-ray absorption. It appears that the left side of the painting was completed at another time using organic pigments of very low x-ray absorption. Courtesy of Eastman Kodak Co.

Table 1.5 The Wealth of NDT methods

Technique	Ref
Acoustic emission	7, 69-71
Dynamic and vibrational analysis	72
Exo-electron emission	74
Fiber-optic sensing	75
Holography in NDT	76-79
Laser techniques	80
Microwave techniques	81
Nuclear magnetic resonance	48, 71, 83-85
Positron annihilation	86
Pressure and leak testing	69, 81
Surface roughness and abrasion	87
Thermography thermal methods	32, 88-92
Thermoelastic effect	93
Thermoluminescence	33
Visual and microscopy	37, 94-100
Xonics electron radiography	101

Note: There are at least 50 other accepted methods of NDT that could be listed here, and many are described in Ref 60. Many recent developments in NDT methods are discussed in Ref 81.

Table 1.6 Selection of NDT process—visual examination is always useful

Defect	Method	Comments
Bursts (wrought metals)	Ultrasonic testing	Internal bursts produce a sharp reflection; able to differentiate types of bursts
	Magnetic particle testing	Surface and near-surface bursts only of ferromagnetic materials
Cold shuts (casts)	Liquid penetrant inspection	Surfaces of most metals; smooth regular line; casts difficult
	X-radiography	Distinct dark line
Fillet cracks (bolts) (wrought metals)	Ultrasonic testing	Extensively used; sharp reflection
	Liquid penetrant inspection	All metals; sharp clear indications; necessary to remove all penetrant subsequent to testing
Grinding cracks	Liquid penetrant inspection	All metals; irregular pattern fine cracks; may require long penetrant dwell times
	Magnetic particle inspection	Ferromagnetic metals only
Convolution cracks nonferrous	X-radiography	Extensively used
Heat-affected zone cracking (HAZ)	Magnetic particle inspection	Ferromagnetic metals only; demagnetization may be difficult; must avoid electric arc from prods
	Liquid penetrant inspection	Nonferrous welds; depends on surface processing
Heat-treat cracks (near areas of stress)	Magnetic particle inspection	Ferromagnetic materials; straight, forked, or curved indications
	Liquid penetrant inspection	Nonferrous metals
Surface shrink cracks	Liquid penetrant inspection	Nonferrous metals; avoid regions such as press fittings
	Magnetic particle inspection	Ferromagnetic materials
	Eddy current	Nonferrous welded piping
Thread cracks (wrought metals)	Liquid penetrant inspection	Use fluorescent penetrant
	Magnetic particle inspection	Ferromagnetic metals; nonrelevant indications from threads
Tubing cracks (nonferrous)	Eddy current	Recommended if tube diameter is less than 1 in. and wall thickness is less than 0.15 in.
	Ultrasonic testing	Suitable for tubing; couplants may affect certain alloys
Hydrogen flake (ferrous)	Ultrasonic testing	Extensively used
	Magnetic particle inspection	Used on finished part; appearance of short discontinuities (hairline cracks)
Hydrogen embrittlement (ferrous)	Magnetic particle inspection	Indications appear as randomly oriented cracks
Inclusions (welds)	X-radiography	Used extensively; sharp well-defined round or other shaped spots; relatively large inclusions
	Eddy current	Thin wall welded tubing
Inclusions (wrought metals)	Ultrasonic testing	Used extensively; large inclusions act as good reflectors; smaller give rise to background "noise"
	Eddy current	Thin wall, small diameter rods; difficult for ferromagnetic metals
	Magnetic particle inspection	On machined surfaces, indicators are straight intermittent or continuous line
Lack of penetration (welds)	X-radiography	Extensively used
	Ultrasonic testing	Used but some geometries difficult
	Eddy current	Nonferrous welded tubing
	Magnetic particle inspection; liquid penetrant inspection	If rear of weld is visible
Laminations (wrought metals)	Ultrasonic testing	Extensively used; sharp signals with loss of rear wall signal
	Magnetic particle inspection; liquid penetrant inspection	Indication straight, broken lines
Laps and seams (rolled metals)	Liquid penetrant inspection; magnetic particle inspection (ferrous)	All metals; fluorescent LPI; indications curved, continuous or broken lines
Laps and seams (wrought metals)	Magnetic particle inspection (ferrous); liquid penetrant inspection	Straight, spiral, or curved indications
	Ultrasonic testing	Extensively used; good signals
	Eddy current	Tubing and piping
Microshrinkage (magnesium castings)	X-radiography	Extensively used; elongated feathery streaks
	Liquid penetrant inspection	Extensively used on finished surfaces where machining opens micropores

(continued)

9

Table 1.6   *(continued)*

Defect	Method	Comments
Gas porosity (welds)	X-radiography	Extensively used; round and elongated spots on radiograph
	Ultrasonic testing	Very sensitive, but depends on grain size
	Eddy current	Thin wall tubing
Unfused porosity (aluminum)	Ultrasonic testing	Extensively used
	Liquid penetrant inspection	Machined article; straight line indications
Stress-corrosion cracking	Liquid penetrant inspection	Extensively used
Hot tears (ferrous coatings)	X-radiography	Extensively used
	Magnetic particle testing	Surface only
Intergranular corrosion (nonferrous)	Liquid penetrant inspection	Extensively used
	X-radiography	Advanced stages of intergranular corrosion

Source: Ref 65

(a)

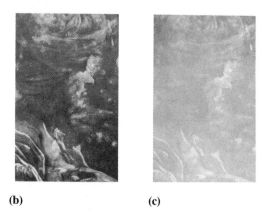

(b)   (c)

Fig. 1.5 *Farmyard Animals* by R.A. Blakelock (Ref 12, p. 10). The radiographs (b) and (c) show a man in a turban, so that the painting is on an older painting. (b) Kodak X-OMAT TL Film, 30 kV, 1200 mA-s. (c) Same film, 70 kV, 200 mA-s. The higher contrast uses a lower voltage, longer exposure. Courtesy of Eastman Kodak Co.

Appendix 1.1: International Organizations of Standardization, Quality Control, and NDT

1. INTERNATIONAL ORGANIZATION FOR STANDARDIZATION (ISO)
   1 rue de Varembé, case postale 56, CH-1211 Genéve 20, Switzerland
   Committee on Nondestructive Testing (ISO/TC135)
2. INTERNATIONAL COMMISSION ON RADIATION UNITS AND MEASUREMENTS (ICRU)
   7910 Woodmont Avenue, Suite 1016, Washington, D.C., USA
3. INTERNATIONAL COMMISSION ON RADIOLOGICAL PROTECTION (ICRP)
   Dr. F. D. Sowby, Clifton Avenue, Sutton SM2 5PU, ENGLAND
4. INTERNATIONAL INSTITUTE OF WELDING (IIW)
   54 Princess Gate, Exhibition Road, London SW7 2PG, ENGLAND
5. EUROPEAN ORGANIZATION FOR QUALITY CONTROL (EOQC)
   P.O. Box 2613, CH-3001, Berne, SWITZERLAND
6. USA NATIONAL COMMISSION ON RADIATION PROTECTION
   P.O. Box 4867, Washington, D.C. 20008, USA
7. AMERICAN SOCIETY FOR TESTING AND MATERIALS (ASTM)
   1916 Race Street, Philadelphia, PA 19103, USA

(Special Technical Publications cover many aspects of NDT)

8. AMERICAN SOCIETY OF MECHANICAL ENGINEERS (ASME)
United Engineering Center, 345 East 47th Street, New York, NY 10017, USA
(Special Publications cover many NDT practices)

9. AMERICAN WELDING SOCIETY (AWS)
550 NW LeJeune Road, Miami, FL 33126, USA

10. AMERICAN SOCIETY FOR NONDE-STRUCTIVE TESTING (ASNT)
3200 Riverside Drive, Columbus, OH 43221, USA
(Guidelines on: Supplement A—Radiographic testing method; Supplement B—Magnetic-particle method; Supplement C—Ultrasonic testing method; Supplement

D—Liquid penetrant testing method; Supplement E—Eddy-current testing method)

11. EUROTEST (International Scientific Association)
Rue du Commerce 20-22 Bte 7, 1040 Brussels, BELGIUM
Secretary General D. Van Elewyck

12. INTERNATIONAL UNION OF TESTING AND RESEARCH LABORATORIES FOR MATERIALS AND STRUCTURES (RILEM)

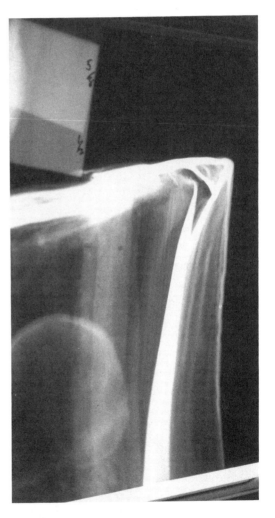

(a)   **(b)**

Fig. 1.6 Lead coffins shown in (a) from the 17th century have been radiographed to decide where to drill air sampling holes without disturbing the remains. A Cobalt-60 isotope source was used with Kodak Industrex AA film; outlines of skulls, wood grain siding, nails, and folds of the half-inch thick lead layers can be distinguished in (b) (Ref 25). The coffin was only partially unearthed at the time providing only limited access to opposite sides of the coffin, which is needed to obtain a radiograph with the best view. Access to opposite sides of the object is essential for radiography. Courtesy of Eastman Kodak Co.

12 rue Brancion, F-75015, Paris, FRANCE
Secretary General Mr. M. Fickelson

Appendix 1.2: Member Societies and Associated Members of the International Committee for Nondestructive Testing

Argentina: Centro Argentino de Ensayos no Destructivos de Materiales (CAEND), Santiago del Estero 250-OF. 31, 1075 Buenos Aires, Argentina

Australia: Australian Institute for Nondestructive Testing, PO Box 334, Box Hill, Victoria 3128, Australia

Austria: Austrian Society for Nondestructive Testing (ÖGZfP), Krügerstrasse 16, A-1015, Vienna, Austria

Belgium: BANT, c/o Katholieke Universiteit Leuven, De Croylaan 2, B-3030 Heverlee, Belgium

Brazil: Associaçáo Brasileira de Ensaios Náo Destrutivos (Brazilian Assn. for NDT] (ABENDE), Rua Luís Góes 2.341, 04043 São Paulo, Brazil

Bulgaria: Bulgarian Society for NDT, Rakovski St. 108, POB431 1000 Sofia, Bulgaria

Burma: Applied Research Institute, Kanbe, Yankin PO, Rangoon, Burma

CEN: European Committee for Standardization, Rue de Stassart, 36-1050 Bruxelles, Belgium

Canada: Canadian Society for Nondestructive Testing (CSNDT), Unit 47, 2400 Lucknow Dr., Mississauga, Ontario L5S1T9, Canada

China, People's Republic of: Chinese Society for Nondestructive Testing, c/o Nondestructive Testing Institute of the

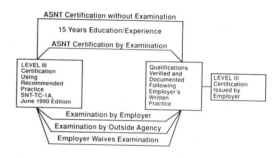

Fig. 1.8 Overall scheme of NDT inspection training prepared by American Society for Nondestructive Testing (ASNT) (Ref 68). See also ASTM standards E 543, E 1212, and E 1359 (Ref 61).

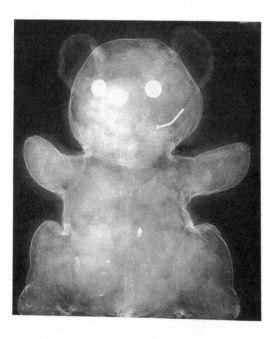

Fig. 1.7 Teddy bear where the filling contains metal scraps. X-radiograph taken at 30 kV, 4 mA, 30 s exposure, at 36 in. on type II film (Ref 10) (see Fig. 2.6a).

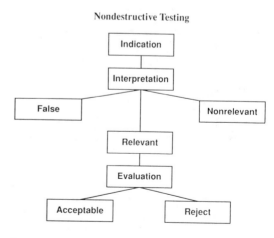

Fig. 1.9 The evaluation of discontinuities detected by NDT

ChineseMechanicalEngineeringSociety, 100 Huihe Rd., Shanghai 200433, China

China, Republic of [Taiwan]: Chinese Society for Nondestructive Testing, 130 Keelung Rd., Section III, Taipei, Taiwan, 10771 ROC

Czechoslovakia: Czechoslovak Society for Nondestructive Testing (CSS NDT), CS 113 12 Praha-1, Opletalova 25, Czechoslovakia

Denmark: FORCE Institutes, Park Alle 345, DK-2605 Broendby, Denmark

Egypt: Egyptian Society for Industrial Inspection, 3A, Moaskar Romani St.-Rushdi, Alexandria, Egypt

Finland: Finnish Society for NDT, Metallimiehenkuja 10, 02150 Espoo, Finland

France: Confédération Française pour les Essais Non Destructifs (COFREND), 32 Boulevard de la Chapelle, 75882 Paris Cedex 18, France

Germany: Deutsche Gesellschaft für Zerstörungsfreie Prüfunge eV (DGZfP) [German Society for Nondestructive Testing], Unter Den Eichen 87, 1000 Berlin 45, Germany

Greece: Hellenic Society for Nondestructive Testing (HSNT), Sofokleous 89 St., 166 73 Voula, Athens, Greece

Hungary: Scientific Society of Mechanical Engineers, Kossuth, Lajos Ter 6-8, Budapest V., Hungary

India: Indian Society for Nondestructive Testing, E-1, 2nd Floor, J. P. Tower, Nungambakkam High Rd., Madras 600 034, India

Iran: The Iranian Institute of Welding and Nondestructive Testing, PO Box 19615-591, Tehran, Iran

Israel: Technion, Israel Institute of Technology, 32000 Haifa, Israel

Italy: Italian Society for Nondestructive Testing, via Foresti 5, 25126 Brescia, Italy

Jamaica: Jamaican Society for Nondestructive Testing (JSNDT), 6 Winchester Rd., Kingston, 10, Jamaica, West Indies

Japan: Japanese Society for Nondestructive Inspection (JSNDI) 5-4-5 Asakusabashi Taitoh-Ku, Tokyo, 111, Japan

Korea, Republic of: Korean Society for Nondestructive Testing, c/o KID 1473-10, Seocho-dong, Seocho-ku, Seoul, 137-070 ROK

Netherlands: KINT Dutch Quality Surveillance and Nondestructive Testing Society, PO Box 390, 3330 AJ Zwijndrecht, Netherlands

New Zealand: New Zealand NDTA, PO Box 12241, Wellington, New Zealand

Norway: Norsk Forening for Ikke-Destruktiv Proving, c/o Robit Training, PO Box 100, N-136 Billingstadsletta, Norway

Philippines: Philippine Society for Nondestructive Testing, Inc., c/o Philippine Nuclear Research Institute, Commonwealth Ave., Diliman, Quezon City, Metro Manila, Philippines

Romania: Romanian Assn. for Nondestructive Testing (ARoENd), Str. Frumoasa 26, Bucharest 78116, Romania

Russia: Soviet Society for Nondestructive Testing and Technical Diagnostics (SSNTTD), 35 St. Usacheva, Moscow, 119048, Russia

Singapore: Singapore Institute of Standards and Industrial Research, 1 Science Park Dr., Singapore 0511, Republic of Singapore

South Africa: South African Assn. of Nondestructive Testing, PO Box 670, Bergvlei 2012, Republic of South Africa

Spain: Associación Española de Ensayos No Destructivos (AEND) (Spanish Assn. for Nondestructive Testing), Zurbano 92, 1°D 28003 Madrid, Spain

Sweden: Swedish Society for NDT, FOP Kansli, Materialnormcentralen, Box 4094, S-102 62 Stockholm, Sweden

Switzerland: Swiss Society for Nondestructive Testing (SSNT) [SVMT/SGZP], c/o EMPA Dübendorf, Überlandstrasse 129, CH-8600, Dübendorf, Switzerland

Thailand: Thai NDT Society, Office of Atomic Energy for Peace, Vibhasadee-Rungsit Rd., Bankehn, Bangkok 10900, Thailand

Tunisia: Permanent National Committee for Nondestructive Testing and Control, Technical Center of Mechanical and Electrical Industries, 1 Rue, 7037 Menzah IV, Tunisia

United Kingdom: British Institute of Nondestructive Testing, 1 Spencer Parade, Northampton, NN1 5AA, UK

Chapter II

Radiography

Part I Introduction to Radiography

Radiographs

Radiography is the technique of obtaining a shadow image of a solid using penetrating radiation such as x-rays or gamma-rays (γ-rays). The image obtained is in projection, with no details of depth within the solid. Images recorded on film are known as radiographs and also less frequently as roentgenograms, skiagrams, roentgenographs, or sciagraphs (Ref 47, 54).

The contrast in a radiograph is due to different degrees of absorption of x-rays in the specimen and depends on variations in specimen thickness, different chemical constituents, nonuniform densities, flaws, discontinuities, or to scattering processes within the specimen. The x-rays need to be generated, projected through the object to give good contrast and definition in the image plane. Access to opposite sides of the object is required. Subsequently, the radiograph needs to be recorded under conditions of safety. Examples of radiographs are given in Fig. 2.5 and 2.49 covering many different applications; see also Table 2.17(a).

The methods of radiography using x-rays and γ-rays are discussed together because both are electromagnetic radiations of comparable wavelengths and energies, where γ-rays are x-rays of high energy emitted by radioactive isotopes. The spectral distributions of x-rays and γ-rays are different, as can be seen in Fig. 2.3(a) and (b), and this leads to some differences in experimental procedures. X-radiography is one of the earliest NDT techniques and its history is reviewed in some detail.

There are several closely related techniques:

- Tomography providing information in three dimensions; that is, details of a selected layer of the object are revealed, so that both the presence and the position of a defect are determined
- Radioscopy (or real-time radiography) where the x-ray photons are converted by one of several methods to be displayed as an analog image on a television monitor
- Xerography, where the latent image is captured on a selenium plate, the plate subjected to a charge, a fine blue powder attracted to charged areas most affected by x-rays, and printed onto white plastic-coated paper. Principal use is in hospitals because of the short exposures required, but it is also used in industry as a coarse sorting inspection
- Using other radiations of neutrons, positrons, protons, or electrons
- Radiometry where the image is recorded by a radiation detector

Only limited discussions are covered of these related techniques.

Definitions given by ASTM document E 1316 (Ref 61) are as follows:

- *Radiology* is the science and application of x-rays, γ-rays, and other penetrating radiations.

- *Radiographic inspection* is the use of x-rays, γ-rays, or other penetrating radiation to detect discontinuities in materials.
- *Radiological examination* has the same definition. The general practice when producing a radiograph to record the image is to use the term radiographic inspection. A glossary of some terms used in radiography is given in Appendix 2.1.

The History of Industrial Radiography

W.C. Roentgen became aware of and can be said to have discovered x-rays 8 November 1895 when he observed a fluorescent glow of crystals on a table near his cathode-ray tube. He characterized very systematically the penetrating radiations emitted by cathode-ray tubes during November and December 1895 (Ref 102), and even sent out as Christmas cards to fellow scientists the x-ray photograph of a human hand as well as a photograph of a set of metal weights inside a closed wooden box. These represent the very first radiographs and several early radiographs taken in 1896 are shown in Fig. 2.1. The signed x-radiograph in Fig. 2.1(e) is of the hand of Lord Kelvin taken in May 1896, showing the ring on his finger. A radiograph of a set of nails on a wooden board is seen in Fig. 2.1(d); this was carried out in March 1896 by scientists in Russia to locate the exact source of the x-rays from a

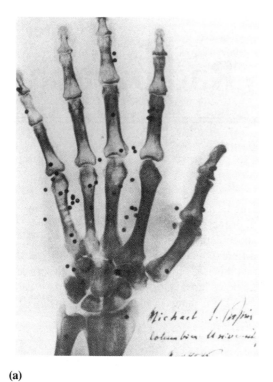

(a)

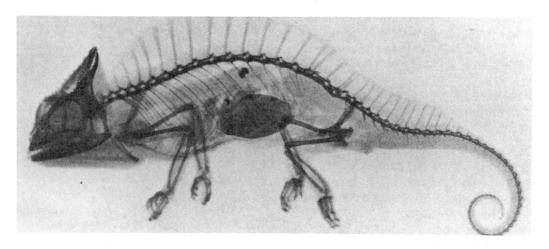

(b)

Fig. 2.1(a) and (b)   The first X-radiographs 1896 (Ref 40). (a) February 1896. Gunshot in hand. Columbia University, New York. (b) February 1896. An African chameleon. Vienna.   *(continued)*

(c)

(d)

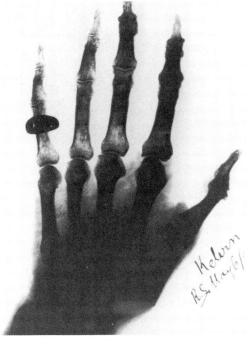

(e)

Fig. 2.1(c), (d), and (e)   The first X-radiographs 1896 (Ref 40). (c) March 1896. Thomas Edison examines the hand of his assistant, C. Dally, using a fluoroscope. (d) March 1896. The source of x-rays from a cathode ray tube, identified by shadows of an array of nails. Imperial Academy of Science, St. Petersburg. (e) The hand of Lord Kelvin, photographed at the Royal Society, London; the radiograph is signed by Lord Kelvin.

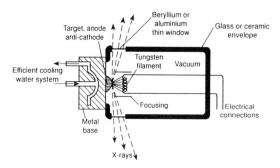

Fig. 2.2 Schematic of a Coolidge x-ray tube. The vacuum is of the order of 10^{-2} Pa. A small voltage is used to pass a current of several amperes through the tungsten filament, which is at negative potential and emits electrons. These electrons are directed by the negatively charged focusing system toward the anode target (anticathode), which is usually at ground potential. The window is of beryllium or aluminum, both metals having low x-ray absorption properties. The target must be carefully cooled since nearly 99% of the energy is converted to heat. The target is made of a metal such as tungsten of high melting point, good thermal conductivity, and high atomic number.

cathode-ray tube (Ref 40). As soon as Roentgen's discovery of x-rays became known, it was easy for scientists everywhere to repeat his experiment because cathode-ray Hittorf-Crookes vacuum tubes were very well known from the mid-19th century onwards and were often treated as drawing-room curiosities. It was thus possible for studies to proceed extremely rapidly, and by early 1896, x-ray fluoroscopes were being used for the inspection of postal packages, examination of porcelains and precious stones, and in medical diagnostic studies (Ref 53, 103). In March 1896, Edison and his assistant C.M. Dally were demonstrating the use of x-rays at the New York National Electrical exhibition; see Fig. 2.1(c). Haplessly, the assistant, Dally, was one of the first persons to die of x-ray burns in 1904. An x-ray photograph of a weld was carried out in January 1896 (Ref 45). In February 1896, the radiograph of a chameleon shown in Fig. 2.1(b) was published with many other radiographs (Ref 40). Also in February 1896, a radiograph was taken of a gunshot hand wound; see Fig. 2.1(a). During 1896, steel objects were examined by x-rays at Pittsburgh Carnegie Steel Works, and x-rays were being used to inspect armaments in the United States, Germany, and Austria (Ref 2, 53).

The earlier gas x-ray tubes were very unreliable, difficult to control, and produced very low x-ray intensities. It was therefore a great advance when the high vacuum x-ray tubes designed by

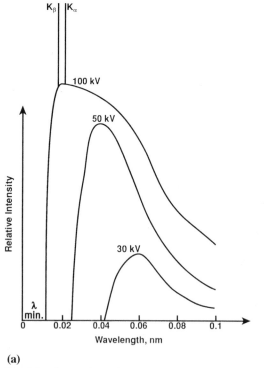

(a)

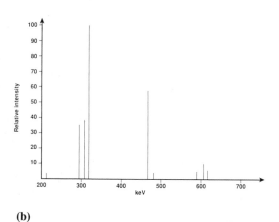

(b)

Fig. 2.3 Comparison of radiation spectra from an x-ray tube and from a radioactive isotope. (a) Spectra from tungsten at x-ray tube excitation voltages of 30, 50, and 100 kV. The characteristic x-ray lines K_α and K_β of tungsten appear at voltages above 69.5 kV, when the tungsten atoms are ionized. The shortest wavelength λ_{min} is given by $\lambda_{min} = (1.24/V_{kilovolts})$ nm. The maximum intensity of the spectra I_{max} occurs at about 1.5 λ_{min}. (b) The γ spectrum of the radioactive isotope Ir^{192} consists of several lines at discrete energies (weaker lines less than 1% relative intensity are not shown).

Coolidge became available in 1913. Industrial radiography now had an intense and reliable x-ray source with energies up to about 100 kV. Many examples of radiographs are given in the book published in 1922 (Ref 104) of armaments, wooden aircraft, defects in wooden constructions, and of various metals.

In 1931, the American Society of Mechanical Engineers permitted x-ray approval of fusion-welded pressure vessels, and in the United Kingdom the Admiralty and Lloyd's Register of Shipping accepted a similar code (Ref 53).

High-voltage x-ray generators (1000 kV) became available in 1931 (General Electric Co.). X-radiography of fast moving objects using intense x-ray beams were undertaken in Germany and the United States in 1938 (Ref 105, 106), in Holland in 1940 (Ref 107), and in the United Kingdom in 1941 (Ref 108). X-rays of energies in the million volts range were obtained in 1940 using the Betatron (Ref 109). Since the 1960s, transportable 15 MeV x-ray sources have provided radiographs of welds in steels up to 30 cm thick (Ref 53). There are presently available 6 MeV betatrons, which can be transported with ease in a small van. These betatrons are tunable and betatrons of 10 MeV are being developed.

Accounts of radiography in the 1930s and 1940s are given in Ref 44 and 110.

Following the discovery of x-rays by Roentgen in 1895, H. Becquerel observed that uranium ores also emitted penetrating radiations (Ref 111). Villard in 1900 was able to show the presence of very penetrating γ-rays from uranium ores and that these γ-rays could pass through 25 cm of lead (Ref 46). Pilon and Laborde (Ref 112) applied these γ-rays to the inspection of metals in 1903. Many of the earlier inspections are described by Pullin (Ref 44). At that time, a typical radioactive source of γ-rays would be about 0.25 g of a radium salt, with an effective source size of about 2 mm. A wide range of γ-ray emitting isotopes are now available from nuclear energy reactors.

X-Rays, Grenz-Rays, and γ-Rays

X-rays are generated when an electron beam impinges on a solid target. A schematic of a typical x-ray tube is shown in Fig. 2.2. The target is cooled very carefully because a very high percentage of the electrical energy is converted into heat, particularly when generating lower energy x-rays. A vacuum of about 10^{-2}Pa (10^{-7} atm) is required, and electrons from a heated filament are accelerated from a high voltage (20 kV to 20 MV) to the anode target, or anticathode, usually at ground potential. A window of a low absorbing foil such as beryllium or aluminum is provided in the vacuum tube to allow the x-rays to exit from the tube.

The anode should be of a high melting point metal, and a typical anode material is tungsten. Spectra from a tungsten anode for a series of excitation voltages are shown in Fig. 2.3(a).

The spectrum contains a range of x-ray energies from the maximum energy of excitation of

Table 2.1(a)   Tabulation relating x-ray wavelength, x-ray tube voltage, and x-ray photon energy

X-ray wavelength (λ) Å	nm	X-ray tube voltage, kV	X-ray photon energy, keV
1	0.1	12.4	12.4
0.1	0.01	124	124
0.01	0.001	1.24 MV	1.24 MeV
0.001	0.0001	12.4 MV	12.4 MeV

The x-ray wavelength, λ, and x-ray tube voltage V, are related by $\lambda\,\text{Å} = 12.4/V$ (kilovolts).

Table 2.1(b)   Tabulation of x-ray wavelengths and energies at maxima of white radiation

X-ray tube voltage, kV	X-ray wavelength at maximum of white radiation [~(1.5 × 12.4)/V], Å	X-ray energy at maximum of white radiation (~0.67 V), keV
25	0.75	17
50	0.37	35
100	0.19	70
200	0.09	130
400	0.05	270
1000	0.02	670

The x-ray wavelength at the maximum of the intensity distribution occurs at a wavelength approximately equal to 1.5 × λ (minimum).

the x-ray tube, corresponding to the minimum in x-ray wavelength λ_{min}, with an energy distribution as shown with a maximum I_{max} at about 1.5 λ_{min}; that is,

$$\lambda \text{ at } I_{max} \cong 1.5 \, \lambda_{min} \qquad \text{(Eq 2.1)}$$

The maximum energy of the x-rays E_{max} corresponds to the minimum value of the wavelength λ_{min}, since

$$E_{max} = Ve = h\nu = \frac{hc}{\lambda_{min}} \qquad \text{(Eq 2.2)}$$

where E is energy (electron volts), V is the applied voltage on the x-ray tube (volts), e is electronic charge, 1.602×10^{-19} C, h is Planck's constant, 6.626×10^{-34} Js, ν is frequency of radiation (Hz), and c is velocity of light, 2.998×10^8 m/s. The relationship between λ and V is a constant, and from Eq 2.2:

$$\lambda_{nm} = \frac{1.24}{V_{kilovolts}} \qquad \text{(Eq 2.2a)}$$

$$\lambda_{\text{Å}} = \frac{12.4}{V_{kilovolts}} \qquad \text{(Eq 2.2b)}$$

A tabulation is given in Table 2.1 relating x-ray wavelength, x-ray tube voltage (V), and x-ray photon energy.

As the incident electron loses its energy in the target, x-rays are emitted of a broad band of energies, known as the continuous, white, or bremsstrahlung radiation. In radiography, it is the broad-band spectrum of x-rays that is employed.

Monochromatic x-radiation, characteristic of the target atoms, are also emitted when the excitation voltage is sufficient to ionize the atoms. One of the highest x-ray photon energies of characteristic lines is ~115 keV for the K-lines of uranium (see Table 2.5) so that at higher energies, bremsstrahlung is the only radiation. X-rays can range in energy from less than 0.1 keV to more than 100 MeV. The lower energy x-rays are described as soft x-rays, and are also known as Grenz-rays (Ref 113).

Gamma-rays are electromagnetic radiation emitted by the disintegration of a radioactive isotope and have energy from about 100 keV to well over 1 MeV, corresponding to about 0.01 to 0.001 Å. The most useful γ-emitting radioactive isotopes for radiological purposes, are found to be cobalt (Co-60), iridium (Ir-192), cesium (Cs-137), ytterbium (Yb-169), and thulium (Tm-170). The essential characteristics of these radioactive isotopes are listed in Table 2.2, and the γ-spectrum from iridium (Ir-192) is presented in Fig. 2.3(b).

Gamma-rays are monochromatic, although several γ-rays of different energies may be emitted by the isotope. As an example, the radioactive disintegration of Co-60 results in the emission of a γ-ray of 0.31 MeV, a γ-ray of 1.17 MeV, another γ-ray of 1.17 MeV, and the atom of cobalt becomes the isotope $^{60}_{28}$Ni.

Table 2.2 Radioactive isotopes used as sources for γ-radiography

Radioactive isotope	τ half-life	γ-rays, MeV	Minimum effective source size diameter, mm	Specific activity, Ci g^{-1}	Radioactive source strength, Ckg^{-1}s^{-1}B$_q$$^{-1}$ $\times 10^{-19}$	Source radiation output, Rhm Ci^{-1}	Dose rate, R/hr · ft^2 Ci	Maximum approximate penetration in steel, cm	Comments
Cobalt-60 $_{27}$Co^{60}	5.3 years	1.17, 1.33	3	20	20	1.3	14.5	23	Requires heavy shielding
Iridium-192 $_{77}$Ir^{192}	74 days	0.31, 0.61	0.5	50	5	0.5	5.9	10	...
Ytterbium-169 $_{70}$Yb^{169}	31 days	0.053, 0.309	0.3	2	2	0.125	...	2	Used for thin welds
Cesium-137 $_{55}$Cs^{137}	30 years	0.66	10	75	75	0.37	4.2	10	...
Thulium-170 $_{69}$Tm^{170}	127 days	0.084, 0.052	3	50	50	0.003	0.03	1.3	...

Other radioactive isotopes are listed in Ref 70.

The number of disintegrations per unit time in a radioactive isotope is observed to decrease exponentially with time, and it is usual to express this activity in the form:

$$n_t = n_0 \exp\left(-\frac{0.693t}{\tau}\right) = n_0 \exp\left(-\Lambda t\right) \qquad \text{(Eq 2.3)}$$

where n_t, n_0 are the number of disintegrations at $t = t$, o respectively, τ is the half-life of the radioactive isotope, and Λ is the decay constant. The number of disintegration in unit time per gram is called:

$$\text{Specific activity} = \frac{0.693}{\tau} \frac{N_0}{A_w} \qquad \text{(Eq 2.4)}$$

where A_w is the atomic weight, and N_0 is Avogadro's Number (Ref 114). The definition of units used in radioactivity is given in the section "Radiation Units" in this chapter.

There are several advantages of using radioactive isotopes as sources of high energy x-rays; the source is compact and easy to transport, independent of electrical and water supplies, of relatively low cost, and monochromatic. One disadvantage of the generally shorter wavelengths of γ-rays is that, in certain applications, the resulting radiograph may lack the contrast possible with a spectrum of x-rays. Other disadvantages are the low intensity giving rise to long exposure times, the fact that the radioactive isotope has to be replaced after a time comparable to the half-life of the isotope, and that the radioactive source cannot be made safe by being switched-off. Gamma radiography is discussed in the section "Gamma Radiography" in this chapter.

Outline of Radiographic Procedures

X-rays travel in straight lines and cannot be focused under normal conditions. This determines the principles of x-radiography which uses a spot source of x-rays or as nearly a point as possible. The size of the source, the focal spot, is important in defining the image, and focal spot sizes range from 1.5 to 5 mm in conventional x-ray tubes, 0.2 to 0.5 mm in some focused x-ray tubes, and 5 to 50 μm in special microfocused x-ray tubes. This x-ray source provides a diverging broad beam through the specimen onto a radiographic film, ionization counter, photon converter or counter, or a fluorescent screen. This is illustrated in Fig. 2.4, where voids may be present with different orientations, the thickness

of the object may vary as well as its composition and density; the x-ray shadow image should be a faithful reflection of the inhomogeneities of the object. The density levels in the x-ray film are represented by the intensity profile cross section. It is necessary to have access to opposite sides of the specimen.

The attenuation of an x-ray beam by a solid is partly by absorption and partly by scattering so that great care must be made in the selection of x-rays of the appropriate energy to obtain a reasonable and optimum contrast in the image. This is one of the most important stages in the produc-

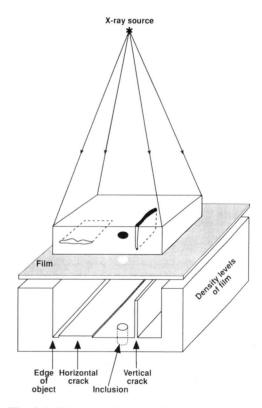

Fig. 2.4 Schematic of an x-radiographic system. A near-point source of x-rays, from the (tungsten) target of the x-ray tube, diverges through the specimen projecting a shadow image onto the film. A dense inclusion, a void, and oriented cracks are illustrated with their resulting effects on the intensity profile of the radiograph. The horizontal crack needs to provide several percent change in attenuation to be visible readily on the film. Example 12 in the section "Calculations for Trial Exposures" is concerned with the radiographic contrast of a crack, 3 x 1 x 0.01 cm in an aluminum block 5 x 5 x 2 cm.

tion of a useful radiograph. As an example of the importance of the selection of x-ray energies, some radiographs are compared in Fig. 2.5. The features that can be seen in the object are quite different depending on the energy of the x-rays used. The crack in a steel plate, 1 in. thick, is recorded in (a) using x-rays from 150 kV generator, and in (b) using γ-rays from Ir-192. The high energy γ-rays give rise to a poor contrast. A gear is recorded in (c) and (d) using two films in the cassette holder. Radiographs (c) and (d) record quite different contrasts. Very often it is necessary to consider other radiations (electron, neutron, proton) that have different absorption characteristics and that might provide a better contrast in the image, or enable the observation

of a specific feature in the specimen under examination. In many cases, nothing useful may be achieved unless the correct radiation of appropriate energy is used.

The steps to be followed in obtaining a useful radiograph are discussed below:

1. Preliminary Visual Examination of the Object. It is important to look carefully at the specimen and to decide on the direction to examine the object considering the probable orientation of defects and the thickness of the specimen in relation to the diverging beam of x-rays.

2. Energy of the X-Rays. The energy (or wavelength) of the x-rays needs to be selected considering the composition of the object, the

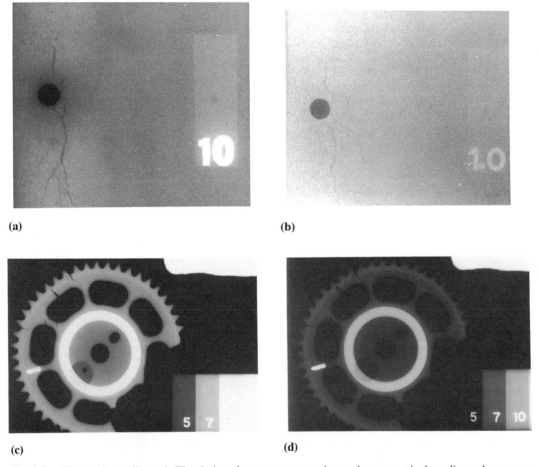

(a)

(b)

(c)

(d)

Fig. 2.5   Contrast in a radiograph. The choice of x-ray energy can change the contrast in the radiograph; an example from ASTM (Ref 61) document E 242 is shown in (a) and (b) of a crack in a steel plate (1 in. thick) examined using (a) 150 kV x-ray generators and (b)Ir[192] radioactive source. The use of two films, one placed behind the other, increases the raiodgrpahic latitude and can improve the contrast; see (c) and (d) of the radiograph of the gear seen in Fig. 2.6(a). Reprinted with permission of ASTM.

path length of x-rays through the object, and the problems of scattered or fluorescent x-rays. Consideration may need to be given to other radiations.

3. Recording the Image. The image can be observed on an image intensifying tube with remote viewing or recorded on film with or without intensifying screens. Grids or blocking materials should be used to reduce scattering effects. The optimum time of exposure will need to be determined by experimental trials using image quality indicator test pieces. A decision is required for the distances to be used between the x-ray source, the object, and the detector plane to provide sufficient magnification, sensitivity, sharpness, and contrast, in a reasonable exposure time. It might be possible to choose a fine-grain film to improve the definition of the image, but such film will be slower, requiring longer exposures. Several examples of calculations for trial exposures are given in the section "Calculations for Trial Exposures" in this chapter.

4. Interpretation of the Radiograph. Radiographs are projections, providing no information about depth within the specimen. To aid in the interpretation, many factors are involved:

- The preliminary selection of x-ray energy (or other radiation) and factors controlling sharpness and contrast of the image
- Use of radiation having the largest possible differences in the relative absorption coefficients of the different compositions present
- Orientation of the object to permit any discontinuity or defect to show maximum contrast
- The radiograph must be examined carefully using a well-illuminated viewing screen under optimum lighting conditions
- Detailed knowledge of the nature of the specimen, the types of defects or features to be expected. Reference radiographs are always advantageous as given in many texts; for example see Ref 61 and 103. These may also be used to provide acceptable defect limitations

- Awareness of artifacts that may be present such as those due to x-ray scattering from high density sharp edges

The basic principles involved in these procedures are discussed in Part II. Practical guidelines and examples of obtaining radiographs are discussed in Part III.

Practical Tips

1. The x-ray target-to-film distance should not be less than 10× the thickness of the specimen.
2. Radiographic films should be exposed to a density D of at least $D = 1.5$ and not more than $D = 3.5$. (Density D is discussed in Part II "Technical Discussions" of this chapter; see Fig. 2.13.)
3. The greatest dimension of the suspected flaw should be parallel to the x-ray beam.
4. Some limitations of x-radiography include: (a) very difficult in the case of complex shapes, (b) access required to both sides of the specimen, and (c) laminar discontinuities are very difficult to observe when not parallel to the x-ray beam.
5. Radiography is relatively expensive.
6. X-rays and γ-rays are dangerous; extensive safety precautions are essential.
7. The unsharpness (penumbra) can be controlled by source-to-specimen distance, specimen-to-film distance, focal spot size, and specimen thickness.
8. The human eye sees an image as sharp with a penumbra up to 0.02 in.
9. A discontinuity of thickness less than 2% of the overall thickness of the specimen is difficult to observe.

Table 2.3 High voltage generators for x-rays

	Voltage range	Tube current, mA	Typical focal spot size, mm	Comments
Iron core transformer	to 400 kV	~50	3.5 × 3.5 at 400 kV	Above 400 kV, generators become too large; see Fig. 2.6(a)
Resonance tuned transformer	250 kV to 4 MV	5	8 × 8 at 1 MV	See Fig. 2.6(b)
Electrostatic van de Graaff	500 kV to 8 MV	1	1 × 1	
Linear electron accelerator	1 MV to 25 MV	...	3 × 3 at 15 MV	X-ray output 100 Grays/min at 1 m. Relatively heavy and bulky, though reliable; see Fig. 2.6(e)

10. A gradual change of photographic density at a boundary is very difficult to observe.
11. A surface crack is placed close to the film to obtain the best contrast.
12. Typical specimen-source distances are: $\frac{1}{8}$ in. thick specimens at ~20 in., 1 in. thick specimens at ~30 in., 10 in. thick specimens at ~100 in.
13. When the presence of a defect is detected, images made from several angles may give information concerning the size and location of the defect.
14. Radiographs are usually viewed as negatives, but positive prints can be prepared for records.

Part II Technical Discussions

The Generation of X-Rays

X-ray generators are very inefficient systems with less than 1% of the energy converted into x-rays at lower energies and approaching only ~25% for x-ray of the highest energy generated. The rest of the electrical energy becomes heat so that an efficient cooling system for the target of the x-ray tube is essential. It is almost always necessary to keep the exposure times to a reasonable minimum, so that the highest x-ray intensity is always being sought. The generated x-ray spectrum covers a wide range of energies, and all of the broad band spectrum is employed in radiography. The x-ray beam is not monochromatic as can be seen in Fig. 2.3(a) where typical spectra from a tungsten target are given as a function of applied voltages for an x-ray tube of design shown schematically in Fig. 2.2.

Some examples of commercial x-ray generators are shown in Fig. 2.6. The high voltages required are generated as listed in Table 2.3. The electrical circuits are either half-wave or full-wave rectified. Details can be found in Ref 67, 69, and 103.

The divergence of the x-ray beam is very large for the iron-core transformers, but become relatively restricted at megavoltages. At 350 keV, a 25° divergence (half-width at half intensity) can be obtained; at 15 MeV, the divergence may only be 5°.

Excitation voltages typically used in radiography are from about 20 kV to 25 MV. The higher energy x-rays are required to penetrate greater thicknesses of objects, whereas lower energies of x-rays provide a different range of contrast in the image; see the section "X-Ray Image Quality" in this chapter.

The x-ray tube targets used for radiography are almost always of tungsten. This is because the overall intensity I obtained from an x-ray tube is given approximately by the empirical relationship

$$I \cong K' i Z_0 V^2 \text{ (Energy/unit area/unit time)} \qquad \text{(Eq 2.5)}$$

where K' is a constant, i and V are the current and applied voltage in the x-ray tube, and Z_0 is the atomic number of the target. The efficiency χ of

(a)

Fig. 2.6(a) Commercial x-ray generators. (a) An x-ray tube for radiography up to 320 kV. The iron-core transformer is on the floor in the background, connected by high-voltage cable to the x-ray tube mounted on an adjustable arm. The size of this unit can be contrasted to the sizes of the higher voltage units shown in (c) and (e) on the next page. See Fig. 1.7 and 2.5(c) and (d) for radiographs taken with this unit. *(continued)*

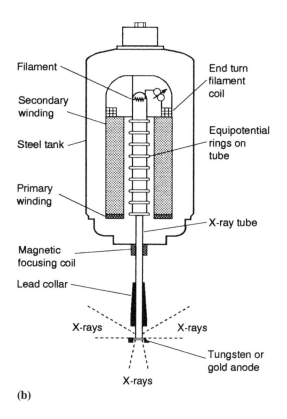

(b)

(c)

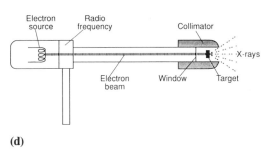

(d)

Fig. 2.6(b), (c), (d), and (e) Commercial x-ray generators. (b) Schematic of a resonance tuned transformer for 500 kV to 4 MV. Reprinted with permission of The American Society for Nondestructive Testing. (c) Commercial unit of the type shown in (b). A typical metal part for examination is seen in front of the operator. (d) Schematic of a linear accelerator 8 MeV. (e) Commercial unit of the type shown in (d). A film is being positioned, held in place using a magnet, for examination of 6 in. steel.

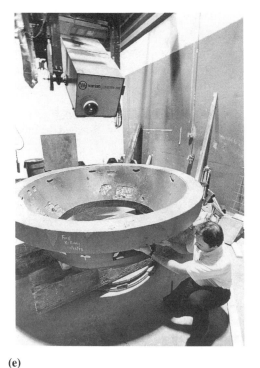

(e)

x-ray production (x-ray energy produced)/(electron energy used), is given by the empirical relationship:

$$\chi \cong 1.4 \times 10^{-9} Z_0 V \qquad \text{(Eq 2.6)}$$

An important factor in the design of an x-ray tube is that a small x-ray source size is essential to obtain good definition in the radiograph. The electron beam is focused onto a small area of the target, the focal point, resulting in severe heating effects so that the target material must have good thermal and mechanical properties. Tungsten is the most suitable target material, because it has a very high melting point, good thermal conductivity, and high atomic number. The x-ray intensity increases with the electron current and with the square of the applied voltage; see Eq 2.5. High energies are required to obtain penetrating x-rays; 2 MeV x-rays decrease in intensity by 50% after approximately 5 mm of steel or 60 mm of concrete. A typical x-ray generator for industrial radiography will therefore be of high energy with a tungsten target.

Absorption and Scattering of X-Rays in Matter

Absorption Mechanisms. Absorption characteristics of materials are important in the development of contrast in a radiograph. Attenuation of x-rays in solids is by several independent mechanisms, some due to absorption, others due to the scattering of the beam. This needs careful consideration because a good radiograph can only be achieved if there is the minimum of x-ray scattering.

The attenuation or absorption coefficient is defined for a narrow well-collimated beam as

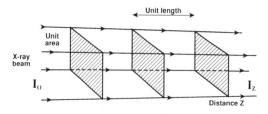

Fig. 2.7 The linear absorption coefficient μ is defined for a narrow, well-collimated, monochromatic x-ray beam. These conditions do not apply in radiography. Scattered x-rays leave the beam and contribute to the decrease in intensity. The mass absorption coefficient $\mu_m = \mu/\rho$ is given by Eq 2.13.

shown in Fig. 2.7. Then the relative decrease in intensity dI/I is proportioned to the distance dz travelled in the medium so that:

$$dI/I = -\mu dz \qquad \text{(Eq 2.7)}$$

where the constant μ is defined as the linear absorption (or attenuation) coefficient, and the negative sign is required because there is a decrease in intensity. Integration from the surface ($z = 0$) of the specimen over the path length (z) leads to:

$$I_z = I_0 \exp -[\mu z] \qquad \text{(Eq 2.8)}$$

where I_z is the intensity of the beam at a distance z, and I_0 is the intensity at the specimen surface. This is known as Lambert's Law (Ref 115) and also as Beer's Law (Ref 116). The linear absorption coefficient μ is dependent on x-ray wavelength.

Equation 2.8 is not appropriate for radiographic studies where we do not have a narrow collimated monochromatic beam, but instead a broad diverging beam of a range of wavelengths. This is illustrated in Fig. 2.8, which is a schematic of an x-ray system for radiography. X-rays scattered by the object F can still attain the x-ray film and so reduce the image contrast. It is necessary to distinguish very carefully between true absorption and scattering effects.

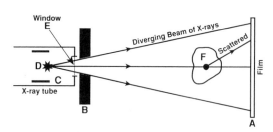

Fig. 2.8 A beam of x-rays leaves the source, D, passes through the window, E, and diverges through the collimator (aperture), B, to the detector, A, which may consist of an image intensifier or x-ray film. Scattered x-rays from the object, F, or from the collimator or air can reach the detector, reducing contrast in the image. The target is frequently tungsten, at a high voltage, from about 20 kV to several MV. The x-ray beam is not monochromatic. In the case of the high voltage equipment using resonance tuned transformers or linear accelerators, seen in Fig. 2.6(b), and 2.6(d) the x-ray beam is emitted from the foil window, or target outside the window.

The linear absorption coefficient μ is the sum of the contributions of various processes and is given by:

$$\mu = Th + ph + j + \zeta + \eta \qquad \text{(Eq 2.9)}$$

where Th is the Thomson scattering, ph is the photoelectric effect, j is the Compton scattering consisting of j_A absorption and j_S scattering, ζ is the pair and triplet formation, and η is the photodisintegration process (Ref 69, 117).

Thomson scattering, also known as coherent or classical scattering, Th, occurs when the x-ray photon interacts with the whole atom so that the photon is scattered with no change in internal energy to the scattering atom, nor to the x-ray photon, and the phase of the scattered x-rays is coherent with the incident radiation. The wavelength of the incident radiation will be comparable to the size of the atom, of the order of 0.1 nm, that is about 10 keV. Thomson scattering is never more than a minor contributor to the absorption coefficient, so that the term Th in Eq 2.9 can be ignored in radiography. Thomson scattering is illustrated in Fig. 2.9(a).

Photoelectric absorption of x-rays occurs when the x-ray photon is absorbed resulting in the ejection of electrons from the outer shell of the atom, resulting in the ionization of the atom. The Einstein photoelectric equation:

$$h\nu = \frac{1}{2}mv^2 + \phi_0 \qquad \text{(Eq 2.10)}$$

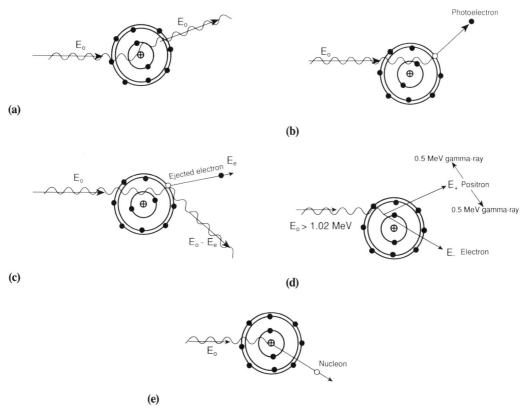

Fig. 2.9 Different mechanisms can reduce the intensity of the incident x-ray beam. These are (a) Thomson (coherent, classical) scattering of an x-ray photon without loss of energy. (b) Photoelectric effect with the absorption of an x-ray photon and the ejection of an electron from the atom, with the subsequent emission of an x-ray photon characteristic of the atom, or by the emission of an Auger electron. (c) Compton (incoherent) scattering with the ejection of an electron and the scattering of an x-ray photon of lower energy. The change in wavelength of the scattered electron is given by $\Delta\lambda = 0.0243(1 - \cos\alpha)$Å, where α is the scattering angle of the scattered photon. (d) Pair-production of an electron and positron from an incident x-ray photon of energy greater than 1.02 MeV. The positron decays with the emission of two x-ray photons as illustrated. This is the basis of positron emission tomography (PET). Different mechanisms can reduce the intensity of the incident x-ray beam. (e) Photodisintegration with the ejection of a particle by the nucleus.

is based on conservation of energy where $h\nu$ is the energy of the x-ray photon, $\frac{1}{2}mv^2$ is the kinetic energy given to the electron, and ϕ_0 is the energy required to remove the electron, that is the work function. Subsequently, the ionized atom returns to the neutral state with the emission of an x-ray characteristic of the atom, or of an Auger electron (Ref 114). Photoelectron absorption is the dominant process for x-ray absorption up to energies of about 0.5 MeV, and for atoms of high atomic number. The photoelectric process is illustrated in Fig. 2.9(b).

Compton scattering also known as incoherent scattering occurs when the incident x-ray photon ejects an electron from an atom and an x-ray photon of lower energy is scattered by the atom. Relativistic energy and momentum are conserved in this process and the scattered x-ray photon has less energy and therefore greater wavelength than the incident photon. This process is illustrated in Fig. 2.9(c).

Pair production can occur when the x-ray photon energy is greater than 1.02 MeV, when an electron and positron are created with the annihilation of the x-ray photon. Positrons have very short lives, disappearing (positron annihilation) with the formation of two photons of 0.5 MeV energy; see the section " Radiographic Techniques—Tomography" in this chapter. Pair production is illustrated in Fig. 2.9(d). The energy balance equation for pair production is:

$$h\nu = m_0^+c^2 + m_0^-c^2 + \text{K.E.} \qquad \text{(Eq 2.11)}$$

where $h\nu$ is the energy of the x-ray photon, which provides the rest masses of the positron $m_0^+c^2$ and electron $m_0^-c^2$, which share kinetic energy K.E. Because the rest mass of each of the electron and positron is 0.51 MeV, the energy equation for pair production is

$$h\nu = 1.02\,\text{MeV} + \text{K.E.} \qquad \text{(Eq 2.12)}$$

Photodisintegration is the process by which the x-ray photon is captured by the nucleus of the atom with the ejection of a particle from the nucleus when all the energy of the x-ray is given to the nucleus. This process may be neglected for the energies of x-rays used in radiography and is illustrated in Fig. 2.9(e); see Ref 69 and 117.

Usually, in industrial radiography, the relative contributions of these factors to the mass attenuation of the beam depend on the photon energy and absorber composition. Below 1.022

MeV, ζ and η are zero; ph is the dominant factor for high atomic number atoms irradiated by low energy photons; Compton scattering is important for low atomic number specimens (Ref 117). Equation 2.9 can be rewritten as:

$$\mu_m = \mu/\rho = \mu_S + \mu_\tau \qquad \text{(Eq 2.13)}$$

where μ_m is the mass absorption coefficient, ρ is the density, μ_S is the scattering component (Compton and Thomson), μ_τ is the photoelectric and other effects where the energy of the x-ray photon is converted into other energy forms. The scattering terms dominate at low atomic number and photon energies up to ~5 MeV, and true absorption terms dominate at high atomic number and low photon energy. The regions of dominant x-ray absorption processes are illustrated in Fig. 2.10(a) for aluminum and in Fig. 2.10(b) for tungsten. The relative importance of the different components contributing to the linear absorption of aluminum, for example, are illustrated in Fig. 2.10(a), showing how these vary with x-ray photon energy. For most purposes, the photoelectric effect and Compton scattering can be considered to be the two dominant absorption processes below about 5 MeV.

Because radiography uses a broad diverging beam, the scattered component can remain within the beam and affect the image. Special attention must be taken when Compton scattering is a dominant absorption process, particularly for low atomic numbers and high voltages. There are several methods to reduce the effects of scattering and these will be discussed in the section "Artifacts and Parasitic Effects—Scattered X-Rays" in this chapter.

Tables of mass absorption coefficients, photon-cross sections, and energy absorption coefficients are listed in many texts, such as Ref 55, 103, and 118. All of these data are for monochromatic x-rays, because absorption is highly dependent on wavelength. A selection of some mass absorption coefficients is given in Tables 2.4(a), (b), and (c) (see the section "Calculations for Trial Exposures" in Part III "Practical Guidelines to Improved Radiography" in this chapter).

For radiographic purposes, it is necessary to have data on the contributions due to Compton scattering, the photoelectric effect, and pair formation. In using these coefficients, careful attention must be made to the method of measurement, whether monochromatic conditions were applied, or whether a filtered broad spectrum x-ray beam was employed.

X-Ray Absorption Coefficients. There are several different ways of describing the attenuation of x-rays by matter, as x-ray absorption coefficients, photon cross-sections, and half-value layers. These units are defined carefully and their interrelationships given (Ref 114, 117-119). The half-value layer is usually determined experimentally.

Linear Absorption Coefficient μ cm^{-1}.

$$I_z = I_0 \exp - [\mu z] \qquad \text{[Eq 2.8]}$$

The linear absorption coefficient is the fractional decrease in intensity on traversing unit length of path (e.g., 1 cm). The linear absorption coefficient cm^{-1} is used in all practical situations.

Mass Absorption Coefficient μ_m cm^2/g.

$$\mu_m = \mu/\rho \text{ cm}^2/\text{g} = -\frac{dI}{I_z} \cdot \frac{1}{\rho dz} \qquad \text{(Eq 2.14)}$$

This is the fractional decrease in intensity in traversing unit mass per unit area. Data tables are always presented in terms of the mass absorption coefficient, because these are independent of the density of the solid.

Gram-Atomic Absorption Coefficient μ_g cm^2/gram-atom.

$$\mu_g = \mu_m \times A = \mu/\rho \times A_w \qquad \text{(Eq 2.15)}$$

where A_w is the atomic weight in grams. This is the fractional decrease in intensity when the x-ray beam traverses 1 gram-atom. This coefficient is frequently used in deriving the absorption coefficient of a compound.

Atomic Absorption Coefficient μ_a cm^2/atom.

$$\mu_a = \frac{\mu_g}{N_0} = \frac{\mu}{\rho} \times \frac{A_w}{N_0} \qquad \text{(Eq 2.16)}$$

where N_0 is Avogadro's Number. This is the fractional decrease in intensity when the beam trav-

Table 2.4(a)   X-ray absorption coefficients for aluminum and iron

Energy	Wavelength, Å	Mass absorption (μ_m), cm^2/g	Cross section in barns, cm$^2 \times 10^{-24}$				
			Total	Coherent	Incoherent	Photoelectric	Pair production
Aluminum							
20 keV	0.620	3.39	151.2	9.145	6.169	135.9	...
40	0.310	0.562	24.96	3.067	6.721	15.17	...
60	0.207	0.278	12.35	1.517	6.663	4.171	...
80	0.155	0.202	9.028	0.897	6.458	1.673	...
100	0.124	0.171	7.644	0.593	6.225	0.827	...
200	0.062	0.123	5.487	0.157	5.234	0.096	...
400	0.031	0.093	4.157	0.040	4.105	0.012	...
600	0.021	0.078	3.495	0.018	3.473	0.004	...
800	0.016	0.068	3.065	0.010	3.053	0.002	...
1 MeV	0.012	0.061	2.753	0.006	2.746	0.001	...
2	0.006	0.043	1.938	0.002	1.905	0	0.031
4	0.003	0.031	1.393	0.001	1.250	0	0.142
Iron							
20 keV	0.620	25.72	2.438	47.57	10.85	2.380	...
40	0.310	3.555	329.7	16.54	12.48	300.7	...
60	0.207	1.176	109.0	8.44	12.45	87.95	...
80	0.155	0.584	54.12	5.15	12.41	36.57	...
100	0.124	0.367	34.05	3.46	12.07	18.52	...
200	0.062	0.147	13.58	0.948	10.34	2.295	...
400	0.031	0.094	8.729	0.247	8.172	0.311	...
600	0.021	0.077	7.143	0.111	6.929	0.103	...
800	0.016	0.067	6.208	0.063	6.096	0.049	...
1 MeV	0.012	0.060	5.554	0.040	5.485	0.028	...
2	0.006	0.043	3.954	0.010	3.809	0.008	0.127
4	0.003	0.033	3.071	0.003	2.500	0.003	0.565

Ref 103

Table 2.4(b) Mass absorption coefficients for x-ray energies from 20 keV to 20 MeV

Energy	Mass absorption coefficients, cm^2/g									
	Hydrogen	Lithium	Carbon	Nitrogen	Oxygen	Magnesium	Aluminum	Silicon	Phosphorus	Sulfur
20 keV	0.369	0.183	0.429	0.596	0.826	2.72	3.39	4.39	5.31	6.66
40	0.346	0.155	0.206	0.229	0.257	0.485	0.562	0.696	0.797	0.968
60	0.326	0.144	0.176	0.182	0.191	0.258	0.278	0.322	0.350	0.404
80	0.309	0.136	0.161	0.164	0.168	0.196	0.202	0.224	0.234	0.259
100	0.294	0.129	0.152	0.153	0.156	0.169	0.171	0.184	0.187	0.202
200	0.243	0.106	0.123	0.123	0.124	0.125	0.123	0.128	0.125	0.130
400	0.189	0.0825	0.0957	0.0957	0.0957	0.0949	0.0927	0.0962	0.0936	0.0966
600	0.160	0.0697	0.0807	0.0805	0.0808	0.0797	0.0780	0.0808	0.0784	0.0810
800	0.140	0.0550	0.0709	0.0708	0.0708	0.0701	0.0684	0.0707	0.0688	0.0709
1 MeV	0.126	0.0383	0.0637	0.0636	0.0637	0.0628	0.0613	0.0635	0.0617	0.0638
2	0.0875	0.0257	0.0445	0.0445	0.0446	0.0442	0.0432	0.0446	0.0436	0.0449
4	0.0581	...	0.0305	0.0307	0.0310	0.0316	0.0311	0.0324	0.0317	0.0329
6	0.0450	...	0.0247	0.0251	0.0255	0.0268	0.0266	0.0279	0.0275	0.0287
8	0.0375	...	0.0216	0.0221	0.0226	0.0244	0.0244	0.0257	0.0255	0.0268
10	0.0325	...	0.0196	0.0202	0.0209	0.0231	0.0231	0.0246	0.0245	0.0258
20	0.0215	...	0.0158	0.0167	0.0177	0.0212	0.0216	0.0233	0.0235	0.0252

Energy	Mass absorption coefficients, cm^2/g									
	Titanium	Vanadium	Chromium	Iron	Nickel	Copper	Zinc	Zirconium	Molybdenum	Silver
20 keV	15.72	17.06	19.83	25.72	32.33	33.86	38.11	71.58	81.05	18.05
40	2.174	2.369	2.767	3.555	4.487	4.779	5.347	11.46	13.09	17.19
60	0.758	0.821	0.947	1.176	1.460	1.562	1.726	3.721	4.331	5.764
80	0.405	0.432	0.489	0.584	0.707	0.752	0.820	1.692	1.993	2.648
100	0.274	0.288	0.319	0.367	0.433	0.455	0.490	0.944	1.116	1.469
200	0.132	0.133	0.139	0.147	0.158	0.157	0.162	0.220	0.249	0.300
400	0.0910	0.0899	0.0924	0.0941	0.0977	0.0944	0.0956	0.1016	0.1059	0.1136
600	0.0753	0.0742	0.0760	0.0770	0.0794	0.0763	0.0770	0.0774	0.0787	0.0814
800	0.0657	0.0646	0.0662	0.0669	0.0688	0.0660	0.0665	0.0658	0.0613	0.0674
1 MeV	0.0589	0.0579	0.0593	0.0599	0.0615	0.0589	0.0594	0.0580	0.0582	0.0589
2	0.0418	0.0411	0.0421	0.0426	0.0438	0.0420	0.0424	0.0414	0.0416	0.0420
4	0.0317	0.0314	0.0324	0.0331	0.0344	0.0322	0.0335	0.0345	0.0350	0.0360
6	0.0287	0.0274	0.0295	0.0304	0.0320	0.0311	0.0316	0.0338	0.0344	0.0361
8	0.0274	0.0271	0.0285	0.0298	0.0315	0.0305	0.0312	0.0343	0.0351	0.0372
10	0.0269	0.0276	0.0281	0.0296	0.0315	0.0306	0.0313	0.0352	0.0360	0.0385
20	0.0282	0.0304	0.0301	0.0319	0.0346	0.0339	0.0348	0.0410	0.0424	0.0459

Energy	Mass absorption coefficients, cm^2/g						Glass			
	Tin	Tungsten	Platinum	Gold	Lead	Concrete	Silica	Pyrex	Polyethylene	Air (normal temperature pressure)
20 keV	21.39	54.4	76.14	78.23	85.91	3.59	2.49	2.25	0.42	0.752
40	19.50	8.09	12.30	12.77	14.00	0.605	0.463	0.431	0.226	0.248
60	6.575	2.681	4.219	4.414	4.863	0.295	0.252	0.242	0.198	0.188
80	3.019	7.70	8.767	2.106	2.331	0.213	0.194	0.190	0.183	0.167
100	1.667	4.36	4.891	5.121	5.461	0.179	0.169	0.166	0.172	0.154
200	0.325	0.747	0.851	0.914	0.991	0.127	0.126	0.125	0.140	0.127
400	0.1156	0.1849	0.2062	0.2133	0.2294	0.0963	0.0959	0.0954	0.109	0.0954
600	0.0809	0.1052	0.1120	0.1150	0.1204	0.0810	0.0808	0.0804	0.0921	0.0805
800	0.0663	0.070	0.0827	0.0840	0.0867	0.0709	0.0707	0.0704	0.0809	0.0707
1 MeV	0.0577	0.0655	0.0676	0.0690	0.0702	0.0636	0.0636	0.0633	0.0727	0.0636
2	0.0410	0.0431	0.0451	0.0455	0.0462	0.0448	0.0447	0.0444	0.0507	0.0445
4	0.0356	0.0406	0.0417	0.0422	0.0427	0.0319	0.0317	0.0314	0.0345	0.0308
6	0.0359	0.0426	0.0438	0.0440	0.0445	0.0270	0.0266	0.0263	0.0277	0.0252
8	0.0372	0.0449	0.0460	0.0464	0.0474	0.0245	0.0241	0.0237	0.0239	0.0223
10	0.0387	0.0478	0.0491	0.0495	0.0503	0.0231	0.0226	0.0222	0.0215	0.0204
20	0.0464	0.0590	0.0611	0.0617	0.0625	0.0210	0.0203	0.0198	0.0166	0.0170

Ref 103 and 189

Table 2.4(c)   Mass absorption coefficients for x-rays of wavelengths λ = 0.56, 0.71, and 1.54 Å

Absorber	Mass absorption coefficient (μ_m), cm^2/g			Absorber	Mass absorption coefficient (μ_m), cm^2/g		
	Ag K_α 0.56 Å	Mo K_α λ = 0.71 Å	Cu K_α λ = 1.54 Å		Ag K_α 0.56 Å	Mo K_α λ = 0.71 Å	Cu K_α λ = 1.54 Å
H	0.371	0.3727	0.435	Zr	58.5	16.10	143
Li	0.187	0.1968	0.716	Nb	61.7	16.96	153
Be	0.229	0.2451	1.50	Mo	64.8	18.44	162
B	0.279	0.3451	2.39	Pd	12.3	24.42	206
C	0.400	0.5348	5.50	Ag	13.1	26.38	218
N	0.544	0.7898	7.52	Cd	14.0	27.73	231
O	0.740	1.147	12.7	In	14.9	29.13	243
F	0.976	1.584	16.4	Sn	15.9	31.18	256
Na	1.67	2.939	30.1	Sb	16.9	33.01	270
Mg	2.12	3.979	38.6	Te	17.9	33.92	282
Al	2.65	5.043	48.6	I	19.0	36.33	294
Si	3.28	6.533	60.6	Cs	21.3	40.44	318
P	4.01	7.870	74.1	Ba	22.5	42.37	358.9
S	4.84	9.625	89.1	La	23.7	45.34	341
Cl	5.77	11.64	106	Ce	25.0	48.56	352
K	8.00	16.20	143	Pr	26.3	50.78	363
Ca	9.28	19.00	162	Nd	27.7	53.28	374
Sc	10.7	21.04	184	Sm	30.6	57.96	397
Ti	12.3	23.25	208	Gd	33.8	62.79	437
V	14.0	25.24	233	Tb	35.5	66.77	273
Cr	15.8	29.25	260	Dy	37.2	68.89	286
Mn	17.7	31.86	285	Er	40.8	75.61	134
Fe	19.7	37.74	308	Yb	44.8	80.23	146
Co	21.8	41.02	313	Hf	48.8	86.33	159
Ni	24.1	47.24	45.7	Ta	50.9	89.51	166
Cu	26.4	49.34	52.9	W	53.0	95.76	172
Zn	28.8	55.46	60.3	Re	55.2	98.74	178
Ga	31.4	56.90	67.9	Os	57.3	100.2	186
Ge	34.1	60.47	75.6	Ir	59.4	103.4	193
As	36.9	65.97	83.4	Pt	61.4	108.6	200
Se	39.8	68.82	91.4	Au	63.1	111.3	208
Rb	48.9	83	117	Hg	64.7	114.7	216
Sr	52.1	88.04	125	Pb	67.7	122.8	232
Y	55.3	97.56	134	Bi	69.1	125.9	240

erses one atom. This is also known as the cross section (CS).

Photon cross section or Cross Section (CS) cm^2/atom. This is the same unit as the atomic absorption coefficient. The unit of cross section is the barn (10^{-24} cm^2). Tables of data are often presented as photon cross sections in the case of high energy x-rays; see Ref 114 and 118.

Electronic Absorption Coefficient μ_E cm^2/electron.

$$\mu_E = \frac{\mu}{\rho N_E} \qquad \text{(Eq 2.17)}$$

where N_E is the number of electrons in one gram.

Half-Value Thickness (HVT) or Half-Value Layer (HVL). This is the thickness of specimen to decrease the intensity of the x-ray by one-half.

Using Eq 2.8:

$$\text{HVL} = \frac{0.693}{\mu} \qquad \text{(Eq 2.81)}$$

Values of HVT (or HVL) are determined experimentally for the filtered spectra used in radiography.

Mean-Free Path θ'cm. The mean free path of an x-ray photon in an absorber is the mean distance traveled by one x-ray photon before undergoing an interaction.

$$\theta = \frac{1}{\mu} = \frac{\text{HVL}}{0.693} \qquad \text{(Eq 2.19)}$$

Other units of absorption in use in radiography and radiology are defined in Ref 55 and 57. All of the absorption coefficients above are for each element at a particular wavelength of radiation.

Absorption Coefficient of a Compound or Mixture. The absorption coefficient of a compound, alloy, mixture, or amorphous material is calculated from the weighted average of the appropriate absorption coefficients of the atoms present.

The gram-molecular absorption coefficient μ_g of compound $X_xY_yZ_z$ is given at a particular wavelength by

$$\mu_g = x\mu_{gx} + y\mu_{gy} + z\mu_{gz} \qquad \text{(Eq 2.20)}$$

where μ_{gx}, μ_{gy}, and μ_{gz} are the gram-atomic absorption coefficient of element, X, Y, Z, respectively (Ref 119). Alternatively, the mass absorption coefficient of a compound $X_x\,Y_y\,Z_z$ is given by

$$\mu_m = w_x\mu_{mx} + w_y\mu_{my} + w_z\mu_{mz} \qquad \text{(Eq 2.21)}$$

where w_x, w_y, and w_z are the weight fractions and μ_{mx}, μ_{my}, and μ_{mz} are the mass absorption coefficients of elements X, Y, Z respectively.

The chemical state does not affect the absorption, so that a mixture, alloy, or chemical compound of the same constitution all have the same absorption. Some examples of calculations of linear absorption coefficients of compounds are given in Part III, Examples 1 to 3.

Absorption Edges. The mass absorption coefficients μ_m vary widely with the energy of the irradiation and with the atomic number of the absorber. Mass absorption coefficients versus photon energy are given in Fig. 2.11 for the elements beryllium, copper, and tungsten. The ordinate scale is logarithmic, and the absorption coefficient is applied in an exponential function (see Eq 2.8), so that the absorption changes very rapidly with photon energy. There are also sharp abrupt changes in absorption at absorption edges (Table 2.5). Absorption edges correspond to the critical energies required to ionize an atom by the ejection of an electron from an energy level.

The presence of absorption edges can give rise to enhanced contrast between different elements (this is discussed under "Contrast" in the section "X-Ray Image Quality—Contrast, Composition, Thickness, Inclusions" in this chapter) or artifact lines in the radiograph (see the section "Artifacts and Parasitic Effects—Absorption Edge Lines" in this chapter).

Filtration of Spectra. The spectrum of x-ray wavelengths emitted by a target is shown in Fig. 2.3(a) and ranges from the short wavelength λ_{min} to very long wavelengths depending on the applied voltage. Moreover, the spectrum of x-rays emitted from the target of the x-ray tube is modified by absorption in the window of the x-ray tube, the air, the specimen, the black paper (or

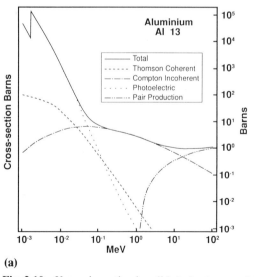

(a)

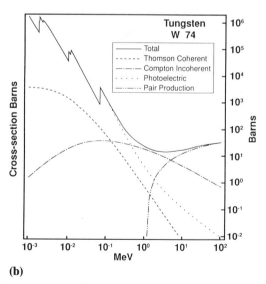

(b)

Fig. 2.10 X-ray absorption in solids is due to several different processes: Thomson (coherent) scattering, Compton (incoherent) scattering, photoelectric effect, pair production. These several processes vary in importance with x-ray energy and atomic number. Regions of dominant x-ray absorption processes are given for (a) aluminum (atomic number 13) and (b) tungsten (atomic number 74).

cassette) covering the film, any screens or foils, as well as being differentially absorbed in the intensifying screen and x-ray film. As a consequence, the x-ray spectrum resulting in the x-ray image is considerably modified by the experimental conditions in each case. The spectrum represented in Fig. 2.3(a) is typical of x-ray beams used in radiographs, and they may not be considered monochromatic. All of the absorption coefficients described in this section and listed in tables such as Table 2.4 are all heavily wavelength dependent. Experimental radiographic studies employ experimental "half-value layers." For calculation purposes, use is made of the wavelength of maximum intensity of the white radiation.

Broad Beam Absorption. There is no simple analytical treatment for absorption in a broad diverging, wide spectrum x-ray beam used in radiography (Ref 119). The absorption characteristics of broad beams need to be determined experimentally in each case. In Fig. 2.12, contours of x-ray intensity are given for a diverging broad x-ray beam. The x-ray intensity distribution depends on the focal spot size, the beam width, degree of divergence, range of x-ray ener-

gies, and the x-ray spectrum. The intensity in a broader beam decreases more slowly than a narrower beam. The normal distance law when radiation is expanding as concentric surfaces of a sphere is given by radiation intensities I decreasing with distance d by

$$\frac{I_1}{I_2} = \frac{d_2^2}{d_1^2} \qquad \text{(Eq 2.22)}$$

This is not the case where secondary scattered x-rays add to the primary beam and are not separated from the incident radiation.

The absorption data to be used in radiography need to be determined experimentally and are usually quoted as half-value layers, HVL (or half-value thickness, HVT), which are thicknesses to reduce the intensity by 50% under those particular experimental conditions. Tenth-value layers, TVL, are also used, where the intensity is decreased to 10% of its original value. The specimen can be placed near the x-ray source (narrow beam) or the specimen can be placed next to the detector (broad beam); see Ref 103. The x-ray spectra will depend on the energy of excitation and on the degree of filtration. Some experimental values of HVL are given in Table 2.6 for several materials, along with some calculated values of HVL.

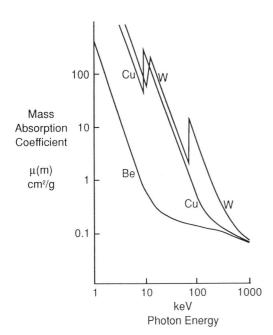

Fig. 2.11 Absorption edges. The dependence of the mass absorption coefficients on x-ray photon energy for several elements. The sharp drop in absorption at the absorption edge corresponds to the critical energy to eject an electron from an energy level of the atom.

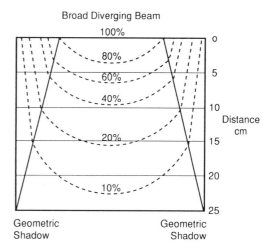

Fig. 2.12 X-ray intensity contours with distance for a diverging broad x-ray beam through a medium. The absorption characteristics of a diverging, broad x-ray beam need to be determined experimentally in each case, because the x-ray intensity contours do not decrease by an inverse square relationship with distance, but depend on the width and the degree of divergence of the x-ray beam.

Table 2.5 Selection of x-ray absorption edges (K$_{ab}$) and characteristic K-lines (Ref 55)

Atomic No.	Element	keV K$_{ab}$	keV K$_{\beta_2}$	keV K$_{\beta_1}$	keV K$_{\alpha_1}$	keV K$_\alpha$	keV K$_{\alpha_2}$
3	Lithium	0.055	...	...	...	0.052	...
4	Beryllium	0.116	...	...	...	0.110	...
5	Boron	0.192	...	...	...	0.185	...
6	Carbon	0.283	...	...	...	0.282	...
7	Nitrogen	0.399	...	...	...	0.392	...
8	Oxygen	0.531	...	...	...	0.523	...
11	Sodium	1.08	...	1.067	...	1.041	...
12	Magnesium	1.303	...	1.297	...	1.254	...
13	Aluminum	1.559	...	1.553	1.487	...	1.486
14	Silicon	1.838	...	1.832	1.740	...	1.739
15	Phosphorus	2.142	...	2.136	2.015	...	2.014
16	Sulfur	2.470	...	2.464	2.308	...	2.306
17	Chlorine	2.819	...	2.815	2.622	...	2.621
19	Potassium	3.607	...	3.589	3.313	...	3.310
20	Calcium	4.038	...	4.012	3.691	...	3.688
21	Scandium	4.496	...	4.460	4.090	...	4.085
22	Titanium	4.964	...	4.931	4.510	...	4.504
23	Vanadium	5.463	...	5.427	4.952	...	4.944
24	Chromium	5.988	...	5.946	5.414	...	5.405
25	Manganese	6.537	...	6.490	5.898	...	5.887
26	Iron	7.111	...	7.057	6.403	...	6.390
27	Cobalt	7.709	...	7.649	6.930	...	6.915
28	Nickel	8.331	8.328	8.264	7.477	...	7.460
29	Copper	8.980	8.976	8.904	8.047	...	8.027
30	Zinc	9.660	9.657	9.571	8.638	...	8.615
31	Gallium	10.368	10.365	10.263	9.251	...	9.234
32	Germanium	11.103	11.100	10.981	9.885	...	9.854
33	Arsenic	11.863	11.863	11.725	10.543	...	10.507
34	Selenium	12.652	12.651	12.495	11.221	...	11.181
35	Bromine	13.475	13.465	13.290	11.923	...	11.877
37	Rubidium	15.201	15.184	14.960	13.394	...	13.335
38	Strontium	16.106	16.083	15.834	14.164	...	14.097
39	Yttrium	17.037	17.011	16.736	14.957	...	14.882
40	Zirconium	17.998	17.969	17.666	15.774	...	15.690
41	Niobium	18.987	18.951	18.621	16.614	...	16.520
42	Molybdenum	20.002	19.964	19.067	17.478	...	17.373
45	Rhodium	23.224	23.169	22.721	20.214	...	20.072
46	Palladium	24.347	24.297	23.816	21.175	...	21.018
47	Silver	25.517	25.454	24.942	22.162	...	21.988
48	Cadmium	26.712	26.641	26.093	23.172	...	22.982
49	Indium	27.928	27.859	27.274	24.207	...	24.000
50	Tin	29.190	29.106	28.483	25.270	...	25.042
51	Antimony	30.486	30.387	29.723	26.357	...	26.109
52	Tellurium	31.809	31.698	30.933	27.471	...	27.200
53	Iodine	33.164	33.016	32.292	28.610	...	28.315
55	Cesium	35.959	35.819	34.984	30.970	...	30.623
56	Barium	37.410	37.255	36.376	32.191	...	31.815
58	Cerium	40.449	40.231	39.355	34.717	...	34.276
60	Neodymium	43.571	43.298	42.269	37.359	...	36.845
72	Hafnium	65.313	64.936	63.209	55.757	...	54.579
73	Tantalum	67.400	66.999	65.210	57.524	...	56.270
74	Tungsten	69.508	69.090	67.233	59.310	...	57.973
75	Rhenium	71.662	71.220	69.298	61.131	...	59.707
76	Osmium	73.860	73.393	71.404	62.991	...	61.477
77	Iridium	76.097	75.605	73.549	64.886	...	63.278
78	Platinum	78.379	77.866	75.736	66.820	...	65.111
79	Gold	80.713	80.165	77.968	68.794	...	66.980
80	Mercury	83.106	82.526	80.258	70.821	...	68.894
81	Thallium	85.517	84.904	82.558	72.860	...	70.820
82	Lead	88.001	87.343	84.922	74.957	...	72.794
83	Bismuth	90.521	89.833	87.335	77.097	...	74.805
90	Thorium	109.630	108.671	105.592	93.334	...	89.942
92	Uranium	115.591	114.549	111.289	98.428	...	94.648

Recording the X-Ray Image

X-rays and γ-rays can be detected using photographic film, ionization chambers, proportional, Geiger and scintillation counters, semiconductor detectors, thermoluminescence, and by calorimetry (Ref 103, 119). Photographic films are used extensively for radiography and this method will be described in detail. Intensifying screens are employed with the photographic film and consist of either thin metal-foils for use with high energy x-rays, or light-emitting inorganic powders used in radioscopy at the front end of an image intensifier.

The Photographic Method. X-radiation affects photographic emulsion in the same way as does visible light, so that x-ray films need to be protected from light at all times. During exposure to x-rays or daylight, the silver halide is reduced to silver (Ag) to form a latent image, which requires subsequent development and fixing. Extensive details are given in Ref 103, 120-122 on the procedures for film and paper photography; see also Ref 67.

The grain size of the silver halide particles in photographic emulsion affects the fine detail in a radiograph. However, fine-grain high-contrast film have lower speeds than fine-grain low-contrast film, and high-speed high-contrast film have larger grains. Inevitably, a compromise must be sought for each individual case, between speed, contrast, and sharpness. The film development process as described by the manufacturer for the films must be adhered to rigorously, otherwise film speed and contrast are adversely affected.

X-ray film consists of silver bromide (AgBr) particles in an organic gelatin matrix. The lightly colored gelatin is almost transparent to x-rays, and it is only the AgBr that will absorb x-rays. However, the quantity of the AgBr particles present is rather small, a few mg/cm^2. Moreover, only a small fraction (about 5%) of the energy absorbed by the AgBr results in the liberation of silver atoms to form a latent image. This fraction is even lower for higher energy x-rays. Photographic emulsion is most sensitive to photons with energy ~45 keV, which corresponds to the energy of the photoelectron interaction in AgBr. The efficiency can be increased by two orders of magnitude by the use of intensifying screens, but this adversely affects the sharpness of the image. The light emission from fluorescent salt screens needs to be matched to the film sensitivity with wavelength.

X-ray film consists of a base of thickness about 0.2 mm thick, and the photographic emulsion layer is about 0.02 mm thick. The base is very often coated on both sides of the base with photographic emulsion. In some cases of oblique incidence of x-rays on doubly coated film, the sharpness of the image may be affected.

Characteristics of X-Ray Film. This is a summary of the properties of x-ray films. Further details can be found in Ref 103, 117, 120, and 121; see Fig. 2.13.

Photographic density, D, is measured with a photographic densitometer in which the intensity of a beam of light is measured with a photodetector without and then with an x-ray film in the beam. If the incident light beam intensity is B_o, and the transmitted light intensity through the darkened

Table 2.6 Experimental (Exp.) and calculated (Calc.) half-value layers with x-ray tube applied voltages

		Half-value layer, cm								
		50 kV	100 kV	200 kV	300 kV	500 kV	1 MV	2 MV	4 MV	10 MV
Tungsten	Exp.	...	...	...	...	...	0.55	0.9	1.15	1.2
	Calc.	...	...	...	...	...	0.57	0.84	0.91	0.78
Lead	Exp.	0.005	0.02	0.05	0.4	0.36	0.75	1.25	1.6	1.7
	Calc.	0.011	...	0.0165	...	0.403	0.87	1.32	1.43	1.22
Steel	Exp.	...	...	...	...	...	1.6	2.0	2.5	3.2
	Calc.	...	...	...	...	...	1.47	2.07	2.67	2.97
Concrete	Exp.	0.51	1.8	2.5	3.1	3.6	4.5	6.2	8.6	11.5
Aluminum	Exp.	...	...	...	...	...	3.9	5.4	7.5	10.0
	Calc.	0.7	1.5	2.1	...	3.0	4.1	6.0	8.3	11.2

The calculated values of half-value layers are given by HVL = 0.693/μ. The experimental half-value layers will depend on the filtered x-ray spectra, and on the geometry of the experiment (Ref 103). See Examples 22 and 23.

developed x-ray film is B_T, then

$$D = \log_{10} \frac{B_o}{B_T} \qquad \text{(Eq 2.23)}$$

This unit D is dimensionless. Photographic density $D = 1$ corresponds to blackened film with transmission 10% of the incident light. $D = 2$ corresponds to 1% transmission of the incident light.

Exposure of film, ε, is the intensity of radiation times time,

$$\varepsilon = I \times t \qquad \text{(Eq 2.24)}$$

The radiographic contrast (C_S) between two neighboring areas of densities D_1 and D_2 is:

$$C_S = D_1 - D_2 = \log \frac{B_o}{B_{T_1}} - \log \frac{B_o}{B_{T_2}} = \log \frac{B_{T_2}}{B_{T_1}}$$

$$\text{(Eq 2.25)}$$

This unit is dimensionless.

X-Ray Film Latitude. A film with a large range of linear response to exposures is said to have a large latitude, so that latitude is inverse with contrast. A film with high latitude has a wide density range of useful exposures.

Film Gradient, G, Film Contrast, or γ. This is the slope of the approximately linear portion of the curve relating D and $\log \varepsilon$; see Fig. 2.13.

$$G = \frac{D_1 - D_2}{\log \varepsilon_1 - \log \varepsilon_2} = \frac{D_1 - D_2}{\log \varepsilon_1/\varepsilon_2} \qquad \text{(Eq 2.26)}$$

Because $\varepsilon_1 = I_1 t$ and $\varepsilon_2 = I_2 t$,

$$G = \frac{D_1 - D_2}{\log I_1/I_2} \qquad \text{(Eq 2.27)}$$

This unit is dimensionless. The radiographic contrast:

$$C_S = D_1 - D_2 = G \log \frac{I_1}{I_2} \qquad \text{(Eq 2.28a)}$$

and so

$$C_S = 0.4343 \, G \ln \frac{I_1}{I_2} \qquad \text{(Eq 2.28b)}$$

Exposures must be within the linear portion of the D versus $\log \varepsilon$. From Fig. 2.13, it can be seen that G is constant within the latitude of the film.

Film Speed, FS. The position along the abscissa of the D versus $\log \varepsilon$ linear portion determines the speed of the film. Film speeds can only be compared for films having the same value of G.

Fog and Base Density. Base density does not vary with time because it is formed during the fabrication of the film. Typical value of $D_{base} \sim$ 0.07. Fog increases with time, with the age of the

film, due to cosmic and other radiations. In fresh film $D_{fog} \sim 0.05$, and $D_{fog+base} \sim 0.12$. A value of $D_{fog+base} > 0.2$ is considered unsatisfactory.

The energy to produce $D = 1$ above fog in typical x-ray film is $\sim 10^{-5}$ J/m^2 ($\sim 2 \times 10^{13}$ photons/m^2) for $\lambda = 436$ nm.

Radiography. Densities range from about 0.25 to 2 (i.e., transmittance $\sim 50\%$ to 1%). X-ray films tend to have high G and low speed. A minimum contrast $C_S = 0.2$ is desired in radiography even though the eye is sensitive to contrast changes as low as 0.05 of photographic film blackening (Ref 116).

Intensifying Screens. Various types of screens are discussed below.

Metal foil intensity screens are used with high energy x-rays above about 100 keV. The metal screens emit electrons that affect the silver halide grains. When using lead screens, 88 keV are required to eject K electrons. The lead screen is placed in front of the screen, nearer the x-ray source, ~ 0.02 to 0.2 mm thick. The electrons

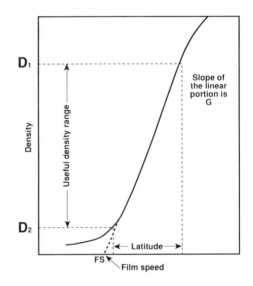

Fig. 2.13 Characteristic density curve of an x-ray film. The density D of an x-ray film is plotted against log of the exposure ε, which equals It, where I is intensity of x-rays and t is time. The radiographic contrast C_S of the film is given by $C_S = G \log I_1/I_2$ where $\varepsilon = I_1 t$ and $\varepsilon_2 = I_2 t$, and G is the gradient (or γ) and is given by the slope of the linear portion of the curve. Because the slope is not truly linear for x-rays, it is necessary to specify the density D for a value of G. The film speed is given by the position FS.

ejected from the lead tend to travel in the same direction as the primary radiation. The lead screen must be in intimate contact with the x-ray film. A second lead screen is added at the back of the film to reduce back-scattered x-rays affecting the film. Lead metal foils are used with x-rays of 100 keV to 2 MeV; other metals, copper, tungsten, and tantalum, are used with higher energy x-rays. Some metal intensifying screens are listed in Table 2.7.

Fluorescent Salt Intensifying Screens. Photographic emulsion is much less sensitive to x-ray photons than to visible light photons. Conse-quently, fluorescent salt screens are used because some inorganic powders emit visible light photons when irradiated by x-rays. The fluorescent efficiency depends on photon energy. The fluorescent salt intensifying screen consists of phosphor particles in a binding matrix, mounted on a white reflecting base. Roentgen discovered x-rays by noting the fluorescence of barium platinocyanide crystals. Several highly efficient phosphors have been developed, such as gadolinium oxygen sulfide (Gd_2O_2S) activated with terbium, lanthanum oxygen bromide (LaOBr) activated with terbium (Tb)

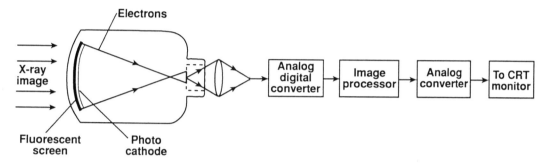

Fig. 2.14 In radioscopy (formerly known as real-time radiography), the x-ray photons are converted by one of several methods and displayed as an analog image on a television monitor.

Table 2.7 Metal foil intensifying screens

Radiation	Screen material	Front screen thickness, mm	Back screen thickness, mm	Resolving power, mm
X-rays				
<120 kV	Lead	None	0.1 (minimum)	0.1
120-250 kV	Lead	0.025-0.05	0.1 (minimum)	0.2
250-400 kV	Lead	0.05-0.15	0.1 (minimum)	0.3
1 MV	Lead	1.5-2.0	1.0 (minimum)	3
5-10 MV	Copper	1.5-2.0	1.5-1.0 (minimum)	3
15-31 MV	Tantalum or tungsten	1.0-1.5	None	3
γ-rays				
Ir[192]	Lead	0.05-0.15	0.15 (minimum)	1
Cs[137]	Lead	0.50-0.15	0.15 (minimum)	1
Co[60]	Iron or copper	0.5-2.0	0.25-1.0 (minimum)	3

Source: Ref 70

Table 2.8 Fluorescent self intensifying screens

Phosphor	Fluorescent wavelength, Å	Mass phosphor per unit area, g/cm^2	Phosphor layer thickness, μm	Phosphor particle size, μm	Resolving power (line pairs/mm)
Calcium tungstate	4300	0.02-0.11	100-300	4-8	9-14
Barium lead sulfate	3700	0.08	200	5	6.5
Zinc cadmium sulfide	5350	0.09-0.15	250-380	3-60	~3

or thulium (Tm). A frequently used phosphor is calcium tungstate, which offers a constant sensitivity to x-rays of energy 60 to 120 kV. Barium lead sulfate is useful for high energy x-rays. In Table 2.8 the characteristics of some frequently used fluorescent salt screens are listed (see Ref 103, 117).

The screens must be in intimate contact with the x-ray film, and it is usual to sandwich the x-ray film between two intensifying screens. The efficiency of the intensifying screen increases with the crystal size of the phosphor, but this causes the sharpness of the image to suffer, so that different intensifying screens are available for different requirements of speed and image sharpness.

Fluorescent salt intensifying screens are rarely used in industrial radiography, but are used extensively in the medical profession. Fluorescent salt screens are used in radioscopy.

Photostimulable Phosphor Plate. A recently developed intensifying screen consists of a phosphor plate, barium fluorohalide activated by europium. After exposure, a latent image is excited by a laser beam. A photomultiplier converts the light output to an analog image, which is converted for digital process and and analysis (Ref 165).

Radioscopy with Image Intensifiers. Direct and immediate viewing of an x-ray image is possible using a detector that converts to an analog image displayed on a television monitor. The image on a fluorescent salt screen is in general of low brightness, so that methods of image intensification are needed, and these are frequently combined with image scanning techniques to provide a further stage in enhancement of the image, which can be observed away from x-ray source and radiographic site.

A radioscopic (fluoroscopic) system of remote instantaneous viewing via a television system is shown schematically in Fig. 2.14 (Ref 103, 123). X-rays enter the tube and strike a thin layer of a fluorescent material on a photocathode, which produces photoelectrons in the cathode. These are accelerated to about 30 kV and focused onto a smaller fluorescent screen area, so that the gain in brightness can be two orders of magnitude. Solid state x-ray image intensifiers with remote viewing are described in Ref 69, 103, 117, and 124 to 128. Radioscopy was formerly known as "real-time radiography."

The unsharpness of a radioscopic-image intensifier-TV system is generally ~0.5 mm (2 line pairs per mm) which is an order of magnitude worse than photographic film. The contrast sensitivity, using image quality indicator (IQI) test pieces, are comparable to x-ray film.

X-Ray Image Quality

Unsharpness, Penumbra. The image qualities of definition (or unsharpness), contrast, and sensitivity (or resolution) are illustrated in Fig. 2.15. The "unsharpness" or penumbra of a radiographic image is defined as the inability to reproduce faithfully the boundary of a given object. The same term unsharpness is also used to describe the minimum distance that can be resolved between two objects (see Fig. 2.16) which is a measure of the sensitivity or the resolution of the image. The unsharpness is the inverse of resolving power, resolving power (in line pairs per mm) = inverse of unsharpness at half-width (in mm).

The focal spot source of x-rays in an x-ray tube is of finite size, which will give rise to penumbra; see Fig. 2.16. This is known as the geometric unsharpness U_g, and is given by simple geometry to be

$$U_g = F \times \frac{l}{L_o} \tag{Eq 2.29}$$

where F is the width (or effective width) of the source, l is the distance of object to screen, and L_o is the distance of the x-ray source to the object; see Ref 69, 103, and 119. Contact radiography is the

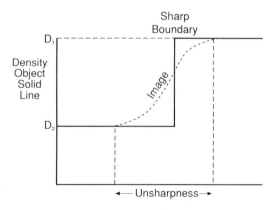

Fig. 2.15 Contrast and unsharpness. A sharp boundary in the object is observed in the image as a smoother boundary, where the radiographic image is unable to follow the abrupt change in density. The unsharpness is as shown. The radiographic contrast C_s is given by $D_1 - D_2$.

case when l is comparable in size to the object thickness Δ for a thin object in contact with the x-ray film cassette. The unsharpness U_g is then as shown in Fig. 2.17(a):

$$U_g = p = \frac{F}{L_o}(\Delta_o + \Delta_F) \qquad \text{(Eq 2.30)}$$

where Δ_F is the film thickness. When l is large, this is known as projection radiography when it is possible to produce an image larger than the object; see Fig. 2.17(b). The image is enlarged with magnification M_g:

$$M_g = \frac{L+l}{l} \qquad \text{(Eq 2.31)}$$

However, the unsharpness is also increased, and at a greater rate than the magnification so that the resolution of the final image is only improved when using a very small x-ray source. When x-ray microsources are employed, this technique is known as x-ray microscopy (Ref 13, 115); see the section "Additional Radiographic Techniques" in this chapter.

There are several other factors adding to the unsharpness. If the object moves, the un-

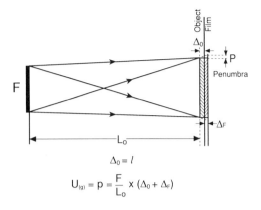

$$\Delta_o = l$$

$$U_{(g)} = p = \frac{F}{L_o} \times (\Delta_o + \Delta_F)$$

(a)

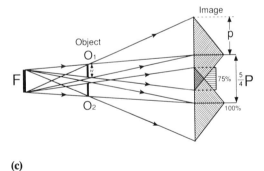

(b)

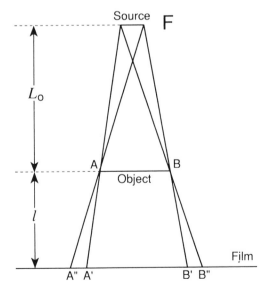

Fig. 2.16 Penumbra, unsharpness and x-ray source size. The penumbra, p, or geometric unsharpness U_g, arises from the finite size of the source, and is given by A'A″ (see Eq 2.30). The distance B'B″ can have a different value from A'A″ when the object and film are not symmetrically arranged relative to the diverging x-ray beam. See Examples 13 and 14.

(c)

Fig. 2.17 Contact and projection radiography. The condition for contact radiography is shown in (a) where the object of width Δ_o is in contact with the film of thickness Δ_F. The unsharpness (See Eq 2.30) is shown. In projection radiography, (b), the image is shown for object of size W, when using an x-ray source of finite size F. The intensity distribution of the umbra S and penumbra p is shown; $p \equiv U_g$, the geometric unsharpness. When $S = 0$, the object size W' and the p' have limiting values, which correspond to the limit of detection. In (c), the resolution limit of the two equal objects is determined by their penumbra formation. The case is shown when the intensity between the objects decreases by 25% of the maximum, when the gap between the objects is a quarter the diameter of the object size.

sharpness of the image is increased by U_m the motion unsharpness. Another factor adding to the unsharpness U_s is due to the use of an intensifying screen, and this is related to the phosphor particle size and to the diffusion of light from the phosphor. There is also an unsharpness contribution due to the x-ray film. When x-rays are absorbed by the AgBr grains in photographic emulsion, secondary electrons are emitted from the AgBr and can affect the adjacent AgBr grains. This leads to an inherent film unsharpness U_f, which depends on the energy of the x-ray photons. Figure 2.18 is an experimental curve relating U_f to x-ray generator voltage (Ref 69). The best value of unsharpness for the most sensitive radiograph will correspond to the case when U_g is equal to U_f.

The total unsharpness U_T is composed of several terms:

$$U_T = U_g + U_m + U_S + U_f \qquad \text{(Eq 2.32)}$$

The assumption that the total unsharpness U_T is the algebraic sum of the various unsharpness factors is not strictly valid. A better approximation is the geometric average, when

$$U_T = \left[\sum_i U_i^2 \right]^{1/2} \qquad \text{(Eq 2.33)}$$

where U_i represents the separate unsharpness factors.

In typical radiographic studies without the use of intensifying screens, and for an object 1 cm thick in contact with the x-ray film cassette, the total unsharpness U_T can be of the order of 0.1 mm (Ref 117). This total unsharpness will increase with the thickness of the object.

Size Limit of Detection of Flaws. The finite size of the x-ray source gives rise to a limit of detection in the image due to the formation of a penumbra which corresponds to the geometric unsharpness. Schematic diagrams are given in Fig. 2.17(a), (b), and (c) of projection and contact radiography with an x-ray source size F. Diffraction effects and Fresnel fringes from edges are negligible when using x-rays of wavelength less than 0.1 nm, so that the x-ray paths are given by simple geometry. In Fig. 2.17(b), x-ray paths of a projection radiograph are given with the intensity distribution across the image plane, from an

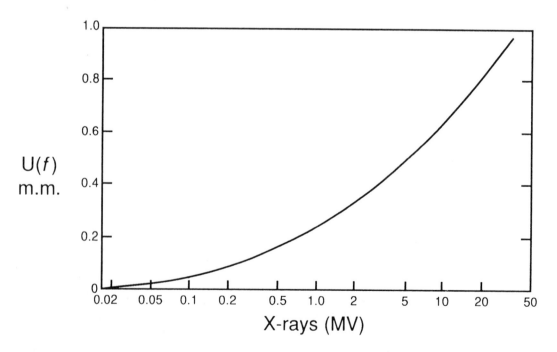

Fig. 2.18 The variation of film unsharpness U_f with x-ray generator voltage. Film unsharpness U_f is due to secondary electrons emitted with kinetic energy from the AgBr grains due to the x-ray photon energy. The most sensitive radiograph is obtained when the geometric unsharpness U_g is equal to U_f. See Examples 13, 14, 16, and 18.

object of size W, showing the penumbra p, and the umbra S.

Then, it can be seen that the magnification is:

$$M_g = \frac{l + L_0}{L_0} \qquad \text{[Eq 2.31]}$$

the penumbra is:

$$p = \frac{F \times l}{L_0} \qquad \text{(Eq 2.34)}$$

the umbra is:

$$S = \frac{(L_0 + l)\, W - Fl}{L_0} \qquad \text{(Eq 2.35)}$$

The reduced penumbra is:

$$p_0 = \frac{p}{M_g} = \frac{Fl}{L_0 + l} \qquad \text{(Eq 2.36)}$$

The limit of detection W' is set equal to the case when the umbra $S = 0$, though it will also depend on the level of contrast. Then,

$$W' = p_0 = \frac{F \times l}{L_0 + l} \qquad \text{(Eq 2.37)}$$

It should be noted that $W' = p_0$ varies with the distances l and L_0. In a thick object, p_0 is smaller for object planes closer to the screen. The object plane nearest to the x-ray source will have the highest magnification, but will be the least sharp. This determines the depth of focus.

In Fig. 2.17(c), the intensity across an image is shown for two objects, O_1 and O_2, of equal width W, separated by a small gap. The critical condition to be able to resolve the two objects is considered to be the case when the intensity falls by 25% between the images as shown. This occurs when the gap between O_1 and O_2 is $W/4$ (by simple geometry). This is the detection limit of the two objects of equal size.

In Fig. 2.17(a), the geometric optics is shown for contact radiography, using a source size F. The penumbra p depends on the ratio of the object thickness Δ_0, to the thickness of the photographic film Δ_F. Now:

$$\frac{p}{\Delta_0 + \Delta_F} = \frac{F}{L_0} \qquad \text{(Eq 2.38)}$$

$$\therefore p = (\Delta_0 + \Delta_F)\left[\frac{F}{L_0}\right] \qquad \text{(Eq 2.39)}$$

This shows that in contact radiography p depends on F/L_0, which is the angular size of the source.

As an example, the penumbra of an object of thickness 1 cm, using an x-ray source size of 3

mm, and source film distance 1 m is given by

$$p \approx \frac{1 \text{ cm} \times 3 \text{ mm}}{1 \text{ m}} \approx 0.03 \text{ mm}$$

and this is the geometric unsharpness U_g.

The effective size of the x-ray source, as illustrated in Fig. 2.19 is the appropriate area projected along the axis of the diverging x-ray beam. The size of the focal-spot can be determined by taking a pinhole photograph using a small hole, a few microns in diameter, in a lead foil (Ref 115).

The shape of the x-ray source will not necessarily be regular in general, and will give rise to different penumbra and unsharpness in different directions. This arises from the divergent nature of the x-ray beam, where the projected image has a shape change. This is illustrated in Fig. 2.20(a) and (b) where the projected image using parallel radiation (a) is compared with the projected image using divergent radiation (b). In Fig. 2.20(c), the image is shown to project with different distortions across the plane of the image.

Image Quality Indicators. Penetrameters or image quality indicators (IQI) are standard test pieces placed on top of or alongside the specimen as a check on the validity of the radiographic

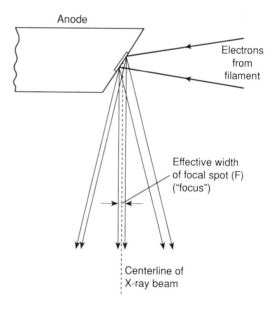

Fig. 2.19 The effective size F of the x-ray source is the area projected along the axis of the diverging x-ray beam, and it is this value of F that determines the geometric unsharpness U_g.

technique. The IQI is usually of the same material as the test object or radiographically similar and in the form of a step-wedge shape plate (plaque) with drilled holes of different sizes, or of a series of wires of different diameters. The wire type IQI is usually placed on the specimen. The plaque type IQI is placed at the periphery of the film; see

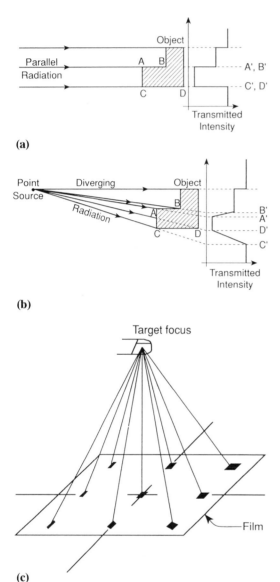

(a)

(b)

(c)

Fig. 2.20 Distortion of the image due to point source projection. A diverging beam of x-rays does not project a faithful image of the object. This can be seen by comparing (a) parallel illumination with (b) diverging beam. (c) Different regions of the image project with different distortions across the image plane.

Fig. 2.49a and b. The IQI is placed on top of the thickest part of the object to verify that the primary incident beam is able to penetrate this part of the object. Use of the IQI is the most common way to evaluate the sensitivity and quality of the radiographic technique. The different types of IQI penetrameters are the American Society of Testing Materials (ASTM) plaque penetrameter, the American Society of Mechanical Engineers (ASME) penetrameter, and the wire penetrameter known as the German DIN, and these are described in Ref 103. Penetrameters are discussed extensively in Ref 129, p 339 to 342.

Quality Level or Standard Sensitivity. One type of penetrameter is shown in Fig. 2.21 with related details in Table 2.9. The thickness T, 0.02 in., is 2% of the 1.0 identification (ID) number, which gives the thickness of the object being radiographed, in this case 1 in. The holes are of diameter 0.02, 0.04, and 0.08 in., as shown. The penetrameter is placed on the source side of the object, and the radiograph should reveal the outline of the penetrameter and of the holes. Standard Sensitivities, Quality Levels, and penetrameter sizes are listed in Table 2.9. In the case where the 2% penetrameter outline is seen in a radiograph as well as the $2T$ hole, the quality level is considered $2\text{-}2T$ (Ref 67).

Shim stock are thin pieces of material radiographically identical to the specimen. These are used when the area to be radiographed is thicker than neighboring areas; see Ref 67, p 6 to 15.

Contrast. Composition contrast, inclusion contrast, and thickness contrast sensitivity are discussed below.

Composition Contrast. If different chemical components are present, it is possible to choose an

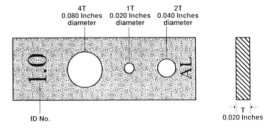

Fig. 2.21 Penetrameter for objects 1 in. thick. The thickness of the penetrameter is 2% of the object being radiographed. In the present case, the penetrameter thickness is 0.02 in., and the identification (ID) No. is 1.0. The material of the penetrameter needs to be identical or very similar in x-ray absorption properties to the test object. The penetrameter is placed on the source-side of the object (Ref 67; see Fig. 2.49a and b).

x-ray wavelength to accentuate their absorption differences and hence augment the contrast due to compositional changes (Ref 115).

Suppose two components of different compositions are present of mass absorption coefficients μ_1/ρ_1 and μ_2/ρ_2, and masses m_1 and m_2 per unit area. Then,

$$I_1 = I_0 \exp - \left[\frac{\mu_1}{\rho_1} \cdot m_1 \right] \qquad \text{(Eq 2.40)}$$

and

$$I_2 = I_0 \exp - \left[\frac{\mu_2}{\rho_2} \cdot m_2 \right] \qquad \text{(Eq 2.41)}$$

where I_0 is the incident intensity, and I_1 and I_2 are the transmitted intensities.

Then

$$\frac{I_1}{I_2} = \exp \left[\frac{\mu_2}{\rho_2} \cdot m_2 - \frac{\mu_1}{\rho_1} \cdot m_1 \right] \qquad \text{(Eq 2.42)}$$

which is the ratio of the transmitted intensities through the two components. Now the radiographic contrast

$$C_S = D_1 - D_2 = 0.4343 \, G \ln \frac{I_1}{I_2} \qquad \text{[Eq 2.28b]}$$

$$= 0.4343 \, G \left[\frac{\mu_2}{\rho_2} \cdot m_2 - \frac{\mu_2}{\rho_1} \cdot m_1 \right] \qquad \text{(Eq 2.43)}$$

and m/ρ can be replaced by the specimen thickness Δ, so that

$$C_S = 0.4343 \, G \, [\mu_1 - \mu_2] \Delta \qquad \text{(Eq 2.44)}$$

Thus, the contrast depends on the difference between the linear absorption coefficients, $\mu_1 - \mu_2$ of the two chemical components.

An example is illustrated in Fig. 2.22(a). The specimen containing copper and platinum is examined using zinc $K\alpha$ radiation which is at a wavelength where there is a large difference in

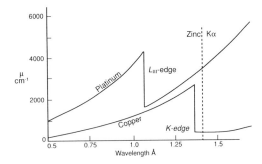

(a)

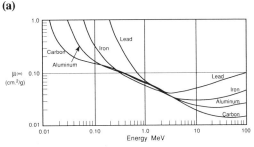

(b)

Fig. 2.22 Composition contrast can be achieved by the careful selection of x-ray wavelength. (a) The use of x-rays of $\lambda = 1.44$ Å is recommended when examining a specimen containing platinum and copper because the difference in the linear absorption coefficients is greatest just above the absorption edge wavelength of copper; see Eq 2.44. This x-ray wavelength can be generated using zinc as target in the x-ray tube. (b) Compositional contrast can be seen at a minimum at x-ray energies of the order of 1 MeV, where the mass absorption coefficients for all materials become very similar.

Table 2.9(a) Standard penetrameter sizes

T_m, in.	Identification (ID) No.	Thickness, T	Hole diameter, in.		
			1T	2T	4T
0.25	25	0.005	0.010	0.020	0.040
0.375	37	0.008	0.010	0.020	0.040
0.5	50	0.010	0.010	0.020	0.040
0.625	62	0.013	0.013	0.025	0.050
0.75	75	0.015	0.015	0.030	0.060
0.875	87	0.018	0.018	0.035	0.070
1.0	1.0	0.020	0.020	0.040	0.080
1.125	1.1	0.023	0.023	0.045	0.090
1.25	1.2	0.025	0.025	0.050	0.100
1.5	1.5	0.030	0.030	0.060	0.120

T_m = ideal thickness of object being radiographed. See also Fig. 2.21 and Ref 67.

Table 2.9(b) Sensitivity and quality levels

Sensitivity, %	Quality level	T/T_M, %	Perceptible hole diameter
0.7	1-1T	1	1T
1.0	1-2T	1	2T
1.4	2-1T	2	1T
2.0	2-2T	2	2T
2.8	2-4T	2	4T
4.0	4-2T	4	2T

See also Fig. 2.21 and Ref 67.

the linear absorption coefficients, due to the Cu $K\alpha$ absorption edge. Another possibility would be to use an x-radiation of wavelength just less than that of the platinum absorption edge. (See Example 3, in the section "Calculations for Trial Exposures" in this chapter.)

Composition contrast is at a minimum at x-ray energies for approximately 1 to 2.5 MeV. This can be seen from Fig. 2.22(b), where the mass absorption coefficients of all materials approach the same value of approximately 0.05 cm²/g.

Inclusion Contrast (Ref 115). Suppose an inclusion is present of thickness Δ_i and linear absorption coefficient μ_1, in a matrix of thickness Δ_M and linear absorption coefficient μ_2. Then the intensity transmitted through the matrix:

$$I_2 = I_0 \exp - [\mu_2 \Delta_M] \qquad \text{[Eq 2.8]}$$

and the intensity transmitted through the matrix and inclusion

$$I_1 = I_0 \exp - \mu_2(\Delta M - \Delta_i) \cdot \exp - \mu_2 \Delta_i \qquad \text{(Eq 2.45)}$$

so that

$$\frac{I_2}{I_1} = \exp - (\mu_2 - \mu_1)\Delta_i \qquad \text{(Eq 2.46)}$$

which is independent of the thickness of the matrix. Then the radiographic contrast

$$C_S = 0.4343 \, G \, (\mu_1 - \mu_2)\Delta_i \qquad \text{(Eq 2.47a)}$$

which is independent of the thickness of the matrix. If the inclusion is a void, $\mu_1 = 0$ then

$$\frac{I_2}{I_1} = \exp - \mu_2 \Delta_i \qquad \text{(Eq 2.47b)}$$

and the contrast

$$C_S = 0.4343G \, \mu_2 \Delta_i \qquad \text{(Eq 2.47c)}$$

To observe an inclusion, the x-radiation must be chosen that can traverse the thickness of the specimen and having the maximum difference in linear absorption coefficients (see Example 12).

The thickness sensitivity depends on the change in the transmitted intensity due to a relative thickness difference $\Delta z/z$; see Fig. 2.23. Equation 2.7 can be expressed as

$$\Delta z = \frac{\Delta I}{\mu I} \qquad \text{(Eq 2.48)}$$

The film density difference $D_2 - D_1$ due to the thickness difference in Fig. 2.23, is the radiographic contrast

$$C_S = D_2 - D_1 = G \log I_2/I_1$$

$$= G \log \left(\frac{I_1 + \Delta I}{I_1}\right) \text{ since } I_2 = I_1 + \Delta I \qquad \text{[Eq 2.28a]}$$

Therefore:

$$D_2 - D_1 \approx G \frac{\Delta I}{I_1} \text{ for } \Delta I \text{ small} \qquad \text{(Eq 2.49)}$$

Therefore

$$\Delta z \approx \frac{D_2 - D_1}{\mu G} \text{ using Eq 2.48} \qquad \text{(Eq 2.50)}$$

The intensity I_T at the film is given by

$$I_T = I_B + I_1$$

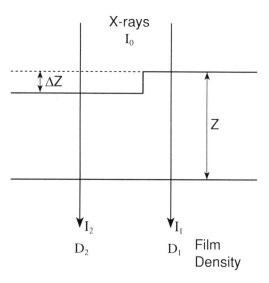

Fig. 2.23 Thickness contrast sensitivity $\Delta z/z$ depends on $D_2 - D_1$ and is inversely proportional to the film gradient G and to the linear absorption coefficient μ; see Eq 2.52.

where I_B is the background intensity at the film (approximately uniform) and I_1 is the intensity from the incident beam traversing the absorber. Equation 2.50 becomes

$$\Delta z = \frac{D_2 - D_1}{\mu G}(1 + \frac{I_B}{I_1}) \qquad \text{(Eq 2.51)}$$

The term $1 + I_B/I_1$ is known as the build-up factor, due to background of scattered x-rays not truly absorbed by the object (Ref 69). This results in the

$$\text{Thickness contrast sensitivity} = \frac{\Delta z}{z} =$$
$$\frac{D_2 - D_1}{z\mu G}\left(1 + \frac{I_B}{I_1}\right) \qquad \text{(Eq 2.52)}$$

depending on $D_2 - D_1$ (minimum density difference visible), $1/G$ (inversely proportional to the film gradient), $1/z$ (inversely proportional to specimen thickness), and

$$\frac{1 + I_B/I_1}{\mu}$$

the build-up factor and linear absorption coefficient, which depend on the nature of the radiation used.

Radiation Units

There are several traditional units as well as the units of Système International (SI) which are used in radiography and radiology. These units are defined and their use discussed. (Further discussion of the units are given in Ref 55, 57, 67, and 103.)

Roentgen, R, is the traditional unit of radiation fluence, dose, quantity, or exposure and is defined in terms of the ionization produced in air. One Roentgen is defined as the amount of radiation that produces ions carrying one electrostatic unit of charge (esu) in 1 cm^3 of dry air at standard temperature and pressure. This corresponds to 2.58×10^{-4} coulombs per kilogram or 83.3×10^{-4} J/kg of air. Because the Roentgen is measured by the energy absorbed by air, it is valid when all the energy is absorbed; consequently this definition is not valid for high energy radiation (above 3 MeV) which has a low absorption in air. The Roentgen applies to ionizing radiations such as x-rays and γ-rays.

The rad is the traditional unit of absorbed dose and is equal to 0.1 J/kg. It should be noted that 1 R in human tissue corresponds to an energy absorption of approximately 0.095 J/kg.

The Gray (Gy) is the SI unit for the radiation dose resulting in the absorption of 1 J/kg. The rad and Gray apply to all radiations.

The relative biological effectiveness (rbe) is the value correcting for various radiation effects of the human body. The rbe is defined as:

$$\text{rbe} = \frac{\text{rads of } 250\,\text{kV x--rays}}{\text{rads of other radiation producing the same effect}}$$
$$\text{(Eq 2.53)}$$

Some accepted rbe values are:

- X-rays, g-rays, electrons (beta particles), rbe = 1
- Thermal neutrons, rbe = 5
- Fast neutrons, rbe = 10
- Alpha particles, rbe = 20

Roentgen Equivalent Man (rem or Sievert). This unit defines the biological effect of radiation on the human body. The rem, Roentgen Equivalent Man, is the absorbed dose where rem = dose in rads × rbe. The Sievert is the SI unit corresponding to the traditional unit rem, and is equal to the radiation dose in Gy × rbe. *Note*: For the human body for x-rays and γ-rays, because 1 R is 0.095 J/kg and rbe = 1, the dose in Roentgen $\cong$ dose in rad $\cong$ dose in rem.

Radioactive Disintegrations (Curie). There are three units of radioactive isotope activity (or decay). The Curie (Ci) is the traditional unit, being equal to 3.7×10^{10} nuclear transformation per second. The Becquerel (Bq) is the SI unit and equals one disintegration per second. The Rutherford (rd) is an obsolete unit and equals 10^6 disintegration per second. *Note*: 1 Ci $\equiv$ 3.7 $\times 10^{10}$ Bq.

The specific activity (S_A) of a radioactive isotope source is the number of disintegration per second per gram of the radioactive element. It is usually quoted in Ci or Bq per gram (or per cc). It can be shown that

$$S_A = \frac{0.693 N_o}{\tau A_w} \qquad \text{[Eq 2.4]}$$

where τ is the half-life of the element and N_o/A_w is the number of atoms per gram.

The activity of a particular radioactive source is the number of disintegration per second and is quoted in Ci or Bq.

The strength of a radioactive source is measured by its radiation intensity and is measured by Roentgen per hour at one meter (Rhm). The radioactive isotope dose rate is Roentgen/hour/Curie/foot^2. Some values for radioactive isotopes are (see Table 2.2):

- Co^{60}, 14.5 (in R/hr ft^2 Ci)
- Ir^{192}, 5.9
- Ce^{137}, 4.2
- Thulium, 0.03

Radiation Intensity. The traditional unit of radiation intensity is Roentgen per hour (R/hr). This is used for x-rays and γ-rays and is only valid for low energy radiations.

Gamma-Radiography

The γ-rays from radioactive isotope sources are of high energy, and consist of several monochromatic lines; see Fig. 2.3(b) for the spectrum of Ir^{192}. The units of radioactive isotope activity are presented in the previous section "Radiation Units." Several γ -ray sources are in use for radiography and are listed in Table 2.2. The effective source sizes range from 0.3 to 10 mm, so that the geometric unsharpness can be derived in any given case. The choice of γ-rays will be determined by the need to penetrate a thick ob-

ject. The half-life of the radioactive isotope is of concern if using a short life isotope such as Ir^{192}, Yb^{169}, or Th^{170}. The strength of the source is always of concern because the radiation intensity levels are lower than from x-ray generators. Figure 2.24 shows a typical exposure chart for the γ-rays from Ir^{192} for several thicknesses of steel and for several film densities. The exposures are given in terms of the specific activity of the radioactive source (in Ci or Bq), allowing for the area of the specimen, which dictates the source-specimen distance. These curves are based on experience and of course depend on the precise x-ray film (with or without screen) being used. An example of a commercial system for γ-rays is given in Fig. 2.25 and is discussed extensively in Ref 67 and 103.

In the section "Calculations for Trial Exposures" in this chapter, Example 22 is concerned with the shielding of a radioactive source. Example 26 derives the optimum exposure using a radioactive source. Safety precautions are discussed in the section "Safety Precautions and Shielding."

Additional Radiographic Techniques

Several useful radiographic methods employ other radiations, produce three-dimensional data, or can observe fast moving objects. Extensive descriptions are available of these methods. However, only very brief descriptions are presented here, and extensive references are provided (Ref 241).

Flash Radiography. High speed flash x-radiography can be carried out using an x-ray pulse of less than 100 ns duration. Peak voltages from a pulse generator are used of 100 kV to over 2 MV. The spectrum from W or Mo targets consists of a short λ component (hard x-rays) with considerable penetration and also longer λ radiation (soft x-rays), which can enhance the contrast of the image. At 100 kV peak voltage, the shortest wavelengths are about 0.1 Å, and at 1 MV is about 0.01 Å. Some examples of flash radiography of ballistics, machinery in motion, projectiles, and phase transformations are described in Ref 130 to 134.

Neutron radiography complements x-radiography because the absorption characteristics of neutrons and x-rays are totally different. While x-rays are more heavily absorbed by elements of high atomic number, neutron absorption varies in

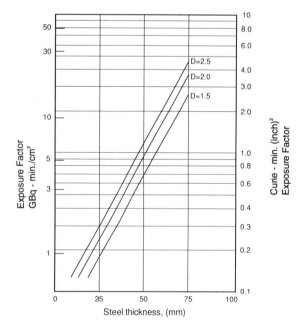

Fig. 2.24 Experimental exposure chart for steel using γ-rays from the radioactive source Ir^{192}. The exposures are given in terms of the source activity and the area of the specimen for different film densities.

a totally different manner with atomic number. As an example, hydrogen has one of the highest neutron absorption coefficients, so that neutron radiography provides much more contrast for organic materials than does x-radiography. An example often quoted is that of a thin wax string encased in a lead block of thickness of 2 in.; the neutron radiograph reveals the presence of the string that would be totally invisible by x-radiography. Neutron radiography also permits the differentiation of elements of similar atomic numbers where x-radiography is not suitable.

Neutrons are usually classified by energy as cold ($<10^{-2}$ eV, $\lambda \sim 0.1$ mm), thermal (<0.3 eV, $\lambda \sim 4$ μm), epithermal ($<10^4$ eV, $\lambda \sim 1$ Å), and fast (10 to 20 MeV, $\lambda \sim 10^{-3}$ Å). Thermal neutrons undergo capture by a nucleus to form a different nucleus; this is the basis of neutron activation analysis. Fast neutrons have had very limited use for radiography to date.

Sources of neutrons are atomic reactors, spallation sources, and radioactive (RA) isotopes. Lists of suitable RA isotopes and their characteristics are listed in Ref 124, Volume 1, Chapter 9. A portable neutron radiography system using californium, Cf^{252} is described in Ref 135, Chapter 7. This system is about the same size as a small portable x-ray system. The half-life of Cf^{252}, is 2.65 years, average neutron energy 2.3 MeV, giving a neutron yield of about 10^{12} neutron/second/gram; γ-rays are also present from the RA source. Thermal neutrons can be detected by a photographic method furnished with a converter screen to change the neutrons to α-, β-, or γ-rays. The converter screen can provide immediate and direct exposures, though γ-rays from the source will be a problem because they affect

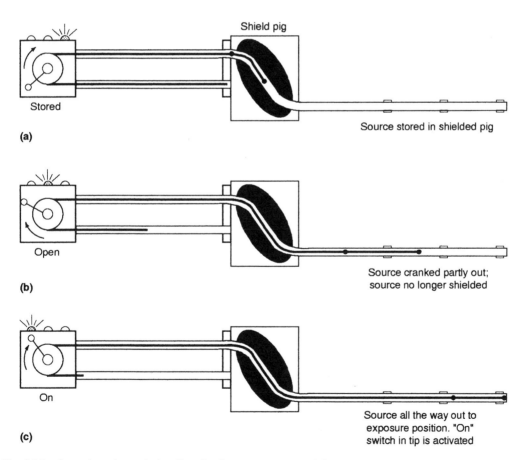

Fig. 2.25 Operation of a typical radioactive isotope camera (Ref 67). (a) Source stored in shielded "pig." (b) Source no longer shielded. (c) Source all the way out to exposure position.

the photographic film. The presence of γ-rays can be overcome by using a transfer exposure method; the converter screen is exposed to the neutron beam, becoming radioactive, and is then transferred to a cassette to expose film (Ref 136).

In a typical direct exposure method, cadmium screens are placed both in front of and behind the film; a recommended combination is a front cadmium screen 250 μm thick, and a back cadmium screen 500 μm thick. A discussion of the merits of different types of converter screens is given in Ref 136. Examples of neutron radiography are described in Ref 20 and 135.

Proton Radiography. Although the use of protons in radiography has been very limited, an extensive discussion of the method is given in Ref 137. The great advantage of this technique is that very small density changes can be detected under suitable conditions, much smaller than by other radiations. Transmission of the monoenergetic proton beam through an object remains approximately constant for about 90% of the trajectory, after which the transmission rapidly drops to zero. It is during this last stage that a small variation in density can have a pronounced effect on transmission. Density changes as little as 0.05% can be detected. The protons can be detected by photographic or polaroid film, and measurements have been made with proton beams of several hundred MeV. Different procedures can be followed using proton absorption, proton scattering, or proton activation autoradiography. Applications include thickness measurements with accuracies possible of $2 \times 10^{-3}\%$; other examples are given in Ref 137.

Activation Analysis. Elemental analysis of an object can be carried out by activation analysis (AA). Thermal neutrons, γ, proton, and deuteron radiations induce nuclear reactions to occur followed by the emission of γ-rays, and it is this spectrum of emitted γ-rays which is characteristic of the solid (Ref 138). The γ-rays can be detected using Ge (Li) semiconductor devices. The emitted γ-spectra, half-life of the radioactivity, and the type of radiation emitted can be used to identify the elements present (Ref 16). In neutron activation analysis (NAA), neutrons are used to induce radiative capture reactions (n,γ), (n,ρ), (n,α), or $(n,2n)$. Typical nuclear reactions are Na^{23} (n,γ) Na^{24}, Al^{27} (n,ρ) Mg^{27}, and Cl^{35} $(n,2n)$ Cl^{34}. Quantitative analysis can be undertaken from the induced radioactivity, usually carried out by comparison with standards.

For archeological purposes, the authenticity of ancient coins has been extensively studied by AA (Ref 6, 16, 18). One example uses the fact that silver coins were contaminated with gold, up to the 6th century AD. Subsequently, with improved methods of refinement, silver coins contain much less gold. The γ-spectra from coins after irradiation by protons of energy 30 MeV permit the gold concentration to be determined, which can be used to date the coins.

The age and provenance of oil paintings can be determined by AA. White lead $2PbCO_3Pb(OH)_2$ has been used in paintings throughout the ages, and the lead purity has improved with the refinement process. During refinement, some of the radium, Ra^{226}, from the uranium decay series is removed, particularly in developments since the 18th century. This means that the relative amounts of radium (Ra^{226}), lead (Pb^{210}), and polonium (Po^{210}) (from the uranium decay series) can be used to determine the age of paint specimens (Ref 17).

It is often important to determine the composition at great depths below the earth's surface. This can be carried out using a drill-hole, typically 4 in. in diameter. A compact source of neutrons is inserted with a particle counter shielded from the source; the source of neutrons may consist of an α-particle emitter mixed with beryllium powder when a (α,n) reaction occurs. The neutrons activate the minerals and the γ-spectra is measured by the counter. Rocks bearing iron have been identified by their characteristic γ-ray spectra (Ref 19).

Forensic studies have been carried out on the bullets fired in the assassination of President J.F. Kennedy (Ref 14). NAA showed similarities in the silver and antimony content of several of the bullets found at the scene.

Shadow Autoradiography. A radioactive specimen in contact with a photographic emulsion will produce an image, and this technique, autoradiography, is used extensively to examine biological specimens, generally deliberately doped with a radioactive isotope. This can be an α-emitter polonium (Pu), or a β-emitter such as iodine (I^{131}), ruthenium (Ru^{106}), strontium (Sr^{90}), and phosphorus (P^{32}). In shadow autoradiography, a preliminary shadowing of the specimen is carried out by a metal coating. Specimens examined by this technique are very diverse and include human skin and sputum, sheep thyroid, lung tissue, plant leaves, as well as radioactive powders (Ref 113, 139, 140).

X-Ray Optical Systems. X-ray microscopy, or micro-focal radiography, is projection x-radiography using an x-ray source having a very fine spot size (Ref 13, 115, 141). These special x-ray generators are reviewed in Ref 13. X-ray focal spot sizes down to 5 μm diameter have been obtained by special focusing techniques at 40 to 100 kV, with a target loading of 1 mA. This is achieved in metal-ceramic x-ray tubes with electrostatic focusing. Various x-ray optical systems are reviewed in Ref 142.

X-ray topography is reviewed in Ref 143 and 144; single crystals can be examined to display dislocations, stacking faults, twinning, and lattice distortions. Polycrystalline aggregates can be examined for strains, surface relief, and texture. Monochromatic x-radiation must be employed, and exposures are long, except when synchrotron radiation and image intensifiers are employed.

Residual stress NDT measurements can be undertaken of polycrystalline surface layers by x-ray diffraction and the technique is reviewed in Ref 145.

Thickness Measurements. Control of thickness during the production of objects is a desirable feature of quality control technology. Thickness can be measured by absorption or by backscattering techniques of x-rays, γ-rays, α- or β-rays, or by x-ray fluorescence (Ref 113, 139). In the x-ray gage, the absorption of a beam of x-rays is used to determine thickness of an object of known composition. A typical example is the use of additives in paper, such as the pigment kaolinite clay ($Al_2O_3 \cdot 2SiO_2 \cdot 2H_2O$), which has been controlled using the attenuation of a monochromatic x-ray beam (Ref 27). Similarly, the porosity of materials can be measured (Ref 146) as well as the observation of steam voids in water (Ref 147).

In the case of radioactive specimens, the absorption of the γ-rays emitted by the specimen itself can be used to control the thickness. An example of this is the loading of U^{235} reactor fuel elements, monitored by the γ-rays (184 keV) emitted by the U^{235} specimen itself (Ref 148). The thickness of thin foils has been measured using a beam of α-particles (Ref 139) using a radioactive source which emits α-particles with a range of about 4 cm in air. The sensitivity can be adjusted to detect small changes in thickness of about 1%. A beam of β-particles has been used to determine the thickness of foils from rolling mills of aluminum and of steel, and in another use, to control the tobacco content of cigarettes (Ref 149); a radioactive isotope strontium Sr^{90} is used, which emits β-rays of energies up to about 1.35 MeV. There is an empirical relationship (Ref 114, 139) relating density, ρ, and specimen thickness, Δ, to the energy of the β-ray E_M MeV, where:

$$\rho\Delta = 0.45\,E_M - 0.16$$

The thickness of coatings on substrates can be measured using the backscattering of x-rays, γ-rays, or β-rays. The sensitivity depends on the difference in atomic number between the substrate and coating materials (Ref 139). Typical measurements are of tin, zinc, chromium, or brass coatings on steel, paint, or lacquer on metallic surfaces, rubber or plastics on calendering rolls, selenium layers on aluminum, barium coatings on photographic paper, plastic coatings on wires, glaze on porcelain. In the case of tin plating on steel, the coatings can be controlled to about 10^{-6} cm, that is about 0.3 mg/cm^2.

X-ray fluorescence (Ref 113) has also been used to estimate surface conditions, and the method is suitable for elements of high atomic number when the fluorescent yield is more important than the production of Auger electrons. Fluorescence is induced in the substrate and the reduction in intensity of the fluorescent beam leads to a calculated value of the thickness of the surface coating. When measurements are carried out in air, the longer fluorescent x-ray wavelengths are absorbed so that the limit in air is the fluorescent radiation from titanium Ti ($Z = 22$ with K_α x-rays, $\lambda = 2.75$ Å), when a path in air of about 10 cm will reduce the x-ray intensity by about 50%. Examples of fluorescent measurements of thickness are silver plating on copper using molydenum radiation (Mo K_α $\lambda = 0.71$ Å), which causes the copper to fluoresce. In deriving the thickness of the silver plating, allowance has to be made for the absorption in the coating of the incoming Mo K_α x-radiation, as well as the attenuation of the CuK_α fluorescent radiation traversing the coating. Another application concerns the thickness of zirconium cladding (about 0.01 cm thick) on uranium core nuclear fuel pins of diameter of about 0.3 cm; x-rays from a tungsten target at 50 kV were used to excite the uranium Lα ($\lambda = 0.911$ Å). This enabled the identification of regions where the zirconium cladding was less than 0.0075 cm thick (Ref 113, 150).

Autoradiography can also be used to measure coating thicknesses, and an example of this is the use of a flat uranium source on which sheets of aluminum, zirconium, or stainless steel can be placed. The β-radiation of energies up to 2 MeV traverse the sheets, leading to accuracies of about 5% in thicknesses of metal sheets of about 0.5 mm (Ref 139, 151).

Tomography is literally, "the picture of a slice." The differences between radiography and tomography are illustrated in Fig. 2.26. In conventional radiography by transmission, the contrast of a defect is due to absorption differences along the total path that the x-ray transverses, as shown in Fig. 2.26(a), and there is no attempt to determine the distance along that path of the position of the defect. Conventional radiography is by shadow projection. The position of the defect is revealed more precisely if Compton scattered x-rays are used as shown in Fig. 2.26(b), where in backscatter imaging, the region under view in the object is defined by the line of vision of the detector. Another procedure uses special

motions in transmission to blur all of the object except the one of interest; see Fig. 2.26(c). This is known as laminography or motion tomography. A further method is reconstructive tomography, see Fig. 2.26(d), where all of the planes are excluded except the one of interest, and mathematical procedures are used to reconstruct the image of the plane. In flashing tomosynthesis, or multiradiograph tomography, the image consists of several overlapping x-radiographs. These are unscrambled optically from the multi x-ray radiograph, see Fig. 2.27.

Tomography can also be carried out by:

- Positron emission tomography (PET); see Ref 86 and 152 to 154. Positron annihilation is illustrated in Fig. 2.9
- Radionuclide tomography; see Ref 154
- Gamma-ray tomography; see Ref 154 to 157
- Nuclear magnetic resonance; see Ref 83, 84, 158 to 160

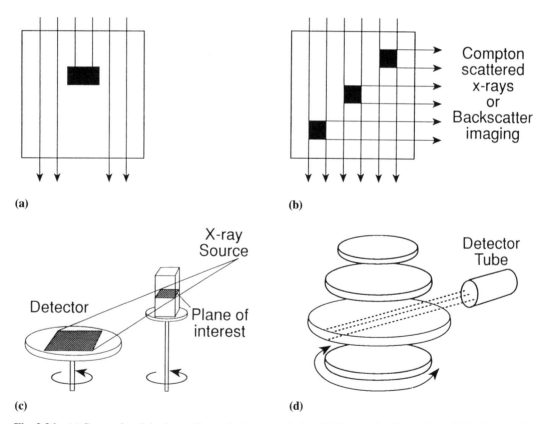

(a)

(b)

(c)

(d)

Fig. 2.26 (a) Conventional shadow radiography by transmission. (b) Scattered radiography at 90°, where the line of vision of the detector defines the volume under view in the object. (c) Laminography, where synchronous rotations of object and detector film result in one plane only being in focus. (d) Reconstructive tomography, where many observations are carried out from different angles within one plane.

- Ultrasonics; see Ref 161 to 164

Industrial computed tomography is discussed extensively in Ref 129, p 358-386. Specialized services of computer tomography for industrial purposes are now available; for example see Ref 165.

Laminography. In laminography, the object and the film are moved simultaneously, so that the projection of one particular layer of the object remains stationary relative to the film: see Fig. 2.26(c). All other layers of the object project as blurred images, because they move relative to the film. The object needs to remain constant throughout the exposure, which may require a high x-radiation dose, so that this procedure is not appropriate for human beings. Different levels of the object can be viewed by adjusting the particular plane of the object in synchronism with the detector plane. Specimens examined include the variations in the wall thickness of a glass bottle, the different levels of a complex circuit board, and the interior of a stopwatch (Ref 29, 113, 166, 167).

Compton (Inelastic) Imaging; Backscatter Imaging. This has been applied to topics as varied as microporosity in plastic explosives, and to medical studies of tumors in human tissue (Ref 168, 169). Inelastic or Compton scattering results from the scattering of x-rays with a loss of energy due to recoil energy transferred to electrons. The change of wavelength $\Delta\lambda$ is given by

$$\Delta\lambda = 0.0243 \, (1 - \cos\alpha)$$

where α is the scattering angle (Ref 114); see Fig. 2.9(c). This provides for a discrimination between the incoming x-rays and the Compton scattered x-rays of lower energy. These Compton scattered x-rays permit the monitoring of the electron density and compositional change, because the image position is defined by the small overlap volume of the incident radiation and the scattered beam; see Fig. 2.26(b).

Compton scattering is of sufficient intensity to be detected and differentiated from the elastically scattered radiation. The use of an industrial-type (tungsten) x-ray tube operating at 200 kV provides sufficient Compton scattered x-rays to record a suitable image of an aluminum casting on film in 2 h (Ref 168). The recording time can be reduced drastically by the use of a standard x-ray image intensifier with a TV monitor so that a suitable image can be obtained in less than 1 s.

The x-ray beam in the form of a diverging fan irradiates one slice of the object; a pin-hole lens permits Compton scattered rays at 90° to enter the x-ray camera and be recorded. With this simple system, a spatial resolution of about 1 mm is obtained of the slice of the object examined. A more advanced system has been constructed known as COMSCAN (Ref 168); the x-ray tube is a standard industrial type and is operated at about 65 kV. Several collimator slits are provided, around the irradiated object, and the scattered x-rays detected by an extended array of scintillation counters of $Bi_4Ge_3O_{12}$. The volume elements of the object examined are $1 \times 1 \times 7$ mm.

In other systems, γ-rays from radioactive sources have been used; e.g., Co^{60} (about 200 TBq) sources, providing γ-rays of energies 1.17 and 1.33 MeV, with NaI(Te) solid-state detectors. The Compton scattering can be observed at any angle, so that measurements can be carried out in an object where there is only restricted access to one side of the object (Ref 168).

The Compton effect is taken advantage of by the techniques known as backscatter image tomography (BIT), where access is only required to one side of the specimen. Applications of Compton scattering tomography include examination of aluminum car engine components, breast cancer and its spread into surrounding tissue, examination of the human skull and jaw, the explosive charge in shells, cooling channels in aero engine turbine blades, location of iron bars behind 4 cm thick concrete walls, examination of the iron base of oil tanks, and the inspection of iron bars and voids in concrete (Ref 168, 169).

Flashing Tomosynthesis. Multiple-radiograph tomography, or flashing tomosynthesis (Ref 170) requires radiographs to be taken of an object from several different angles. The reconstruction of the image of a particular layer is carried out optically, and the principle of the method is illustrated in Fig. 2.27. The first step is the formation of a multiple image taken by flashing x-radiographs of the object using a series of x-ray sources set in the same plane, and subtending different angles at the object. A multiple image is formed, which is then decoded by projecting light through the image from a series of lenses in a pattern related to that of the x-ray sources. In Fig. 2.27, an object is shown consisting of a square and a circle in different levels of the object, and illumination by three sources gives rise to a complex multiple image. Three sources of light are now projected through this complex image in such a way that one feature of the object, here the square, superimposes in the final image; see Fig. 2.26(b). There is a background in the image due to the noncoinciding parts of the image and this noise level can be minimized by careful distribution of the x-ray sources used, as well as the diameter of the object examined. Objects up to 60 mm in diameter have been examined, giving good images using an array of 25 x-ray tubes, when depths of 15 cm can be portrayed in planar slices; see also Ref 155 and 171. Good images have been obtained of a simulated human head, skull, and neck (Ref 170).

Reconstructive Tomography. Mathematical procedures can be used to reconstruct the desired image plane from transmission data taken along different paths within one plane of the object; see Fig. 2.28. This is known as reconstructive tomography (RT), computed tomography (CT), or computer-assisted (axial) tomography (CAT) (Ref 154, 164, 172-175). The various mathematical procedures are known as iterative least square technique (ILST), algebraic reconstructive technique (ART), and simultaneous iterative reconstruction technique (SIRT). The data used can be x-ray transmission attenuation, ultrasonic (Ref 175), nuclear magnetic resonance (Ref 176), neutron transmission (Ref 175, 177, 178), or positron annihilation (Ref 154).

The principle of these methods is illustrated in Fig. 2.28 (Ref 172). The object is considered as subdivided into an array of cells as in Fig. 2.28(a). The contributions of the ith cell in the jth array are considered in the observed sinogram, so that the object is replaced by an array of cells, in each of which it is required to determine the relative density of matter. Any observation will provide the effect of the sum of the ith cells along the direction of the jth ray. The process to do this is illustrated in Fig. 2.28(b) and (c) as a simplified case of how the observations can be unraveled to arrive at the original object densities. The original object and ray measurements are shown in Fig. 2.28(b). One method of procedure is presented schematically using additive corrections. Starting from zero in all cells, one can correct each ray sum to approach the observed values.

Practical examples of reconstructive tomography are body scans (Ref 173), mammography (by ultrasonic transmission) (Ref 175, 179), nuclear fuel bundles inside a reactor by neutron tomography using 20 neutron radiographs (Ref 175), aerospace structures by x-ray tomography (Ref 171), and turbine blades and vanes (Ref 155).

Part III Practical Guidelines to Improved Radiography

Useful Radiographs

The preparations needed to obtain useful radiographs include several calculations, as well as trial exposures. The technical discussions in Part II provide the background to these considerations. It is always necessary to commence by examining the object visually and to consider what is known of the probable location of any defects. The following steps need to be considered (Ref 169).

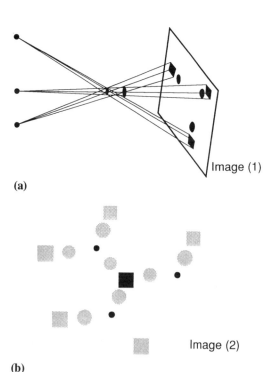

Fig. 2.27 Flashing tomosynthesis, or multiple radiograph tomography. As an illustration, an object consisting of a square and a circle at different levels are illuminated by three sources resulting in multiple image in (a). Three light sources project image 1 along the same directions as used by the x-rays so that the square part of the object is superimposed in (b) image 2. The remainder of the image gives rise to background noise.

Selection of X-Ray Tube Voltage. An estimate is required of the x-ray energy to penetrate the thickness of the specimen. This can be undertaken in various ways:

- Using tables and figures of the variation of half-value layers (HVL) and tenth-value layers (TVL) with x-ray energy; see Tables 2.6 and 2.10. The HVL and TVL have been determined experimentally for broad beam x-ray filtered spectra in particular experimental settings
- Using tables of maximum thickness of specimens suggested for examination with x-ray energies; see Table 2.11. These are

based on experimental data
- Using plots of x-ray intensity versus distance into the solid. Several such plots are given in Ref 55 for steel, lead, concrete, and other materials

Once an estimate has been made of the voltage required for a radiograph, a series of trial exposures can be carried out using IQI test pieces that are placed on the source side of the object. Such experiments provide the final selection of the x-ray tube voltage to obtain the desired penetration and contrast. IQI test pieces are reviewed in Ref 69; see also the section "X-Ray Image Quality—Image Quality Indicators" in this chapter.

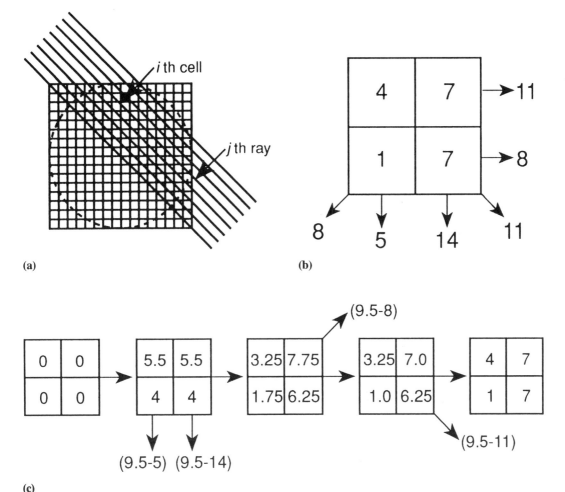

(a)

(b)

(c)

Fig. 2.28 Iterative reconstruction tomography. (a) The object is considered as an array of small volumes (voxel). The jth ray sums the effects due to the voxels along its path. (b) Object of 4 voxel (pixel in two dimensions). The ray sums are given. (c) Additive correction scheme.

Considerable data are available to aid in the taking of a radiograph of steel; see for example Ref 69 and 103. Other materials are compared to steel in tables of approximate equivalent thickness factors as given in Table 2.12. As an example, a radiograph of magnesium of 18 mm thickness at a voltage of 150 kV is equivalent to (18 × 0.06) = 1.08 mm of steel or $18 \times 0.06 \times 1.3 = 1.40$ mm of copper.

Exposure Times. Charts are available providing an estimate of the exposure in terms of x-ray tube current multiplied by time (mA-s), for radiographs of aluminum or steel, depending on

the thickness of the specimen and on the applied voltage; see Fig. 2.29 and 2.30. Such charts are usually provided for a particular x-ray generator, using specified film-source distance, x-ray film type, film developing-processing, for a film density D, usually 2.0. Exposures for other materials can be derived using Table 2.12 (equivalent thickness factors). Another useful experiment exposure chart is given in Fig. 2.31, for x-rays of energies from 100 to 350 kV, and for some γ-ray

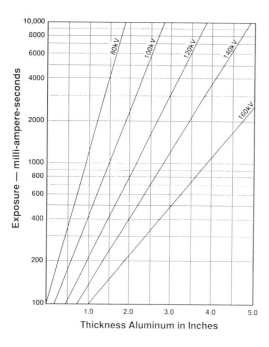

Fig. 2.29 X-ray exposure for specified conditions of screens, target-to-film distance and film (Ref 67). Film density is 2.0; target-to-film distance is 36 in.; film is type II (see Fig. 2.45); lead screens: front, 0.005 in.; back, 0.010 in. See Example 25.

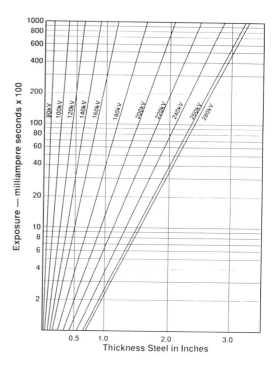

Fig. 2.30 X-ray exposure chart. Data provided for a particular x-ray generator and film (Ref 67). Film is type II (see Fig. 2.45); lead screens: front, 0.005 in.; back, 0.010 in. Target-to-film distance is 36 in. Film density is 2; Example 1 in. of steel, at 160 kV, requires 30,000 mA-s. See Example 27.

Table 2.10   Half-value layers and tenth-value layers for some radioactive isotopes

		Co^{60}	Ir^{192}	Ce^{137}
HVL, in.	Lead	0.49	0.19	0.25
	Steel	0.87	0.61	0.68
	Concrete	2.6	1.9	2.1
	Aluminum	2.6	1.9	2.1
TVL, in.	Lead	1.62	0.64	0.84
	Steel	2.9	2.0	2.25
	Concrete	8.6	6.2	7.1
	Aluminum	8.6	6.2	7.1

radioactive isotope sources. Allowance can be made for a range of film densities D; see Ref 67, 69, and 103.

The exposure times in radiography tend to be long, due to the low efficiency of x-ray film, so intensifying screens are used to reduce exposures that can be less than one-tenth that required in the absence of an intensifying screen; see the section "Recording the X-Ray Image—Intensifying Screen" in this chapter. However, the unsharpness of the radiograph is seriously affected. The unsharpness of film is of the order of 0.01 mm, but those of the intensifying screens are much greater, approaching 1 mm; see Table 2.7.

Therefore, a decision will be required concerning exposure time and unsharpness. It may be useful to take radiographs with and without intensifying screens to compare contrast, unsharpness, and exposure times.

X-Ray Target-to-Film Distance. The distance of the x-ray source to the film and the size of the x-ray source, with the specimen close to the film, determine the unsharpness of the image due to the penumbra. The optimum source-film distance will depend on the voltage applied, because this gives rise to an inherent film unsharpness U_f, as discussed in the section "X-Ray Image Quality—Unsharpness, Penumbra" in this chapter. After obtaining an estimate of the x-ray tube voltage to be used, the value of U_f can be

obtained from Fig. 2.18, which is an experimentally determined curve relating the inherent film unsharpness U_f to the applied voltage V. The value of the geometric unsharpness U_g is set equal to U_f to obtain a highly-sensitive radiograph. Then,

$$U_g = U_f$$

and because:

$$U_g = \frac{F \times l}{L_o} \qquad \text{(from Fig. 2.16 and Eq 2.29)}$$

then

$$U_f = U_g = \frac{F \times l}{L_o} \qquad \text{(Eq 2.54)}$$

For contact radiography, the specimen thickness $\Delta_o = l$. Table 2.13 provides source-to-film ($L + l$) distances for a range of x-ray (and γ-ray) energies, and effective focal spot size, F.

Source-to-Film Distance (γ-Rays). Three factors are important in determining the source-to-film distance, and they are:

- Source size, F
- Specimen thickness, Δ_o

Table 2.11 Guide to the penetration of x-rays in steel and aluminum

Radiation	Steel, mm	Aluminum, mm
X-rays (kV)		
50	1	20
75	3	50
100	10	100
150	15	130
200	25	160
300	40	200
400	75	300
1,000	125	…
2,000	200	…
8,000	300	…
30,000	325	…
γ-rays		
Ir^{192}	60	…
Co^{60}	125	…
Yb^{169}	8	30

These thicknesses are the maximum suggested for radiography at the energies listed (Ref 69).

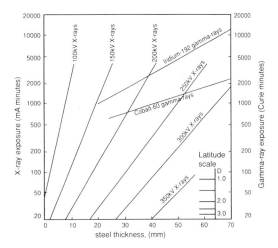

Fig. 2.31 Experimental exposure charts for steel with x-ray energy. These charts are for a particular x-ray film, using lead screens, with source-to-film distance at 1 m, with film density of 2. The inset scale is to allow for other film densities; see Ref 69. These exposures are only estimates and should only be used as a guide to a first trial. For example, at 100 kV, 5 mm steel require an exposure of 500 mA-min at 1 m, that is 10 min at 50 mA. See Examples 8, 13, 14, 16, and 18.

• Source-to-film distance, L_s

There are distortions that arise where the γ-rays penetrate the specimen at different angles. The optimum geometric sharpness is obtained when the radiation source is small, the source-to-film distance is great, and the specimen-to-film distance is small. However, the human eye limit in observing an unsharpness (penumbra) is 0.02 in. so that the following expression is valid (see Ref 67):

$$L_S = \Delta_o \left[\frac{F + 0.02}{0.02} \right] \text{in.} \qquad \text{(Eq 2.55)}$$

where Δ_o is the distance in inches from the upper surface of the specimen to the film. This is the same as the specimen thickness when the specimen is resting on the film; see Fig. 2.17(a). F is the diameter of the source in inches. As an example, for a specimen thickness $\Delta_o = 2$ in., source diameter $F = 1$ in., the source-to-film distance $L_S = 102$ in. *Note*: The rule-of-thumb is that the source-to-film distance should be ~10× the specimen thickness. However, for thin specimens, a distance of 40 in. should be used.

In the case of a steel specimen, and using the inverse square law, the exposure can be estimated using the following empirical relationship:

$$t = \frac{\text{EF}}{S_R} \times \frac{L_s^2}{144} \qquad \text{(Eq 2.56)}$$

where t is the exposure time in minutes for a film density, $D = 2$ and EF is the exposure factor from the exposure chart for Ir^{192} and the film indicated; see Fig. 2.32. S_R is the source strength in Curies. Example 26 utilizes Eq 2.56 (Ref 67).

Artifacts and Parasitic Effects

There are many observations on radiographs that are not real, but are artifacts due to the processing of the film, the fixing solution, stray light effects, film not properly wetted by the developer, even errors in the original preparation of the photographic emulsion on the base. These

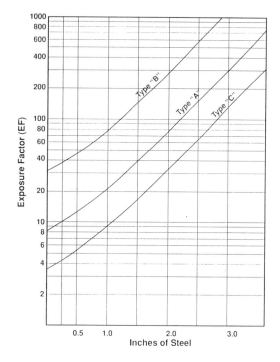

Fig. 2.32 Ir^{192} exposure chart (Ref 67). The exposure time for film density is given by Eq 2.56. See Example 26.

Table 2.12   Approximate equivalent thickness factors

Material	\multicolumn{7}{c}{X-rays, kV}							γ-rays				
	50	100	150	200	250	300	400	4 to 15 MeV	Ir^{192}	Cs^{137}	Co^{60}	Radium
Magnesium	0.03	0.05	0.06	0.07	...	...	...	...	0.22	0.22	0.22	...
Aluminum	0.06	0.08	0.12	0.16	0.17	0.19	0.22	...	0.35	0.35	0.35	0.40
Titanium	...	...	0.45	0.35	0.71	...	0.71	0.9	0.9	0.9	0.9	...
Steel	1.0	1.0	1.0	1.0	1.0	1.0	1.0	1.0	1.0	1.0	1.0	1.0
Copper	1.3	1.5	1.3	1.4	1.45	1.5	1.7	1.3	1.1	1.1	1.1	1.1
Zinc	...	...	1.4	1.3	1.3	1.3	...	1.2	1.1	1.0	1.0	1.0
Brass	...	...	1.4	1.3	1.3	1.3	1.2	1.2	1.1	1.1	1.0	1.1
Zirconium	...	...	2.3	2.0	...	...	1.0	...	...	...	...	...
Lead	5.3	8.3	11	13	15	17	...	3.0	4.0	3.2	2.3	2.0

This table provides approximate conversion factors for the transmission of x-rays through different metal absorbers (Ref 69, 70, 103). See Examples 8, 14, 16, 18, 25, and 26. For example, 3 in. of aluminum at 150 kV is equivalent to 3 × 0.12 in. = 0.36 in. steel.

effects are well documented and discussed in texts such as Ref 103 and 120 and are usually readily discernible from experience.

Several subtle artifacts are discussed below.

X-Ray Fluorescent Radiation. The incident x-ray beam may cause the specimen itself to generate and radiate x-rays in all directions. This fluorescence radiation is the characteristic radiation of the specimen and occurs if the incoming x-radiation is of sufficient energy to ionize the atoms of the specimen. Subsequently, the ionized atoms emit x-radiation of lower energy, known as fluorescence radiation, as the atoms return to the ground state. The condition for fluorescence radiation to occur is when the energy of the x-rays from the x-ray source is just greater than the absorption edge energy of the specimen.

This radiation is a nuisance and must be avoided if at all possible, or reduced to a minimum. A well-known example is that of Cu Kα characteristic radiation ($\lambda = 0.154$ nm) which causes iron, absorption edge ($\lambda = 0.174$ nm), to fluoresce emitting x-rays of $\lambda = 0.193$ nm. A thin metallic screen can be used because this will absorb more of the longer wavelength x-rays (in the absence of an absorption edge). The fluores-

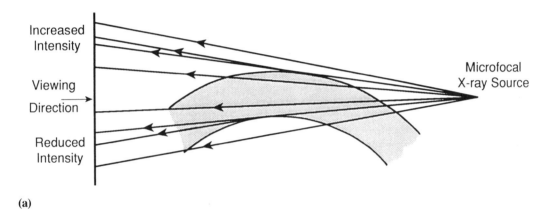

(a)

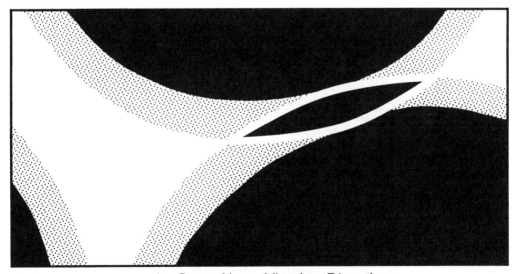

As Seen Along Viewing Direction

(b)

Fig. 2.33 An artifact may be observed in a radiograph due to total external reflection. This can occur when a diverging x-ray beam is incident on a smooth surface (a). A fringe can arise in the image plane (b), particularly if a microfocal x-ray source is being used.

cence radiates uniformly in all directions.

Surface Reflections. X-rays have a refractive index in solids of less than 1 (Ref 13, 180, 181) so that when the angle of incidence at a denser boundary is less than the critical angle (which is very small), total external reflection will occur. An example of this is given in Fig. 2.33 where a diverging beam of x-rays is incident on a smooth surface. The result recorded in the image is a line of increased intensity following the edge on the outside of the object, and having the appearance of a "fringe." There is also a line of decreased intensity following the edge on the inside of the object. This inside fringe may be very difficult to observe, but the outer fringe is frequently visible. Such a fringe has no structural meaning of any sort.

Absorption Edge Lines. The x-ray spectrum used in radiography is modified by the absorption edges of silver and bromine in the photographic emulsion. The absorption of x-rays in photographic emulsion changes dramatically at the absorption edges, and this is frequently visible in the blackening of the film in the form of a line. Lines occur where the wavelength of the broad band spectra of x-rays used in radiography corresponds exactly to the absorption edges of silver and bromine. Such lines are in no way due to the structure of the object under examination and are artifacts.

Effects comparable to this can occur when using intensifying screens, due to the absorption edges in the barium lead sulfate or calcium tungstate (Ref 117). However, because the unsharpness is considerably greater in intensifying screens than in films, the absorption edge effect is probably not detectable when using screens.

Scattered X-Rays. X-rays scattered by the specimen (secondary x-rays) give rise to a general fogging of the film, which reduces the sharpness, clarity, contrast, and resolution; see Fig. 2.8. The extent of the secondary x-rays in radiography depend on the thickness of the specimen and on x-ray energy and the elements present. Scattering can also occur in a parasitic fashion from all material which may be within the direct primary beam; the film cassette, walls of the room, edges of diaphragms; see Fig. 2.34. Care must be taken to restrict the field of the x-ray beam to prevent all unnecessary scattering. Masks and diaphragms may be useful to eliminate any such extra scattering or low atomic number materials (e.g., wood) and can be used to cover concrete and steel surfaces. X-ray diffraction effects can also occur.

There are several methods of reducing the effects of scattering on the x-ray film (Ref 103).

Object-Film Distance. When the object is at a distance from the film, as in projection radiography, fewer scattered rays reach the image.

X-Ray Grids, Roentgen Grids, Scattered Ray Grids, Potter-Bucky Diaphragms. A system of

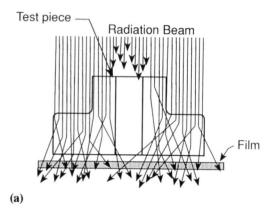

(a)

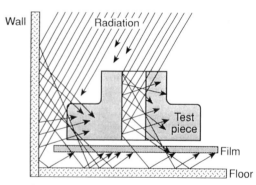

(b)

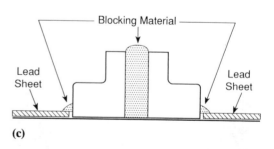

(c)

Fig. 2.34 When taking a radiograph of a specimen, the scattered x-radiation must be restricted from affecting the x-ray film. The scattering can arise from the testpiece itself (a), from surrounding walls, or by backscattering from the floor, as illustrated in (b). Blocking materials are also used to avoid the direct beam exposing unnecessarily the x-ray film, as illustrated in (c).

grids can be used to reduce the amount of scattered (secondary) x-rays arriving at the film. The grids are generally constructed of lead strips alternating with an x-ray transparent medium (e.g., plastic). The grids are arranged to be consistent with the diverging nature of the primary x-ray beam from the source to the film; see Fig. 2.35. The grids can be oscillated parallel to the film so that the shadows of the grid bars disappear into the general background. When used correctly, there can be a significant reduction in the fogging due to the scattered x-rays. The grids are used in particular in medical radiology, and when dealing with low atomic number materials such as Be (Ref 117).

Filters. The scattered x-rays and fluorescent x-rays are of lower energy than the incident radiation, so that a thin metal screen in front of the film can act as a filter reducing the scattered rays more than the primary beam. Lead screens are useful in the MeV range, and screens of aluminum are useful at lower energies. The usefulness of screens needs to be determined in each case experimentally by trial and error, because the increase in exposure time will need to be taken into account.

Blocking Materials. The edges of the specimen can scatter x-rays very badly, fogging the film unnecessarily. Similarly, holes in the specimen itself can result in overexposure of parts of the film,

when the film darkening may spread well beyond the size of the hole. Blocking materials, masks, or diaphragms can be used to reduce this problem. The blocking material should have the same x-ray absorption as the specimen; see Fig. 2.34(c) (Ref 103).

Calculations for Trial Exposures

A range of examples and calculations are presented to aid in the determination of trial exposures to provide sharp radiographs with good contrast. The calculations tend to be based on the use of monochromatic x-radiation but give good estimates for white radiation using the wavelength of maximum intensity of the spectrum. The calculations should be considered as being accurate to within about 10%. The examples are as follows.

Use of Absorption Coefficients. These determine the x-ray intensity transmitted by the object.

- Example 1. Absorption Coefficients
  1A. Absorption Coefficient of an Inorganic Compound
  1B. Conversion of Absorption Coefficients

Table 2.13 Source-to-film distances

Radiation	Source diameter (F), mm	Specimen thickness ($\Delta_0 = l$), cm	Minimum source-to-film distances ($L + l$), cm
X-rays (kV)			
100	1.0	0.5	10.5
150	2.0	2.0	60.0
150	2.0	10.0	300.0
250	4.0	2.5	92.0
400	5.0	7.0	240.0
1000	7.0	10.0	294.0
8000	2.0	25.0	125.0
γ-rays			
Ir^{192}	0.5	1.0	5.0
	2.0	5.0	82.0
Co^{60}	2.0	10.0	67.0
	4.0	10.0	123.0
Yb^{169}	0.5	0.5	4.5
	0.3	0.2	1.4

The geometric unsharpness U_g is set equal to the film unsharpness U_f when $U_g = U_f = (F \times l)/L_o$; see Fig. 2.16 and Example 13. For contact radiography, the specimen thickness $\Delta_0 = l$.

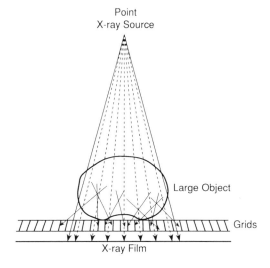

Fig. 2.35 A grid system is used to reduce the amount of scattered x-rays that can attain the film and so reduce the film clarity and contrast. The grids can be oscillated parallel to the x-ray film to avoid the formation of shadows from the grids on the film. The grids are known as roentgen grids, scattered ray grids, and Potter-Bucky diaphragms.

Metal Plates with Inclusions or Voids. Choice of x-ray generator voltage and exposure times is illustrated to obtain sharp radiographs.

Detailed Calculations

Nonmetallic Materials

Thickness Measurements. Monochromatic x-rays are needed for accurate thickness measurements. Balanced filters can be used to give an approximate monochromatic beam. Further details on obtaining monochromatic x-ray beams are given in Ref 119, 182-184.

Protective Shielding

Exposure Charts

Example 1A. Absorption Coefficient of an Inorganic Compound. The calculation of the linear absorption coefficient for barium carbonate for x-rays of $\lambda = 0.154$ nm, is given here as an example. The mass absorption coefficients for $\lambda = 0.154$ nm are given in Table 2.4(c) for barium 358.9 cm²/g, carbon 5.5 cm²/g, and oxygen 12.7 cm²/g. The density of barium carbonate, $BaCO_3$, is 4.43 g/cm³, and the molecular weight is 197.4. (Atomic weights are barium 137.34, carbon 12, oxygen 16.) Then, using

$$\mu_m = w_x\mu_{mx} + w_y\mu_{my} + w_z\,\mu_{mz} \qquad \text{[see Eq 2.21]}$$

$$\mu_m = \frac{137.4}{197.4} \times 358.9 + \frac{12}{197.4} \times 5.5 + \frac{3 \times 16}{197.4} \times 12.7$$

$$= 254 \text{ cm}^2/\text{g}$$

The linear absorption coefficient μ for $BaCO_3$ is μ_m × density so that $\mu = 254 \times 4.43 = 1125$ cm⁻¹ at $\lambda = 0.154$ nm.

Example 1B. Conversion of Absorption Coefficients. The conversion of cross section to mass absorption for aluminum is as follows. The cross section at 100 kV for aluminum is 7.644 Barns from Table 2.4(a). Now:

$$\mu_m = \frac{N_o}{A_w}(CS) = \frac{6.022 \times 10^{23}}{26.981} \times 7.644 \times 10^{-24} \text{ cm}^2$$

$$= 0.1706 \text{ cm}^2/\text{g}$$

Example 2. Absorption in Air of X-Rays. Calculate the absorption coefficients of dry air at room temperature and pressure for x-rays of $\lambda = 0.154$ nm and x-rays from a generator at 120 kV.

Density dry air, $\rho_{air}, = 1.2 \times 10^{-3}$ g/cm³. Air weight fractions are w(oxygen) 0.2, w(nitrogen) 0.8

For $\lambda = 0.154$ nm:

μ_m(oxygen) = 11.5 cm^2/g, μ_m(nitrogen) = 7.52 cm^2/g
[see Table 2.4c]

μ_m(air) = $0.2 \times 11.5 + 0.8 \times 7.52 = 8.32$ cm^2/g

μ_{air} (linear absorption coefficient) = $\mu_m \times \rho$
= $8.32 \times 1.2 \times 10^{-3} = 9.98 \times 10^{-3}$ cm^{-1}

For λ (120 kV generator):

Maximum in x-ray spectrum corresponds to ~80 kV

μ_m(oxygen) = 0.168 cm^2/g, μ_m(nitrogen)
= 0.164 cm^2/g   [see Table 2.4c]

Then:

μ_m(air) = $0.2 \times 0.168 + 0.8 \times 0.164 = 0.165$cm^2/g

μ_{air} (linear absorption coefficient)
= $1.2 \times 10^{-3} \times 0.165 = 0.198 \times 10^{-3}$ cm^{-1}

Note: The intensity transmitted by 1 m of air will be:

For $\lambda = 0.154$ nm, exp $- [9.98 \times 10^{-3} \times 10^2] = 37\%$

For x-rays from a 120 kV generator, exp $- [0.198 \times 10^{-3} \times 10^2] = 98\%$.

Example 3. Linear Absorption Coefficient of a Titanium-Nickel Alloy; Segregation Effects. Calculate the linear absorption coefficient of an alloy containing 51 wt% titanium (Ti), and 49 wt% nickel (Ni). Select an x-ray wavelength most able to reveal segregation effects in this

alloy, and determine the minimum size of segregated regions that can be identified.

Data: Absorption edges: Ti, 0.25 nm; Ni, 0.1489 nm. These values are obtained from Table 2.5, where λ_{nm} of absorption edge is given by $1.24/E_{keV}$ (see Eq 2.2a). Select Cu Kα $\lambda = 0.154$ nm just above the absorption edge of nickel. (This is discussed in the section "X-Ray Image Quality—Contrast" in this chapter.) Densities: Ti, 4.54 g/cm^3, Ni, 8.85 g/cm^3. Mass absorption coefficients: Ni, 45.7 cm^2/g; Ti, 208 cm^2/g (see Table 2.4c).

The density of the alloy
= $0.51 \times 4.54 + 0.49 \times 8.85 = 6.65$g/cm^3

The linear absorption coefficient of the alloy $\mu_{alloy} = 6.65 \times (0.51 \times 208 + 0.49 \times 45.7) = 854.4$ cm^{-1}. The difference between the linear absorption coefficients of titanium and nickel is

$[\mu_m\rho]_{Ti} - [\mu_m\rho]_{Ni} = 4.54 \times 208 - 8.85 \times 45.7$
$= 540$ cm^{-1}

Using Eq 2.44 for $G = 4$ at $D = 2$, the radiographic contrast

$C_S = 0.4343\ G(\mu_1 - \mu_2)\Delta = 0.4343 \times 4 \times 540\ \Delta = 0.2$

where Δ is the minimum size of segregated regions. The minimum acceptable contrast C_S is 0.2; see the section "Recording the X-Ray Image." Hence, $\Delta = 2.1 \times 10^{-4}$ cm.

Example 4. Detection of a Razor Blade in an Apple. An x-radiographic method is to be developed capable of detecting a razor blade in an apple. X-rays are available of wavelength

Table 2.14   Calculations for x-ray detection of a razor blade in an apple (Example 4)

Derivation	X-ray wavelengths		
	0.154 nm	**0.071 nm**	**(80 kV)**
1. Mass absorption coefficients for carbon (Table 2.4c), cm^2/g	5.5	0.53	0.161
2. Mass absorption coefficients for iron, cm^2/g	308	36.7	0.584
3. Fraction x-ray intensity transmitted through apple only: exp $- (0.25\ \mu_m \times 2 \times 2.54)$	9.3×10^{-4}	0.51	0.815
4. Fraction x-ray intensity transmitted through razor blade in apple: exp $- (7.87 \times \mu_m \times 0.025)$	~0	6×10^{-4}	0.89
5. Observations	Virtually no x-rays traverse the apple	~50% intensity through apple. Almost none through razor blade	Comparable intensities of x-rays traverse the apple and the razor blade
Conclusion	Use x-rays of $\lambda = 0.071$ nm		

0.154 nm (characteristic Cu Kα), 0.071 nm (characteristic Mo Kα), or x-rays from an x-ray generator using a tungsten target at 120 kV. The razor blade can be assumed to be 0.25 mm thick, of density 7.87 g/cm^3, in an apple of 2 in. diameter and of density 0.25 g/cm^3 consisting of carbon only; see Fig. 2.36. The maximum intensity of the white x-radiation from a generator at 120 kV will be at ~120/1.5 kV = 80 kV. The appropriate calculations are given in Table 2.14.

Example 5. Examination of a Tungsten Inclusion in an Aluminum Plate. A welded joint in an aluminum plate contains an inclusion of tungsten from the welding. The tungsten inclusion is 4×10^{-2} cm in diameter, and the aluminum plate is 10 cm thick (Fig. 2.37). What is a convenient x-ray tube voltage to employ? (The density of tungsten is 19.32 g/cm^3.) The radiographic contrast

$$C_S = D_1 - D_2 = 0.4343\, G \ln \frac{I_1}{I_2} \qquad \text{[see Eq 2.28b]}$$

Assume $G = 4$ (at $D = 2$); we require at least a minimum value of $C_S = 0.2$ so that

$$0.4343 \times 4 \times \ln \frac{I_1}{I_2} = 0.2$$

hence,

$$\frac{I_1}{I_2} = 1.122$$

From Fig. 2.37:

$$I_1 = I_0 \exp - (\mu_{Al} 10)$$

$$I_2 = I_0 \exp - \mu_{Al}[10 - 4 \times 10^{-2}]\, \exp- (\mu_w \times 4 \times 10^{-2})$$

so

$$\frac{I_1}{I_2} = \frac{1}{\exp - (\mu_w \times 4 \times 10^2)} = 1.122$$

Therefore:

$$\mu_w = 2.878 \text{ cm}^{-1}, \text{ hence, } \mu_w(m) = \frac{2.878}{19.32}$$
$$= 0.149 \text{ cm}^2/\text{g}$$

From Table 2.4(b), tungsten has mass absorption coefficients at 200, 400, and 600 kV of 0.747, 0.185, and 0.105 cm^2/g, respectively, so that 500 kV should be employed. These x-rays should correspond to I_{max} of the intensity distribution, so that the voltage on the x-ray tube voltage = 500×1.5 kV = 750 kV.

Fig. 2.36 Example 4. Detection of a razor blade in an apple using x-radiography.

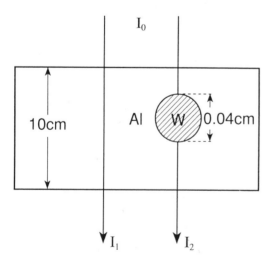

Fig. 2.37 Example 5. Aluminum with tungsten inclusion.

kV	Mass absorption coefficient (μ_m), cm^2/g (Table 2.4)	$\mu = \mu_m \cdot \rho$, cm^{-1}
20	3.375	9.146
100	1.706×10^{-1}	0.462

Example 6. Transmission through an Aluminum Rod. An aluminum rod, 1 cm in diameter, is to be radiographed using x-rays from:

- 30 kV generator (soft x-rays): maximum x-ray intensity corresponds to ~20 kV

- 150 kV generator (hard x-rays): maximum x-ray intensity corresponds to ~100 kV

Compare the transmitted x-ray intensities for aluminum at a density of 2.71 g/cm^3:

$$I_1 = I_0 \exp - [\mu x] \text{ and } x = 1 \text{ cm}$$

At 20 kV, $I_1/I_0 = \exp - [9.146 \times 1] = 1.1 \times 10^{-4}$. At 100 kV, $I_1/I_0 = \exp - [0.462 \times 1] = 0.63$.

Conclusion. At 30 kV almost no x-rays traverse the aluminum rod which appears opaque. However, at 150 kV over half of the x-rays traverse so that the aluminum rod is not totally opaque, and density differences or defects may become visible. The terms "soft" and "hard" x-rays are relative terms only.

Example 7. X-Ray Generator Voltage for Transmission across a Steel Plate. An x-ray beam is required to transmit 5% intensity through 1 in. of steel. Using Eq 2.8:

$$\frac{I_z}{I_0} = 0.05 = \exp(-\mu z) = \exp(-\mu \times 1 \times 2.54)$$

then

$$\mu_m = \frac{\mu}{\rho} = \frac{-\ln 0.05}{2.54 \times 7.87} = 0.15 \text{ cm}^2/\text{g}$$

where the density ρ of Fe is 7.87 g/cm^3.

From Table 2.4(b) the mass absorption coefficient μ_m of Fe has a value of 0.147 cm^2 at 200 kV. Because the maximum intensity in the x-ray spectrum occurs at approximately $1.5 \times$ excitation voltage, the required x-ray tube voltage will be 300 kV.

Example 8. Exposure Times for Metal Plates.

Steel Plate. The exposure is required for a steel plate, 1 in. thick using 200 kV x-rays. This is given by Fig. 2.31 to be 350 mA-min, so that an exposure of 10 min would be required at 35 mA at a source-to-film distance of one meter.

Copper Plate. A radiograph is to be taken at 250 kV of a copper plate, of thickness $\frac{1}{2}$ in. Using Table 2.12 the equivalent steel thickness is 0.5×1.45; i.e., 0.73 in. (18.4 mm) of steel. The exposure from Fig. 2.31 for 18.4 mm steel, at 250 kV, is 25 mA-s, so that if the x-ray tube current is 25 mA, the duration of the exposure will be ~1 s.

Example 9. Aluminum Plate with Voids. A plate of aluminum, 4 cm in thickness, contains a void of size 0.2 cm. A radiograph is taken of the aluminum plate using x-rays from a generator at 120 kV. Will the void be detected in film of gradient 5 (see Fig. 2.38)?

Case 1. The mass absorption coefficient of aluminum, using data from Table 2.4(b) is $\mu_m = 0.202 \text{ cm}^2/\text{g}$ and the density ρ of aluminum is 2.7 g/cm^3. (The maximum intensity of the x-ray spectrum is taken to be at $120/1.5$ kV = 80 kV.) The linear absorption $\mu = \mu_m\rho = 0.545 \text{ cm}^{-1}$. Then, from Fig. 2.38, $I_1 > I_2$.

$$\frac{I_2}{I_0} = \exp - [\mu\Delta_m] \text{ and } \frac{I_1}{I_0} = \exp - [\mu(\Delta_m - \Delta_i)] \qquad [\text{see Eq 2.8}]$$

so that

$$\frac{I_2}{I_1} = \exp - \mu\Delta_i \qquad [\text{see Eq 2.47b}]$$

$$\frac{I_1}{I_2} = \exp 0.545 \times 0.2$$

The contrast, using Eq 2.28(b)

$$C_S = 0.4343 \, G \ln \frac{I_1}{I_2} = 0.4343 \times 5 \times 0.545 \times 0.2 = 0.24$$

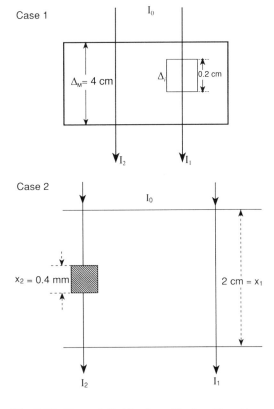

Fig. 2.38 Example 9. Aluminum blocks with voids.

This is greater than the minimum contrast that can be detected in x-ray films (see the section " Recording the X-Ray Image" in this chapter) so that the void is visible. *Note*: $I_1/I_0 = \exp - (0.545 \times 4) = 0.11$ so that sufficient x-rays traverse the specimen.

Case 2. Consider an aluminum plate of 2 cm in thickness containing an internal void 4×10^{-2} cm in size. If this plate is radiographed, what is the ratio of the intensity passing through the whole plate compared to that passing through the plate with the internal void (see Fig. 2.38)? Consider the case of:

- (a) X-rays from a generator at 120 kV with white radiation maximum at approximately 80 kV
- (b) Neutrons of $\lambda = 1.08$ Å
- (c) Gamma rays of 1 MeV

Now

$$\frac{I_1}{I_2} = \exp - \mu\Delta_i \qquad \text{[see Eq 2.47b]}$$

- (a) X-rays (80 kV) $\mu_m = 0.202$ cm²/g for Al (see Table 2.4b) $\rho_{Al} = 2.7$ g/cm³, therefore $\mu = 0.545$ cm⁻¹ and $I_1/I_2 = 0.978$
- (b) Neutrons $\mu_m = 0.036$ cm²/g. *Note*: Neutron mass absorption coefficients are listed in Ref 136. Therefore $\mu = 0.97$ cm⁻¹ and $I_1/I_2 = 0.996$
- (c) X-rays $\mu_m = 0.0613$ cm²/g, therefore $\mu = 0.166$ cm⁻¹ and $I_1/I_2 = 0.993$

Example 10. Void in Iron Plate. An iron plate, 0.5 cm thick, contains a void of diameter 5% of the plate thickness. How much will the transmitted x-ray intensity be increased by the presence of the void? Assume x-rays from a generator at 120 kV, with maximum in the white radiation at about 80 kV. The mass absorption coefficient of iron is 0.584 cm²/g for these x-rays and density of iron is 7.87 g/cm³.

$$I = I_0 \exp [-\mu z]$$

Therefore, the intensity transmitted by the iron plate

$$\frac{I_1}{I_0} = \exp - [7.87 \times 0.584 \times 0.5] = 0.100$$

The intensity transmitted in presence of void

$$\frac{I_2}{I_0} = \exp - [7.87 \times 0.584 \times 0.5 \times 0.95] = 0.113$$

Therefore, increase of intensity in presence of void

$$\frac{I_2}{I_1} = \frac{0.113}{0.100} \approx 11.3\% \text{ increase.}$$

Example 11. Copper-Titanium Bonded Plate. A plate of thickness 1 in. consists of copper bonded to titanium. An x-ray beam with maximum white radiation at approximately 100 kV is reduced in intensity by 99% on traversing the plate. What are the thicknesses of copper and titanium present (Fig. 2.39)?

Data	cm²/g μ_m at 100 kV,	Density (ρ),	$\mu_{cm}^{-1} = \mu_m\rho$
Copper	0.455	8.96	4.077
Titanium	0.274	4.54	1.244

See Table 2.4(b).

Now, $I/I_0 = 0.01 = \exp - [\mu_m\rho x)_{Cu} + (\mu_m\rho(2.54 - x])_{Ti}]$. Therefore, $x = 0.51$ cm copper; $2.54 - x = 2.03$ cm titanium. *Note*: These values are approximate because the x-ray beam is not monochromatic.

These results can be verified as follows:

- Copper of thickness 0.51 cm will reduce the x-ray beam by $e^{-4.077 \times 0.51} = 0.125$
- Titanium of thickness 2.03 cm will reduce the x-ray beam by $e^{-1.244 \times 2.03} = 0.08$

The two methods together will reduce the intensity by $0.125 \times 0.08 = 0.01$, that is, 99%.

Example 12. The Orientation of Flaws. The orientation of a fine crack in the x-ray beam can

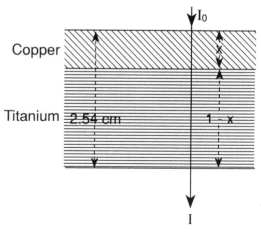

Fig. 2.39 Example 11. Copper-titanium bonded plate.

make all the difference between detection and nondetection, where the contrast can be either nearly zero, or quite high. To illustrate the importance of orientation, the two cracks illustrated in Fig. 2.4 will be considered. One crack is approximately parallel to the x-rays, the other nearly perpendicular to the beam. The contrast to be expected in each case can be calculated, using Eq 2.47(a), (b), and (c) for an inclusion of thickness Δ_i, of linear absorption coefficient μ_{inc} in a matrix of linear absorption coefficient μ. The radiographic contrast:

$$C_S = 0.4343\, G\, (\mu - \mu_{inc})\, \Delta_i \qquad \text{[see Eq 2.47a]}$$

In the case of a crack, there is a void so $\mu_{inc} = 0$ and

$$C_S = 0.4343\, G\mu\Delta_i \qquad \text{[see Eq 2.47c]}$$

so that contrast is directly proportional to the size of the crack.

Contrast from a Crack. A crack 3 cm long, 1 cm deep, and 0.01 cm wide is present in a $5 \times 5 \times 2$ cm block of aluminum (see Fig. 2.4). Determine the radiographic contrast in x-ray film of gradient $G = 4$ at film density $D = 2$, if the x-rays are parallel to the length of the crack and repeat for the case where the x-rays are perpendicular to the crack, being parallel to the crack width. Assume:

$$\mu_{void} \text{ in crack} = 0$$

$\mu_m = 0.202$ cm^2/g for aluminum using x-rays from a generator at 120 kV, with maximum intensity at approximately 80 kV. (See Table 2.4b).

Because the density of aluminum = 2.7 g/cm^3, μ will be $0.202 \times 2.7 = 0.545$ cm^{-1}.

$$\text{Contrast } C_S = 0.4343\, G\mu\Delta_i \qquad \text{[see Eq 2.47(c)]}$$

The contrast with the crack parallel to the x-ray beam

$$C_{S\parallel} = 0.4343 \times 0.545 \text{ cm}^{-1} \times 3 \text{ cm} \sim 2.34$$

The contrast with the crack perpendicular to the x-ray beam

$$C_{S\perp} = 0.4343 \times 0.545 \text{ cm}^{-1} \times 0.01 \text{ cm} = 0.009$$

A contrast of 0.009 is below the minimum for an observation (usually taken as 0.2). The image of the crack parallel to the x-ray beam will be easily

visible in the radiography, but not detectable when the x-ray beam is perpendicular to the crack length.

Example 13. Exposure of a Steel Plate with Flaws. A block of steel, 2.5 in. thick is to be examined for flaws of sizes down to 2 mm × 0.1 mm. An x-ray generator is available capable of energies up to 1 MeV, of effective source size 2.5 mm, and 20 mA current. Select the x-ray tube voltage, the x-ray tube current, and the source-to-film distance. The energy of x-rays to traverse 2.5 in. (63.5 mm) of steel will need to be about 350 kV (see Table 2.11), and these give rise to an unsharpness $U_g = U_f \approx 0.15$ mm (see Fig. 2.18).

The contrast parallel and perpendicular to the crack can be calculated as in Example 12.

$$C_S = 0.4343\, G\mu\Delta_i = 0.2 \text{ (minimum)}$$

Assume $G = 5$ at film density $D = 2$; for steel μ_m for "400 kV" x-rays is 0.0941 cm^2/g (see Table 2.4b), and density is 7.87 g/cm^3. Therefore

$$\mu = \mu_m\rho = 0.0941 \times 787 = 0.74 \text{ cm}^{-1}$$

Then

$$\Delta_i \text{ (minimum)} = \frac{0.2}{0.4343 G\mu}$$

$$= \frac{0.2}{0.4343 \times 5 \times 0.74 \text{ cm}^{-1}} = 0.12 \text{ cm}$$

This flaw will be visible only when the x-rays are parallel to the length (2 mm) and not visible when perpendicular to the length (0.1 mm).

Calculation for the most sensitive radiograph with $U_g = U_f$ is as follows:

$$U_f = U_g = (F \times l)/L \qquad \text{[see Fig. 2.16 and Table 2.13]}$$

so that, with l set equal to specimen thickness

$$\frac{U_g}{l} = \frac{F}{L_o} \qquad \text{[see Eq 2.29]}$$

$$\frac{0.15 \text{ mm}}{63.5 \text{ mm}} = \frac{2.5 \text{ mm}}{\text{Source–object distance}}$$

Source-object distance = 105.8 cm, and the source-film distance $(105.8 + 6.35) \sim 112$ cm. The exposure (see Fig. 2.31) for 63.5 mm of steel at 350 kV requires ~200 mA-min for a film density $D = 2$ at 1 m. At 112 cm, the exposure will be increased by a factor inversely proportional to the distance squared (see Eq 2.22); that is $\times (1.12)^2 = 250$ mA-min. Because the x-ray tube is capable of 20 mA, an exposure of about

12.5 min is required. *Note*: This is only a first estimate of exposure required.

Example 14. Examination of a Zinc Casting. An x-ray beam is to be used to examine a zinc casting, 2.5 cm thick, containing a spherical void 0.2 cm in diameter. Estimate exposure time, x-ray source-to-film distance for contact radiography, using an x-ray generator at 300 kV and 10 mA, with an effective source size of 3 mm, for film of gradient 4 (at $D = 2$).

At 300 kV, the maximum intensity of the x-ray spectrum is taken to correspond to $(300/1.5) = 200$ kV. The mass absorption coefficient for zinc from Table 2.4(b) is $\mu_m = 0.162$ cm^2/g, and the density $\rho = 7.140$ g/cm^3. The intensity traversing the whole zinc cast is

$$\frac{I_1}{I_0} = \exp - [\mu_m \rho z] = \exp - [0.162 \times 7.140 \times 2.5]$$
$$= 0.0555$$

The fraction of the intensity traversing the zinc cast at the void (assumed to contain a vacuum) is

$$\frac{I_2}{I_0} = \exp - [0.162 \times 7.140 \times 2.3] = 0.0699$$

so that

$$= \frac{I_2}{I_1} = 1.26$$

The radiographic contrast C_S can be derived using Eq 2.28(b):

$$C_S = 0.4343 \, G \ln \frac{I_2}{I_1} = 4 \times 0.4343 \ln 1.26 = 0.4$$

so that the void is clearly visible. At 300 kV, the value of U_f is ~0.01 cm (Fig. 2.18). Setting U_g equal to U_f, we have the minimum useful source-to-film distance, using Eq 2.29 with $\Delta_0 = l = 2.5$ cm, and $F = 0.4$ cm, then

$$L_0 = \frac{F \times l}{U_g} = \frac{0.4 \times 2.5}{0.01} = 100 \text{ cm} \qquad \text{[see Fig. 2.16]}$$

so that the source-to-film = $L + \Delta_0 = 102.5$ cm.

The exposure can be estimated as follows. The steel equivalence of zinc is ×1.3 (see Table 2.12), so that the 2.5 cm thick zinc casting is equivalent to 2.5×1.3 cm = 3.25 cm steel. Using Fig. 2.31, an estimate of the exposure at 300 kV for 3.25 cm steel for density, $D = 2$ at 1 m, is ~50 mA-min. At 10 mA, the exposure time would be ~5 min at 1 m. At a greater distance of 102.5 cm, the exposure is increased by ×(1.025)^2 to ~5.25 min.

Example 15. Shrinkage Voids in Cast Iron. The thickness of an iron casting and the size of void are to be determined by the relative absorption of x-rays. Observations are made using x-rays from a generator with a tungsten target at 120 kV at a distance of 2 m. The relative inten-

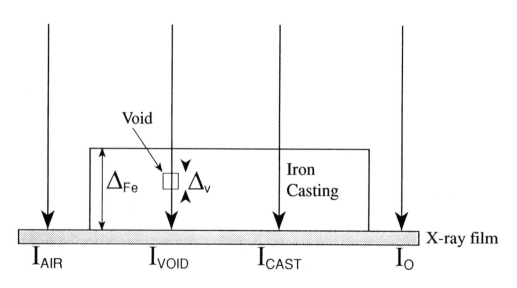

Fig. 2.40 Data for Example 15. X-ray source at 2 m. For I_0 through vacuum path, I_{air} through air path, I_{cast} through iron casting path and I_{void} through casting with void: $I_{air}/I_{void} = 190$, $I_{air}/I_{cast} = 200$. Δ_V is the size void; Δ_{Fe} is the thickness casting.

sities of the transmitted x-ray beams through air to that through the casting is 200, and the relative intensity of x-rays through air compared to the casting with a void is 190. This is illustrated in Fig. 2.40.

The mass absorption coefficients should correspond to the maximum x-ray intensities in the spectrum, 120/1.5 kV = 80 kV. Then,

- $\mu_m(\text{air}) = 0.165 \text{ cm}^2/\text{g}$ (see Example 2)
- $\mu_m(\text{iron}) = 0.367 \text{ cm}^2/\text{g}$ (see Table 2.4b)
- Density air $\rho_1 = 1.2 \times 10^{-3} \text{ g/cm}^3$ and iron $\rho_2 = 7.87 \text{ g/cm}^3$

Then, from Fig. 2.40

$$\frac{I_{\text{air}}}{I_0} = \exp - [\mu_m(\text{air}) \times \rho_1 \times 200]$$
$$= \exp - [0.165 \times 1.2 \times 10^{-3} \times 200] = 0.961$$

$$\frac{I_{\text{cast}}}{I_0} = \exp - [\mu_m(\text{iron}) \times \rho_2 \times \Delta_{\text{Fe}}]$$
$$= \exp - [0.367 \times 7.87 \times \Delta_{\text{Fe}}] = \exp - [2.89 \Delta_{\text{Fe}}]$$

$$\frac{I_{\text{void}}}{I_0} = \exp - [\mu_m(\text{iron}) \times \rho_2 \times (\Delta_{\text{Fe}} - \Delta_{\text{v}})]$$
$$= \exp - [2.89(\Delta_{\text{Fe}} - \Delta_{\text{v}})]$$

Therefore

$$\frac{I_{\text{air}}}{I_{\text{cast}}} = 200 = \frac{0.961}{\exp - 2.89\Delta_{\text{Fe}}}$$

so that $\Delta_{\text{Fe}} = 1.847$ cm.

$$\frac{I_{\text{air}}}{I_{\text{void}}} = 190 = \frac{0.961}{2.89(\Delta_{\text{Fe}} - \Delta_{\text{v}})}$$

so that $\Delta_{\text{v}} = 0.018$ cm. *Note*: These sizes are only approximate because the x-ray beam is not monochromatic.

Example 16. Copper Plate with Lead Shot.
A copper metal plate 2 cm thick is covered by a fine powder of lead shot. An x-ray generator is available with effective source size 2 mm capable of 400 kV and 10 mA and x-ray film of γ 5 at a film density of 2. What conditions should be used to obtain the sharpest radiograph in less than 10 min, and what is the minimum size of lead powder that can be detected? (See Fig. 2.41).

The sharpest radiograph is obtained using the minimum energy x-rays to give the minimum unsharpness. The steps in the calculations are given in Table 2.15 where the different columns are as follows:

- *Column 1*: The range of kV available on x-ray generator
- *Column 2*: The film unsharpness U_f dependence on the x-ray energy; see Fig. 2.18

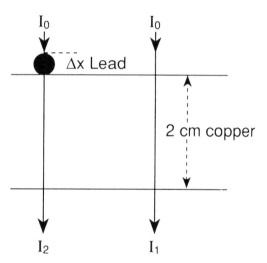

Fig. 2.41 Example 16. Copper plate with lead shot.

Table 2.15 Data for Example 16

1 X-ray generator, kV	2 U_f, mm	3 Copper equivalence steel, mm	4 Source- to-film $l(1 + F/U_f)$, cm	5 mA-min at 1 m	6 At source-to- film distance mA-min	7 Exposure min at 10 mA
400	0.15	11.8	14.3	<20	...	...
300	0.10	13.4	21	<20	...	...
250	0.09	13.8	23.2	~20	...	...
200	0.08	14.2	26	80	5.4	0.54
150	0.065	15.4	32	300	30.72	3.07
100	0.05	13.4	41	~20,000	3360	336
	From Fig. 2.18	From Table 2.12	From Fig. 2.16	From Fig. 2.31	Using Eq 2.22	...

Note: $l = 1$ cm, $F = 2$ mm

- *Column 3*: The steel equivalent of 1 cm copper plate; see Table 2.12
- *Column 4*: The source to film distance is $L_0 + l = l\,[1 + \dfrac{F}{U_f}]$ where $l = 1$ cm and $F = 2$ mm;
  see Fig. 2.16 and Eq 2.34 where $U_f = p$
- *Column 5*: Exposure in mA-min at 1 m distance for equivalent steel thickness; see Fig. 2.31
- *Column 6*: Exposure in mA-min at actual source to film distance from Column 5, using inverse square law; see Eq 2.22
- *Column 7*: The exposure in minutes using 10 mA on x-ray generator

The minimum energy giving an exposure in less than 10 min is 150 kV. At 150 kV, U_f is ~0.065 mm. Estimate size detectable as $2 \times U_g = 2 \times U_f \sim 0.13$ mm.

Example 17. Segregation in Copper-Zinc Brass Alloys. Part A estimates the degree of segregation that can be observed in a plate 1 cm thick of a copper-zinc brass alloy using x-radiation from a generator operating at 150 kV (maximum intensity at 100 kV). Part B suggests alternative x-radiation to measure the smallest degree of segregation in this alloy.

The x-ray examination of the Cu-Zn alloy is illustrated in Fig. 2.42. Assume $G = 4$ at $D = 2$.

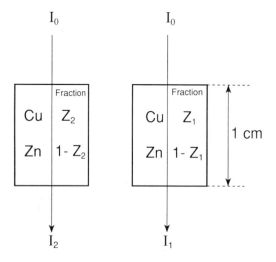

Fig. 2.42 Illustration for Example 17. The segregation can be considered as different path lengths in copper and zinc for the x-ray beam traversing the specimen. The volume fractions are treated as specimen thicknesses of z for copper and $(1 - z)$ for zinc.

Part A. From Table 2.4, μ_m Cu at 100 kV is 0.455 cm^2/g; μ_m Zn at 100 kV is 0.490 cm^2/g. The degree of segregation is given by $Z_2 - Z_1$ (see Fig. 2.42).

$$\frac{I_1}{I_0} = \exp - \mu_{Cu}\,Z_1 \exp - \mu_{Zn}(1 - Z_1)$$
$$= \exp - [\mu_{Cu}Z_1 + \mu_{Zn}(1 - Z_1)]$$

and

$$\frac{I_2}{I_0} = \exp - \mu_{Cu}\,Z_2 \exp - \mu_{Zn}(1 - Z_2)$$
$$= \exp - [\mu_{Cu}Z_2 + \mu_{Zn}(1 - Z_2)]$$

so that

$$\ln \frac{I_1}{I_2} = (\mu_{Cu} - \mu_{Zn})(Z_2 - Z_1)$$

and

$$Z_2 - Z_1 = \frac{\ln I_1/I_2}{\mu_{Cu} - \mu_{Zn}}$$

Now

$$\mu_{Cu} - \mu_{Zn} = [\mu_m\rho]_{Cu} - [\mu_m\rho]_{Zn}$$
$$= 8.9 \times 0.455 - 7.13 \times 0.4 = 0.554 \text{ cm}^{-1}$$

For minimum contrast

$$C_S = 0.2 = 0.4343\,G\,\ln\frac{I_1}{I_2} \qquad \text{[see Eq 2.28b]}$$

Therefore,

$$\ln \frac{I_1}{I_2} = 0.115$$

Therefore, the minimum degree of segregation that can be observed is

$$Z_2 - Z_1 = \frac{\ln I_1/I_2}{\mu_{Cu} - \mu_{Zn}} = \frac{0.115}{0.554} = 0.208 \ (\approx 20\%)$$

The minimum volume concentration differences for contrast is 20.8% with x-ray generator at 150 kV.

Part B (use of absorption edges). The difference $Z_2 - Z_1$ for minimum contrast decreases with increase of $\Delta\mu = \mu_{Cu} - \mu_{Zn}$. This value $\Delta\mu$ is increased using an x-radiation between absorption edges; see the section " X-Ray Image Quality" in this chapter.

The absorption edge for copper is 0.1381 nm and for zinc is 0.1284 nm (see Table 2.5). Use of radiation such as $\lambda = 0.1295$ nm will provide a large value of $\Delta\mu$. (*Note*: $\lambda = 0.1295$ nm can be generated using a zinc target in an x-ray tube with a monochromator.) The mass absorption coefficients μ_m are, for $\lambda = 0.1295$ nm:

- Copper: 246 cm^2/g (density, $\rho = 8.90$ g/cm^3)
- Zinc: 36.3 cm^2/g (density, $\rho = 7.133$ g/cm^3) (from Table 2.4)

This gives

$$\Delta\mu = [\mu_m\rho]_{Cu} - [\mu_m\rho]_{Zn} = 1936 \text{ cm}^{-1}$$

Thus, the minimum degree of segregation detectable becomes

$$Z_2 - Z_1 = \frac{0.115}{\Delta\mu} = \frac{0.115}{1936} = 5.9 \times 10^{-5} \ (\approx 0.01\%)$$

It is to be noted how the careful choice of x-ray wavelength changes the detection of degree of separation by three orders of magnitude.

Example 18. X-Radiograph of a Composite Metal (Detailed Calculations). A radiograph is to be taken of a composite copper-aluminum-lead plate, consisting of 3.6 mm copper, 1.5 mm aluminum, and 0.9 mm lead. An x-ray generator is available, with effective source size of 3 mm, capable of up to 0.3 MV and 8 mA. What source to film distance and other conditions should be used to obtain the sharpest radiograph in under 2 min?

The following steps in the derivation are detailed in Table 2.16.

- *Step 1*: It is necessary to determine the equivalent thickness of steel. This can be obtained using the data from Table 2.12 at each possible voltage. These are presented in Table 2.16(a); for example at 300 kV, the equivalent thickness of steel is 21 mm, and at 100 kV it is 13 mm of steel
- *Step 2*: The exposures at 1 m are determined using Fig. 2.31 using the equivalent thickness of steel, and are recorded in Table 2.16(b), column 2. The exposure is in mA-min
- *Step 3*: The value of U_f at each voltage is obtained from Fig. 2.18 and listed in Table 2.16(b), column 3. The sharpest radiograph corresponds to $U_g = U_f$
- *Step 4*: The distance from source to object, L_o, is now calculated using the fact that $L_o = Fl/U_f$, and these values of L_o are recorded in Table 2.16(b), column 4
- *Step 5*: The distance of the x-ray source to film is $(L_o + l)$, and these are listed in Table 2.16(b), column 5. l = thickness specimen = 6 mm
- *Step 6*: Finally, the exposure in mA-min is obtained, making allowance for the distance being different from 1 m and using the fact that the intensity decreases with the square

Table 2.16(a)   Calculations for Example 18 (using Table 2.12)

kV	Copper	Aluminum	Lead	Equivalence thickness steel, mm
300	[3.6 × 1.5	+ 1.5 × 0.19	+ 0.9 × 17]	21
250	[3.6 × 1.45	+ 1.5 × 0.17	+ 0.9 × 15]	19
200	[3.6 × 1.4	+ 1.5 × 0.16	+ 0.9 × 13]	17
150	[3.6 × 1.3	+ 1.5 × 0.12	+ 0.9 × 11]	15
100	[3.6 × 1.5	+ 1.5 × 0.08	+ 0.9 × 8.3]	13

Table 2.16(b)   Calculations for Example 18

kV	Required exposure at 1 m (Fig. 2.31) mA-min	Using Fig. 2.18 U_f, mm	(Fig. 2.16) Distance x-ray source to object, m	Distance x-ray to film, m	Exposure correction for distance mA-min	Exposure minutes (using 8 mA)
1	2	3	4	5	6	7
300	10	0.12	0.150	0.156	10 × (0.156)^2	0.03
250	35	0.10	0.180	0.186	35 × (0.186)^2	0.15
200	90	0.08	0.225	0.231	90 × (0.231)^2	0.60
150	350	0.06	0.300	0.306	350 × (0.306)^2	4.1
100	>5000	0.04	0.450	0.456	5000) × (0.45)^2	>127

$F = 3$ mm; $l = 6$ mm

of the distance; see Table 2.16(b), column 6. These exposures are recorded in column 7 using 8 mA on the x-ray tube

The conditions for the radiograph required in under two minutes are 200 kV at 0.231 m source-film distance, with an exposure of 0.6 min. The lowest voltage possible is chosen to reduce unsharpness.

Example 19. Radiograph of a Polymethylmethacrylate Plate Containing Internal Voids. What size (dimension along the beam) of a hole can be observed in a radiograph with contrast equal to 0.2 in a plate of polymethylmethacrylate (PMMA) of thickness 1 in., using x-rays from a generator at 120 kV or γ-rays of 1MeV?

- When the hole is empty?
- When the hole is full of ice?

Assume an x-ray film is to be used with gradient G = 5 at D = 2.

PMMA is the polymer polymethylmethacrylate for which the mer is –CH$_3$C$_2$O$_2$CH$_3$CH$_3$–, of density ρ_1 = 1.2 g/cm^3. The mer is of molecular weight 100, and the weight fractions are: W_H for hydrogen, 8%; W_O for oxygen, 32%; W_C for carbon, 60%. The density of ice (ρ)$_2$ is approximately 0.95 g/cm^3. The weight fractions in ice are approximately W_H = 0.11; W_O = 0.89. The PMMA plate is illustrated in Fig. 2.43. The maximum in the x-ray spectrum occurs at approximately ⅔ energy of the x-ray generator, so that the mass absorption coefficients for PMMA and for ice are calculated assuming 80 kV using Eq 2.21.

Calculation of Absorption Coefficients for PMMA. Using x-rays (80 kV): μ_m for hydrogen is 0.309 cm^2/g; for carbon, 0.161 cm^2/g; for oxygen, 0.168 cm^2/g (from Table 2.4b)

$$\mu_m(PMMA)$$
$$= 0.08 \times 0.309 + 0.6 \times 0.161 + 0.32 \times 0.168$$
$$= 0.176 \text{ cm}^2/\text{g}$$

$$\mu_1 = \mu_m(PMMA) \times \rho_1 = 0.176 \times 1.2 = 0.211 \text{ cm}^{-1}$$

Using γ-rays (1 MeV): μ_m for hydrogen is 0.126 cm^2/g for carbon, 0.064 cm^2/g; for oxygen, 0.064 cm^2/g (from Table 2.4b)

$$\mu_m(PMMA)$$
$$= 0.08 \times 0.126 + 0.6 \times 0.064 + 0.32 \times 0.064$$
$$= 0.069 \text{ cm}^2/\text{g}$$

$$\mu_2 = \mu_m(PMMA) \times \rho_1 = 0.069 \times 1.2 = 0.086 \text{ cm}^{-1}$$

Absorption Coefficients of Ice. Using x-rays (80 kV):

$$\mu_m(H_2O) = 0.11 \times 0.309 + 0.89 \times 0.168$$
$$= 0.184 \text{ cm}^2/\text{g}$$

$$\mu_3 = \mu_m(H_2O) \times \rho_2 = 0.184 \times 0.95 = 0.175 \text{ cm}^{-1}$$

Using γ-rays (1 MeV):

$$\mu_m(H_2O) = 0.11 \times 0.26 + 0.89 \times 0.064$$
$$= 0.071 \text{ cm}^2/\text{g}$$

$$\mu_4 = \mu_m(H_2O) \times \rho_2 = 0.071 \times 0.95 = 0.067 \text{ cm}^{-1}$$

To summarize the linear absorption coefficients:

	X-rays at 80 kV, cm^{-1}	Gamma rays at 1 MeV, cm^{-1}
PMMA	μ_1 0.211	μ_2 0.086
Ice	μ_3 0.175	μ_4 0.067

The required minimum contrast

$$C_S = 0.2 = 0.4343 \ G \ln \frac{I_1}{I_2} \quad \text{[see Eq 2.28b]}$$

$$= 0.4343 \text{ x } 5 \text{ x } \ln \frac{I_1}{I_2}, \text{ so that } \frac{I_1}{I_2} = 1.10$$

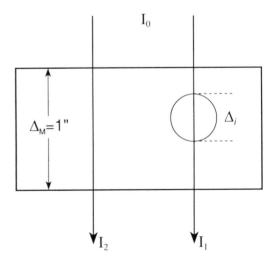

Fig. 2.43 Example 19. Hole in a block of polymethylmethacrylate. The hole may be empty or full of ice. A spherical hole will have less sharply defined edges.

From Fig. 2.43:

$$I_1 = I_o \exp\left[-\mu_m(\Delta_m - \Delta_i)\right]\exp - \mu_i\Delta_i$$
$$= I_0 \exp\left[-\mu_m\Delta_m + \Delta_i(\mu_m - \mu_i)\right]$$

$$I_2 = I_0 \exp - [\mu_m\,\Delta_m]$$

Therefore:

$$\frac{I_1}{I_2} = \exp\left[\Delta_i(\mu_m - \mu_i)\right] = 1.10\ \text{for minimum contrast}$$

Then

$$\Delta_i = \frac{\ln 1.10}{\mu_{PMMA} - \mu_i} = \frac{0.095}{\mu_{PMMA} - \mu_i}$$

Hole Empty:

$$\mu_i = 0,\ \Delta_i = \frac{0.095}{\mu_{PMMA}}$$

For x-ray (80 kV):

$$\Delta_i = \frac{0.095}{0.211} = 0.45\ \text{cm}$$

For γ-rays (1 MeV):

$$\Delta_i = \frac{0.095}{0.086} = 1.10\ \text{cm}$$

Hole Full of Ice: For x-ray (80 kV):

$$\Delta_i = \frac{0.095}{0.211 - 0.175} = 2.64\ \text{cm}$$

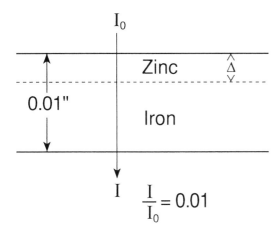

Fig. 2.44   Illustration for Example 20.

For γ-rays (1 MeV):

$$\Delta_i = \frac{0.095}{0.086 - 0.067} = 5.0\ \text{cm}$$

The calculated size of hole full of ice is more than 100% of the PMMA thickness. In conclusion, an empty hole in the PMMA plate can be observed, but a hole full of ice will not be in contrast.

Example 20. Zinc-Coated Iron Foil. A zinc-coated iron foil of total thickness 0.01 in. reduces monochromatic x-rays of λ = 0.056 nm (characteristic K_α x-rays from a silver target) by 99%. Determine the thickness of the zinc coating; see Fig. 2.44.

Data: mass absorption coefficients (Table 2.4c): iron, 19.7 cm²/g; zinc, 55.46 cm²/g. Densities: iron, 7.87 g/cm³; zinc, 7.133 g/cm³. The linear absorption coefficients are:

$$\mu_{Fe} = 19.7 \times 7.879 = 155\ \text{cm}^{-1}$$

$$\mu_{Zn} = 55.46 \times 7.133 = 396\ \text{cm}^{-1}$$

so that

$$\mu_{Zn} - \mu_{Fe} = 396 - 155 = 241\ \text{cm}^{-1}$$

If the thickness of zinc is Δ cm, the iron thickness is (0.0254 – Δ) cm, then

$$\frac{I}{I_0} = 0.01 = \exp - [\mu_{Zn}\Delta]\exp - [\mu_{Fe}(0.0254 - \Delta)]$$

Therefore:

$$\ln 0.01 = -\mu_{Zn}\Delta - \mu_{Fe}(0.0254 - \Delta)$$

so that:

$$\Delta = \frac{4.6 - \mu_{Fe} \times 0.0254}{\mu_{Zn} - \mu_{Fe}}$$

Hence, Δ = 0.003 cm = 0.0014 in. Therefore, the thickness of iron = 0.0086 in.

Example 21. Thickness of Titanium Plate. What thickness variation can be detected by x-rays, of maximum intensity of white radiation corresponding to 80 kV, in a titanium plate of specification thickness 1 cm? It can be assumed that a change of (a) 2%, (b) 5%, in the transmitted intensity is the minimum difference that can be detected.

Data:

- μ_m, mass absorption coefficient of titanium is 0.405 cm^2/g (see Table 2.4b)
- Density of titanium is 4.54 g/cm^3

The specified thickness $\Delta_1 = 1$ cm gives rise to:

$$\frac{I_1}{I_0} = \exp - [\mu_m\Delta_1] = \exp - [0.405 \times 4.54 \times 1]$$
$$= 0.159$$

The thickness Δ_2 corresponding to a change in transmitted intensity of 2% is given by

$$1.02\,\frac{I_1}{I_0} = 1.02 \times 0.159 = \exp - [0.405 \times 4.54 \times \Delta_2]$$

Therefore:

$$\Delta_2 = 0.989 \text{ cm}$$

This represents a variation of 0.011 cm from the specified thickness of 1 cm. Similarly for 5% change, $\Delta_3 = 0.974$ cm. Hence, the thickness variations detectable are at (a) 2%, 0.011 cm; (b) 5%, 0.026 cm in 1 cm.

Example 22. Shielding for a Radioactive Source. A technician will be working at a distance of 10 feet from an Ir^{192} radioactive source of 12 Ci activity. What lead protection should be installed?

This Ir^{192} source will give rise to 12×5.9 R/hour at 1 foot; see Table 2.2 and the section "Radiation Units" in this chapter. At 10 feet, the radiation intensity will be, by the inverse square law of distance, $[12 \times 5.9\text{R}/10^2]$/h, that is 0.708 R/hour, which corresponds to 28.32 R/week of 40 working hours. For biological effects on human beings, 1 R corresponds to 1 rem, so that the dose per week is 28.32 rem. The permitted dose is 0.1 rem per week (see the section "Safety Precautions and Shielding" in this chapter), so the radiation level needs to be reduced by a factor $0.1/28.32 = 3.53 \times 10^{-3}$.

The tenth value layer (TVL) for lead for Ir^{192} is 0.64 in.; see Table 2.10. Four such layers will reduce the intensity by 10^{-4}; that is, 4×0.64 in. = 2.56 in. lead will be needed. *Note*: The shielding should give a greater reduction than estimated.

Example 23. Shielding for an X-Ray Generator. What shielding must be provided so that an x-ray beam generated at 300 kV can be reduced by a factor of 1000?

From Table 2.6, at 300 kV, HVL for lead is 0.16 in. and for concrete is 1.2 in. Ten thicknesses will reduce the intensity by $2^{10} = 1024$, that is 16 in. of lead, or 12 in. of concrete. *Note*: These are approximate values.

Example 24. X-Radiograph Film Density of Steel Plate. A steel plate requires an exposure of 200 mA-s at 300 kV to obtain a film density D = 1.5 on type II film. What exposure is required to obtain $D = 2.5$?

At $D = 1.5$, the log relative exposure is 1.8. $D = 2.5$ under the same conditions requires a log relative exposure of 2.05; see Fig. 2.45. Therefore,

$$\text{Required exposure} = 200 \text{ mA-s}_x 10^{(2.05 - 1.80)}$$
$$= 356 \text{ mA-s}$$

Example 25. X-Radiograph of an Aluminum Plate with 2% Sensitivity. An x-radiograph is to be taken of an aluminum plate, 3 in. thick. A sensitivity of 2% is required with a film density D of 1.5 on type II film, using lead screens (as given on Fig. 2.29).

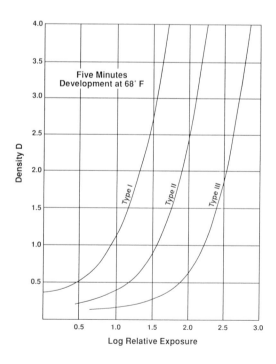

Fig. 2.45 Film characteristic curves. H & D (Hurter & Driffield) Curves, provided by the manufacturer. Type I is a slower film providing higher contrast. See Examples 24, 25, and 27 (Ref 67, p 6-37).

The rule-of-thumb suggests a target-to-film distance of 10×3 in. = 30 in.

1. Using Fig. 2.29, a 3 in. thick specimen at 36 in. and $D = 2$ requires exposures of either 120 kV, 3200 mA-s; or 140 kV, 1200 mA-s; or 160 kV, 500 mA-s
2. Using Table 2.12, the steel equivalent of 3 in. aluminum at 150 kV is 3×0.12 in., i.e. 0.36 in.
3. Using Fig. 2.46, a 2% sensitivity for 0.36 in. requires the voltage to be between 220 and 135 kV. Select 160 kV because this will give the shortest exposure
4. Using Fig. 2.45 for $D = 2$ on type II film, the log ε (relative exposure) is 1.9, and for $D = 1.5$, log $\varepsilon = 1.75$. This means the exposure is changed by $\times 10^{(1.75 - 1.9)} = \times 0.69$
5. The exposure at 160 kV is now 500 mA-s $\times$ 0.69 = 345 mA-s
6. This exposure is for a specimen at 36 in. A distance of 30 in. requires a shorter exposure

(inverse square law with distance) of

$$\times \left(\frac{30}{36}\right)^2, \text{ i.e. } 345 \times \left(\frac{30}{36}\right)^2 = 240 \text{ mA-s}$$

Example 26. Gamma Radiograph of Aluminum Plate with 2% Sensitivity. The same specimen considered in Example 25 is to be radiographed using an Ir^{192} radioactive source. The decay curve of the radioactive source is shown in Fig. 2.47, which is provided with the source. The Ir^{192} source is to be used when 100 days old, and so has an activity of 20 Ci.

1. The steel equivalent (see Table 2.12) of aluminum for Ir^{192} radiation is 0.35, so that 3 in. of aluminum becomes 3×0.35 in. = 1.05 in. equivalent steel.
2. From Fig. 2.46(a), Ir^{192} can be used with 2% sensitivity to examine this thickness of steel equivalence
3. Using Fig. 2.32, the exposure factor EF for type A film for 1.05 in. steel equivalence EF = 22
4. Using Eq 2.56

$$t_{mins} = \frac{EF \times L_s^2}{S_R \times 144}$$

where S_R = 20 Ci, L_s = 30 in., and EF = 22; so that $t = 6.88$ min

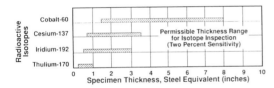

(a)

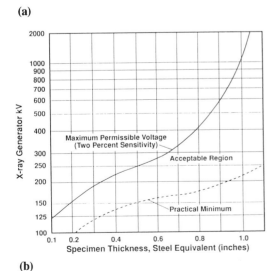

(b)

Fig. 2.46 Gamma-rays and x-ray generator voltages to obtain a sensitivity of 2% (Ref 67). See Examples 25 to 27.

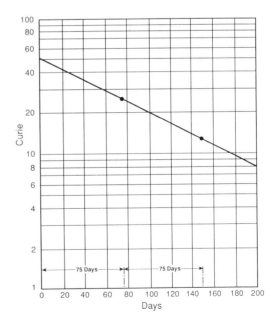

Fig. 2.47 Ir^{192} decay curve (provided with the radioactive source).

5. To change to $D = 1.5$, using the correction factor determined for Example 25 of 0.69

Exposure (min) = $6.88 \times 0.69 \cong 4.75$ min

Note: These calculations are accurate to within approximately 10%.

Example 27. Radiograph of a Plate of Nonuniform Thickness. A radiograph is required of a steel plate varying in thickness from 0.3 to 0.4 in. A sensitivity of 2% is required and film densities are acceptable between $D = 1.5$ and 3.0.

1. Using Fig. 2.46(b), exposures for 2% sensitivity: for 0.3 in. are between 120 and 190 kV, and for 0.4 in. are between 135 and 225 kV. Select 160 kV.
2. Using Fig. 2.30 with $D = 2$, type film II, at 36 in. target-to-film with lead screens then 0.3 in. steel is exposed to 220 mA-s at 160 kV, and 0.4 in. steel is exposed to 400 mA-s at 160 kV
3. Using Fig. 2.45 for type II film, the log relative exposure at $D = 1.5$ is 1.75; at $D = 2$, 1.9; and at $D = 3$, 2.1. Hence the exposures are changed by $\times 10^{(1.75 - 1.9)} = 0.7$ at $D = 1.5$ and $\times 10^{(2.1 - 1.9)} = 1.6$ at $D = 3.0$
4. Thus, the range of exposures at 160 kV are

Steel plate thickness, in.	Film density, D	Exposure, mA · s
0.3	1.5	$220 \times 0.7 = 154$
	3.0	$220 \times 1.6 = 352$
0.4	1.5	$400 \times 0.7 = 280$
	3.0	$400 \times 1.6 = 1640$

Hence, at 160 kV, and D between 1.5 and 3.0, permitted exposures can range from 280 to 352 mA-s (suggest 300 mA-s).

Safety Precautions and Shielding

Exposure to x-rays and γ-rays represents a potentially severe health hazard so that care and intelligence must be used when working with these radiations. The danger of x-ray burns was observed very early when Dally, the assistant of Edison, who demonstrated x-rays at an exhibition in 1896, died in 1904 of x-ray burns. It is said that Edison discontinued all further work with x-rays (Ref 40). A review of the biological effects of x-radiation is given in Ref 55. A very clear description of safety and protection is given in Chapter 5 of Ref 67.

Safety precautions are prescribed by several organizations such as the International Commission on Radiation Protection (ICRP); see also Ref 184 to 188. There are regulations covering the maximum permissible levels of absorbed radiation dose. The radiographic and radiological units of exposure dose, absorbed dose, radiation intensity, radioactive isotope source intensity, specific activity of a radioactive source, and related units (rad, rem, rbe, Roentgen, Curie) are defined in the section "Radiation Units" in this chapter.

The maximum permissible dose accumulated by the whole body recommended by ICRP is

$$H = 5 (u - 18) \text{ rems} \qquad \text{(Eq 2.57)}$$

where H is the accumulated dose in rem, and u is the age of the individual in years. This assumes no exposure before the age of 18, and an average absorbed dose of 0.1 rem per working week, and the exposure in any one year must not exceed 12 rem. An "accumulated dose" concept is used, and this is illustrated in Fig. 2.48. (In the section "Radiation Units," it is pointed out that for the human body, a dose in Roentgen equals the dose in rem.)

Shielding. Useful empirical formulas (Ref 55) for a radioactive source are:

Roentgen-hour at one foot = $5.6 A_S \cdot E_g$ (Eq 2.58)

Danger range of an unshielded radioactive source D_R is

$$D_R = \frac{5000 A_S \cdot E_g}{d^2} \text{ cm} \qquad \text{(Eq 2.59)}$$

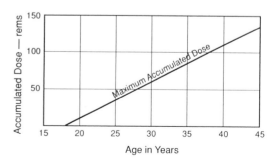

Fig. 2.48 The permitted accumulated dose is $5(u - 18)$ rems where u is the age of the operator in years.

where A_S is the activity of the radioactive source in Ci, E_g is the total gamma radiation energy, in MeV, and d is distance in cm.

Shielding must be used to protect from x-rays and γ-rays, and thicknesses of lead or concrete are used to reduce the intensity by a large factor (perhaps 10^{-4}). A TVL reduction of 90% in intensity as given in some tables, for example Table 2.10, is probably suitable for scattered x-rays, but not for the direct x-ray beam. Typical calculations of shielding are given in Examples 22 and 23. Data concerning x-ray shielding design, dose build-up factors are given in Ref 55, 56, 58, 67, and 184-189.

Intensifying screens are not sufficient in themselves to stop the x-ray beam. Observations must be carried out only in reflection, using remote viewing system, or by using a radiation barrier window made of lead glass, lead-containing plastic, or a lead-containing liquid of minimum density 6 g/cm^3.

It is important to have available a convenient x-ray radiation detector, and this should be the first item of equipment to have available (and in working condition). There are several good radiation detectors on the market:

- Victoreen 440; see Ref 103, p 740
- Baird Atomics 904-416CP, see Ref 103, p 741
- Eberline E-112B and PNR-4, see Ref 103, p 742
- Mini-Instruments, Model X, Southampton Street, London, WC2, U.K.

- Rad Alert, Perspective Scientific Ltd., Baker Street, London W1M 1LA, U.K.
- Monitor 4, S.E. International Instrument Division, 156, Drakes Lane, Summertown, TN, USA

Details for the first three units are given in Ref 103. The last two units are very compact in size, permitting the examination of small spaces that are sometimes inaccessible to the larger units. Government regulations require periodic calibration of radiation monitors when used in connection with licensed radioactive materials.

Personnel Monitoring. All personnel working in the vicinity of x-ray generators or radioactive isotopes should be required to carry

Table 2.17(a)   ASTM Reference radiographs

Standard No.	Description
E 155	Aluminum and Magnesium Castings (1991)
E 186	Steel Castings Heavy-Walled (2 to $4\frac{1}{2}$ in.) (1991)
E 192	Investment Steel Castings for Aerospace Applications (1991)
E 242	Radiographic Images as Certain Parameters are Changed (1991)
E 272	High-Strength Copper-Based and Nickel-Copper Alloy Castings (1991)
E 280	Steel Castings Heavy-Walled ($4\frac{1}{2}$ to 12 in.) (1991)
E 310	Tin Bronze Castings (1991)
E 390	Steel Fusion Welds (1991)
E 446	Steel Castings up to 2 in. Thickness (1991)
E 505	Aluminum and Magnesium Die Castings (1991)
E 689	Ductile Iron Castings (1990)
E 802	Gray Iron Castings up to $4\frac{1}{2}$ in. Thickness (1991)
E 1320	Titanium Castings (1991)

Table 2.17(b)   ASTM standards for radiography

Standard No.	Description
E 94	Radiographic Testing (1991)
E 142	Controlling Quality of Radiographic Testing (1992)
E 431	Radiographs of Semiconductors (1992)
E 545	Thermal Neutron Radiographic Examination (1991)
E 592	Penetrameter Sensitivity for Radiography of Steel Plates (1989)
E 746	Radiographic Film (1987)
E 747	Radiographic Examination Using Wire Penetrameters (1990)
E 748	Thermal Neutron Radiography
E 801	Radiographic Testing of Electronic Devices (1991)
E 803	Neutron Radiography Beams (1991)
E 999	Industrial Radiographic Film Processing (1990)
E 1000	Radioscopy (1992)
E 1025	Hole-Type Image Quality Indicators (1989)
E 1030	Radiographic Examination of Castings (1991)
E 1032	Radiographic Examination of Weldments (1992)
E 1079	Transmission Densitometers (1991)
E 1114	Iridium-192 Industrial Radiographic Sources (1986)
E 1161	Radiographic Testing of Semiconductors (1987)
E 1165	Focal Spots of Industrial X-Ray Tubes by Pinhole Imaging (1992)
E 1254	Radiographs and Industrial Radiographic Films (1988)
E 1255	Radioscopy (1992)
E 1390	Illuminators Used for Viewing Industrial Radiographs (1990)
E 1411	Qualification of Radioscopic Systems (1991)
E 1416	Radioscopic Examination of Weldments (1991)
E 1441	Computed Tomography Imaging (1992)
E 1453	Storage of Media that contains Analog or Digital Radioscopic Data (1992)
E 1475	Computerized Transfer of Digital Radiological Test Data (1992)

individual monitoring devices. These can be:

- *Film Badges*: The film needs to be developed to compare with reference standards.
- *Ionization Chamber Pocket Dosimeters*: These have the advantage of being easy to read and instantaneous
- *Thermoluminescent Dosimeters*

These devices are described further in Ref 103, p 736 to 745. Regulations governing the use of personnel monitoring devices are described in several handbooks published by the National Bureau of Standards (Ref 58), and in the *Radiological Health Handbook* (Ref 55). Other useful references are 186 to 188.

ASTM Reference Radiographs

A most important and essential service is provided by ASTM in developing a series of codes

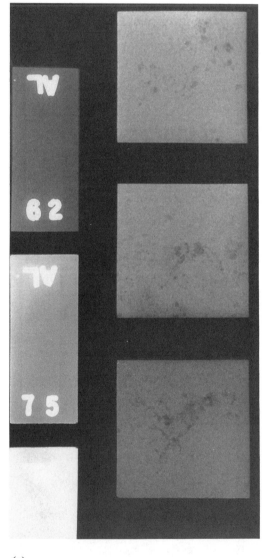

(a)

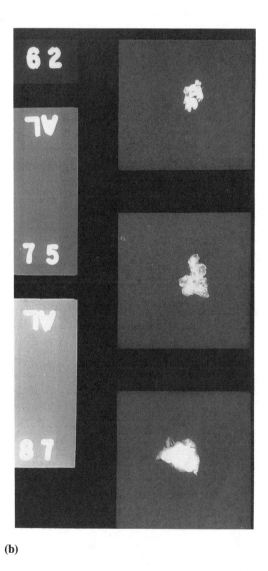

(b)

Fig. 2.49 (a) and (b) Reference radiographs from the ASTM Series (Ref 61). (a) Low-density inclusions (probably sand from cast in an aluminum plate; from E 155. Penetrators alongside are for ⅝ in. and ¾ in. thickness. (b) Tungsten inclusions in an aluminum plate; from E 155. Penetrameters alongside are MIL Standard 453C ¾ in. and ⅞ in.

(continued)

and standards covering all aspects of radiography. These are listed in Table 2.17(a) and (b). In particular, the reference radiographs play a key role in all examinations by radiography; see Table 2.17(a).

Some photographs shown in this text, Fig. 2.5 and 2.49 are taken from these ASTM reference data (with the permission of ASTM). A glossary of nomenclature is also available in ASTM document E 1316.

Appendix 2.1: Glossary of Some Terms used in Radiography (Ref 55, 61, 190)

absorbed dose. The energy imparted by ionizing radiation per unit mass of irradiated material; measured by "rad" where 1 rad = 0.01 J/kg. The SI unit is "Gray" Gy where 1 Gy = 1 J/kg.

absorbed dose rate. Absorbed dose per unit time; rad/s. The SI unit is Gy/s.

absorption (attenuation) coefficient, linear. The fractional attenuation of x-rays per unit length of material.

absorption (attenuation) coefficient, mass. The fractional attenuation of x-rays per unit mass of material.

absorption edge. A discontinuity in the absorption curve where, as the energy of an x-ray is increased, the absorption in the

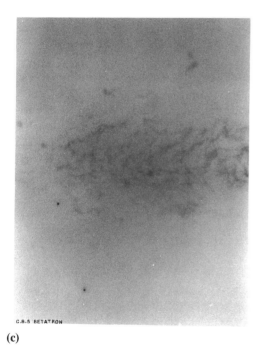

(c)

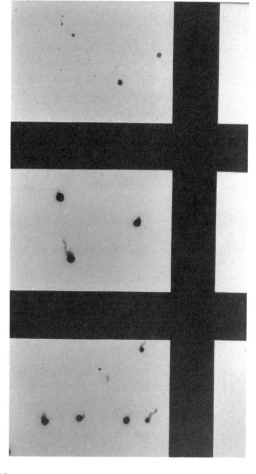

(d)

Fig. 2.49 (c) and (d) Reference radiographs from the ASTM Series (Ref 61). (c) Shrinkage in a steel plate; from E 280. Steel thicknesses of 4½ in. to 12 in. can be examined by 4 to 30 MeV. (d) Gas holes in a steel investment casting; from E 192. Reprinted with permission of ASTM.

solid increases abruptly at levels where an electron is ejected from the atom.

activity. The number of nuclear disintegrations that occur in unit time in a quantity of radioactive material.

angstrom (Å). Measurement of length, 1 Å = 10^{-10} m.

antiscatter grid. A lead-slatted grid between object and film orientated so that only radiation in the direction of the primary beam reaches the film.

autoradiography. A substance, rendered radioactive, is placed in contact with a film so that a map of the radioactive areas appears on the film.

barn. A unit of area. One barn = 10^{-28} m^2.

becquerel (Bq). The SI unit of activity of a radioactive isotope; one nuclear transformation per second. 3.7×10^{10} Bq = 1 Curie (Ci).

betatron. An electron accelerator in an evacuated toroid, with a special magnetic field constraining the electrons to a circular orbit. This type of equipment usually operates at energies between 5 and 30 MeV.

bremsstrahlung (white radiation). The electromagnetic radiation resulting from the retardation of charged particles, usually electrons.

broad beam attenuation (absorption). The x-ray attenuation when contributions from all sources, including secondary radiation, are included.

build-up factor. Detector output due to all x-rays, including secondaries, compared to the output when only primary radiation is considered.

cathode rays. A stream of electrons emitted from the cathode when an electric discharge takes place in a gas at low pressures ($\sim 10^{-2}$ Pa).

characteristic curve (H & D, Hurter & Driffield) of photographic film. The plot of density versus log of exposure or of relative exposure; see Fig. 2.45.

characteristic x-radiation. Each element, when bombarded with electrons, emits x-rays of a characteristic wavelength depending on the wave number.

collimator. A device of radiation absorbent material defining the direction and angular divergence of the radiation beam.

Compton scattering. When a photon collides with an electron it may not lose all its energy, and a lower energy photon will then be emitted from the atom at an angle to the incident photon path; see Fig. 2.9(c).

contrast sensitivity. The minimum change in an object that produces a perceptible density change in the radiograph.

cumulative dose. The total amount of radiation received in a specified time.

Curie (Ci). That quantity of a radioactive isotope which decays at the rate of 3.7×10^{10} disintegration per second.

decay. The gradual decrease in the activity of a radioactive source.

density (film). The quantitative measure of film blackening when light is transmitted.

$$D = \log \frac{B_0}{B_T}$$

where: D = density, B_0 = light intensity incident on the film, and B_T = intensity transmitted.

density gradient (γ) G. The slope of the curve of density against log exposure for a film; see Fig. 2.13.

diffraction. Interference effects giving rise to illumination beyond the geometrical shadow. This effect becomes important when the dimensions of the apertures or obstructions are comparable to the wavelength of the radiation.

dosemeter (dosimeter). A device to measure the x-ray dose to which an object, or person, has been subjected.

electron volt (eV). 1 eV is equal to the kinetic energy acquired by an electron when accelerated through a potential difference of 1 volt.

equivalent activity. The activity of a point source of the same radioisotope giving the same exposure rate at the same distance from the center of the source.

exposure. (1) The quantity of ionization produced in unit volume of air. The SI unit is the coulomb/kilogram. (1 C/kg = 3876 Roentgen). (2) The dose received by a film, expressed by (tube current) × (time of exposure).

exposure range (latitude). The range of exposures over which a film can be employed usefully; see Fig. 2.13.

film badge. A film carried or worn by radiographic personnel indicating the exposure to radiation by the film density after a given time.

film contrast. The slope of the characteristic curve of a photographic material that is related to the density difference resulting from a given exposure difference.

film speed. The time required to achieve a given density under given conditions; see Fig. 2.13.

flash radiography. Very short, high intensity x-ray pulses to examine rapidly moving objects.

fluoroscopy. Observation of an x-ray image on a fluorescent screen.

fog. A term used to denote any increase in density of a processed photographic emulsion caused by (for example):

- *Aging*: deterioration due to film stored for too long a period of time, or under incorrect conditions
- *Base*: the uniform density inherent in a processed emulsion
- *Chemical*: unwanted reactions during processing

Fraunhofer diffraction. The light source and detector are effectively at an infinite distance from the scatterer.

Fresnel diffraction. The light source or detector are at a finite distance from the scatterer.

Fresnel fringe. A fringe formed under *Fresnel diffraction* conditions.

gamma-ray (γ-ray). Electromagnetic radiation, comparable to x-rays, but occurring as a result of a radioactive event.

geometric unsharpness U_g (penumbra). The penumbral shadow in a radiograph.

graininess. The impression of irregularity of silver deposit in a radiograph.

Gray (Gy). The SI unit of absorbed dose. 1 Gy = 1 Joule/kg = 100 rad.

half-life. The time for one-half of a given number of radioactive atoms to decay.

half-value layer (HVL) [half-value thickness (HVT)]. The thickness of a substance which, introduced into the path of a beam of radiation, reduces its intensity by one-half.

Image quality indicator (IQI). The minimum discernible image as described by the designated hole in a plaque-type, or the designated wire image in the wire type IQI.

image quality indicator (IQI) (penetrameter). A device whose image provides visual and qualitative data of quality and sensitivity in the radiograph.

intensifying screen. A screen that converts radiographic energy into light or electrons and reduces the exposure time required to produce a radiograph. The two processes most commonly used are:

- *Metal screen*: consisting of a dense metal (usually lead) that emits primary electrons when exposed to x- or γ-rays
- *Fluorescent salt screen*: consisting of a coating of phosphors that fluoresces when exposed to x- or γ-radiation

laminography. The specimen and the detector move in step so that one image plane only of the specimen remains relatively sharp on the detector; see Fig. 2.26(c).

latent image. An image in a detecting receptor that requires further processing to convert into a visible image; e.g., film developing.

linear accelerator (linac). A high energy x-ray source in which the electrons are accelerated in a straight line by passing through resonant cavities in the correct phase before impinging on a target; see Fig. 2.6(d)and (e).

line pairs per millimeter. A measure of the spatial resolution of an image consisting of a test pattern of pairs of equal width, high contrast lines and spaces.

non-screen type film (direct-type film). X-ray film for use with or without metal screens, but not for use with fluorescent salt screens.

nuclear activity. The number of disintegrations in a specific quantity of material per unit of time. This is specified in Curie (Ci) where $1\ Ci = 3.7 \times 10^{10}$ disintegrations per second.

pair production. An incident photon with energy greater than 1.02 MeV is converted into an electron-positron pair. Subsequent annihilation of the positron results in the production of two 0.511 MeV γ photons; see Fig. 2.9(d).

penumbra. See *geometric unsharpness*.

photodisintegration. The capture of an x-ray photon by an atomic nucleus with the subsequent emission of a particle; see Fig. 2.9(e).

photoelectric effect. A photon transfers its energy to an electron to eject it from the atom; see Fig. 2.9(b).

quality factor. See *relative biological effectiveness.*

rad. Unit of absorbed dose of radiation equal to 0.01 J/kg in the medium.

radiographic contrast. The density difference between an image and its surroundings on a radiograph.

radiographic equivalence factor. By which the thickness of a material must be multiplied to determine the thickness of a standard material having the same absorption.

radiographic exposure. The subjection of a recording medium to radiation for the purpose of producing an image.

radiographic filter. Use of metal foils to absorb the longer softer x-ray wavelengths leaving a higher proportion of harder x-rays in a beam.

radioscopy (real-time radioscopy). The electronic production of a radiological image.

relative biological effectiveness. A factor used to compare the biological effects of radiation doses of different types of ionizing radiations. It is the experimentally determined ratio of the radiation dose compared to a reference radiation dose to produce identical biological effects; see Ref 55. Also known as "quality factor."

Roentgen (Röntgen). Unit of exposure. One roentgen (R) is the quantity of x- or γ-radiation that produces, in dry air at normal temperature and pressure, ions carrying one electrostatic unit of quantity of electricity of either sign.

Roentgen Equivalent Man (rem). The unit of dose used in radiation protection. It is given by the absorbed rad multiplied by a quality factor, Q.

specific activity. The activity of a radioactive isotope per gram of solid.

tenth-value-layer (TVL). The thickness of a layer of material that reduces the intensity when introduced into the path of a beam of radiation by a factor of ten.

tomography ("the picture of a slice"). A procedure to obtain an image of a cross section at one level of the specimen.

unsharpness (film). The blurring of a radiographic image due to the inherent resolution of the film.

Chapter III

Ultrasonic Testing

Part I General Discussion

Introduction

Sound travels by the vibrations of the atoms and molecules present, traveling with a velocity depending on the mechanical properties of the medium. Imperfections and inclusions in solids cause sound waves to be scattered, resulting in echoes, reverberations, and a general dampening of the sound wave. It is well known that the state of perfection of a solid can be deduced from the ringing of china cups, of crystal glassware, and the tapping of train wheels (Ref 191). In this type of test, the resonance of the object in the audible range is used, and that gives a good idea of the state of the object. In liquids, sound can travel great distances; Leonardo da Vinci described in 1490 a simple hydrophone whereby the ear, placed over a long tube inserted into the sea, enables one to hear ships at great distances; indeed, recently a series of sound pulses near Perth, Australia, could be detected at Bermuda, Jamaica (Ref 192, 193). However, the method of da Vinci does not indicate any direction of the sound, and directions are not readily obtainable by simple methods. More precise techniques are necessary to locate the sources of the sound waves and this requires the use of a narrow beam of sound acting as a "searchlight" in the solid. The use of sound waves of short wavelength (high frequency) is illustrated in Fig. 3.1, for the case when the transducer diameter is much greater than the sound wavelength.

Nondestructive testing is carried out using ultrasonic waves of high frequency above the audible range, that is, above ~20 kHz. Sound waves of frequency above 20 kHz are also

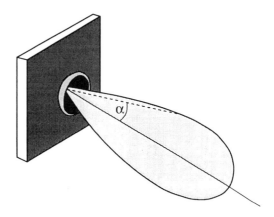

Fig. 3.1 Direction of sound radiating from a transducer. Sound from a transducer is concentrated in a half-angle α, determined by the direction of the first minimum in the sound diffraction pattern. This applies for the far-field beam away from the immediate vicinity of the transducer, and when the transducer diameter Δ is much greater than the sound wavelength λ. As an example, in the case of steel ultrasonic waves of frequency 1 MHz, and using a source of 3 cm diameter will give rise to a divergent angle $2\alpha \cong 14°$.

known as ultrasound or ultrasonics. Measurements by sound waves at sea are known as SONAR or ASDIC observations (Ref 30, 194).

The acoustic spectra of various mechanical waves are shown in Table 3.1. The normal audible range is from about 20 to 20,000 Hz. The velocity (V) of sound waves depends on the medium and varies from about 300 to 6000 m/s. The range of frequencies used in ultrasonic testing is from less than 0.1 to greater than 15 MHz, and typical values of wavelengths in ultrasonic testing are from 1 to 10 mm (Ref 163, 194-196).

Historical Background

Echo ranging detection of objects at sea was proposed by Richardson in a patent in 1912, subsequent to the sinking of the ocean liner Titanic. Fessenden designed a transducer for a source of high-frequency sound waves for submarine signaling and echo ranging, and he was able to detect an iceberg at a distance of 2 miles in 1914, using 1000 Hz sound waves. During World War I, Chilowsky and Langevin experimented in France with ultrasonic waves, and Langevin developed a source of ultrasonic waves using the piezoelectric effect with quartz crystals between steel plates. In 1918, Langevin was able to detect echoes from a submarine at distances of 1.5 km (Ref 30, 197, 198).

Table 3.1 Acoustic spectra of mechanical waves

Frequencies (ν), Hz cycles/s	Common sources of sound	Comments
10^{-2}	Atmospheric thunder	
10^{-1}	Ship's propeller	Infrasound
10^0	Heavy trucks	
10^1	Low notes	
10^2		
10^3	kHz	Normal audible 20-20,000 Hz
10^4	Noise of waves at sea	Underwater sonar 20-500 kHz
10^5		
10^6	MHz	Tool on a grindstone
10^7		Ultrasonic NDT 0.1-15 MHz
10^8		
10^9	GHz	Highest frequencies attained

Source: Ref 163, 194-196

Other experiments were carried out by Boyle and by Wood and Loomis in 1927 (Ref 199) using quartz piezoelectric transducers. Pierce (1928) developed the magnetostriction device as an ultrasonic oscillator (Ref 200). ASDIC (World War I) and SONAR (World War II) are terms used to describe the underwater echo techniques of ranging objects (e.g., submarines, shoals of fish) and measuring sea-depth. Behun in 1921 reported a method of measuring the depth of the sea bed by an ultrasonic resonance method, using standing waves (Ref 195, 201). Nicholson carried out ultrasonic studies using the piezoelectric effect of Rochelle salt, potassium sodium tartrate ($KNaC_4H_4O_6 \cdot 4H_2O$) (Ref 202).

In 1929 and 1935, Sokolov discussed the use of ultrasonic waves in detecting defects in metal objects (Ref 203, 204), and Mulhauser in 1931 obtained a patent using ultrasonic waves to detect flaws in solids, using a transmission mode operating with two transducers—one to transmit, the other to receive the ultrasonic waves (Ref 191, 195, 205). Sokolov carried out many pioneering experiments in ultrasonic testing: investigating coupling liquids, matching impedances, and developing methods of obtaining an image of a defect by a scanning technique; his patent was the first issued in the United States for ultrasonic testing (Ref 206). An instrument based on his work was marketed under the tradename Ultrasonel (Ref 191, 207). Schraiber in 1940 developed methods of continuous ultrasonic wave testing (Ref 1), and Erwin (Ref 208) reported in 1945 on thickness measurement by ultrasonic methods.

A very important development came with the studies of Firestone in 1940 and Simmons in 1945 of pulsed ultrasonic testing, using the echo principle (Ref 209-211). This method was shown to be relatively simple and quick. Moreover, the pulsed method avoided many of the problems associated with standing-wave formation. Present day methods of ultrasonic testing have evolved from this method. Firestone designed an instrument, called the "Reflectoscope," in which a pulse of ultrasonic waves was transmitted into the body under examination, and the time interval measured for the pulse to return.

The pulse echo method in which the same transducer transmits and receives the ultrasonic pulse has become the most generally used system. The basic system developed by Firestone (1940), Sokolov (1941) (Ref 210), and Sproule (1945) (Ref 212) is essentially the same as that described by Hueter and Bolt in 1955 (Ref 213). This method is discussed in the section "Outline of Pulse Echo and Transmission Methods of NDT" in this chapter.

The development of ultrasonic cameras to provide an image of a defect has been attempted by Pohlman (Ref 214) and by Sokolov (Ref 215) and more successfully recently by Sandhu and Thomas (Ref 216).

A review of ultrasonic measurements of the position of defects in solids is reported by Doyle and Scala (Ref 217) and the development of time-of-flight measurements in Ref 218. A listing of the history of biomedical ultrasonics is given by Ref 166. A short outline of the development of pulse transit-time (pulse echo) method is given in Ref 191 and 219 and the development of underwater detection is reviewed in Ref 220.

Ultrasonic Waves

When a disturbance occurs at one end of a solid, it travels through the solid in a finite time as a sound wave by the vibrations of the molecules, atoms, or particles present. These vibrations lead to a propagated wave traveling through the medium with wavelength λ ranging from as long as 10,000 m to very short λ of 10^{-5} m. The very long λ correspond to the sound emitted by a ship's propeller in water, and the very short λ are used in some medical and industrial research applications. Examples of the different frequencies ν of the mechanical vibration spectrum is given in Table 3.1 (Ref 195) [λ and ν are related to the velocity V of the waves by $V = \lambda\nu$].

Audible sound corresponds to the range of frequencies of approximately 20 to 20,000 Hz (cycles per second). Ultrasound, ultrasonics, or sonics are the terms used to describe mechanical vibration waves above the audible range (frequency greater than 20,000 Hz). (Supersonics refers to the motion of objects or particles faster than the velocity of sound in air, or in the medium in question.)

Sounds travel at different velocities through different media (see Tables 3.3 and 3.6); sound velocities vary very little with frequency in most metals. The wavelengths, frequencies, and wave velocities used in nondestructive testing are generally:

- *Wavelengths*: 1 to 10 mm
- *Frequencies*: 0.1 to over 15 MHz; most applications use below 10 MHz

- *Sound velocities*: 1 to 10 km/s

A simple illustration of the vibrations in a solid is shown in Fig. 3.2, where a blow at one end of a bar can result in longitudinal vibrations or in transverse vibrations, so that a pulse of waves traverse the bar. Waves propagate in solids and depend on the resistance of the atoms of the solid to vibrate when a force is applied; that is, the impedance Z. In nondestructive testing, cracks, boundaries, or inclusions are detected by the change in Z between the different media, when scattering and reflection of sound waves occur.

Outline of Pulse Echo and Through-Transmission Methods

Ultrasonic waves are transmitted through solids over distances of several meters in fine-grain

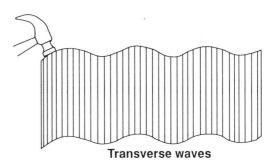

Transverse waves

(a)

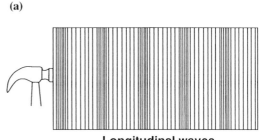

Longitudinal waves

(b)

Fig. 3.2 Transverse (a) and longitudinal (b) vibrations and waves in a bulk specimen. The ability of a solid to vibrate under an applied force is characterized by the specific acoustic impedance Z. Impedance (the ratio of cause to effect) can be considered as the resistance of the material to the passage of sound waves. Sound waves do not propagate in vacua because the vibration of particles is essential.

steels, but only about 10 cm in some cast irons. Discontinuities or defects cause scattering and reflection of the waves, and the detection of the reflected or transmitted waves permits the defect to be located. The general dispositions of the various methods are illustrated in Fig. 3.3. The arrangements can use one transducer only, as in the pulse echo (PE) (A-scan) method in Fig. 3.3(a), or two transducers, as in (b), (c), and (d). Transmission methods require access to both sides of the specimen, while back reflection

methods can manage when access is restricted to one side only of the specimen. The coupling transducer-specimen surface must be very good. This transmission method is used when small defects are present which do not give adequate reflection signals in the pulse echo mode. The transmission signal is reduced in intensity if small defects are present. The pulsed through-transmission testing is quite common and is often used for thin sheet metals where pulse-echo testing is impractical due to the dead zone effect. It

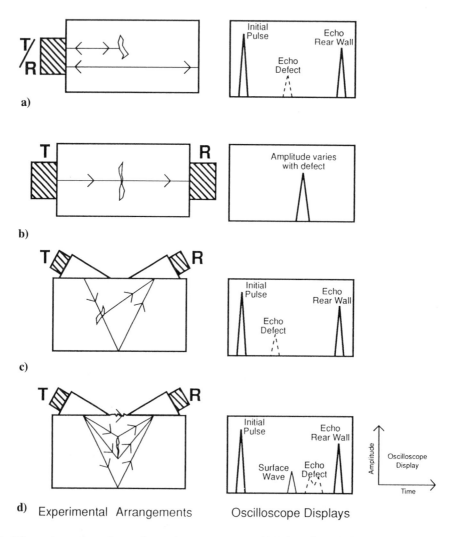

d) Experimental Arrangements   Oscilloscope Displays

Fig. 3.3 Ultrasonic wave nondestructive testing arrangements. (a) Pulse echo techniques (PE) (pulse transit time). (b) The transmission method, pulsed through-transmission testing, requires two transducers with access to both sides of the specimen. (c) Reflection method requires two transducers (pitch-catch) but access to one side only of the specimen. (d) An alternative transmission-reflection crack tip diffraction technique compares the time taken by the surface wave to the time taken by waves diffracted from cracks within the specimen. Note: T, transmitter transducer; R, receiver transducer; T/R, transmitter and receiver transducer.

is also used for the inspection of composites for large flaws. Methods are available where reflections from the far side of the specimen, or from flaws within the specimen, are picked up by a second transducer, as shown in Fig. 3.3(c) and (d). This method of angular transmission permits the selection of the shear mode only to traverse and be reflected by the specimen. A continuous wave CW or a pulse echo PE method can be employed. The distance between transmitter and detector transducers is kept constant with good surface coupling. When a pulse technique is used, an oscilloscope is required to detect the time as well as the amplitude of the transmitted pulse. The angle-beam pulse re-

flections method (Fig. 3.3d) is known as crack-tip diffraction. In this method, the time for the surface wave traveling to the receiver is compared to other signals scattered from the far surface or diffracted from a crack as illustrated. The size of the crack can be derived if the scattered (diffracted) beams for the two ends of the crack can be time resolved in the detector system (Ref 218).

The pulse echo method is used extensively and is shown in Fig. 3.4. A transducer touches the surface of the specimen through a coupling liquid, sending a pulse of ultrasonic waves traveling through the medium. Alternatively, the specimen may be immersed in water, with the probe immersed as well. The pulse is reflected by a discontinuity or from the rear surface and is detected by the same transducer. The time required for the pulse to travel out and back can be displayed on an oscilloscope, and it is common to use a regular train of ultrasonic pulses so that the oscilloscope signal is more easily observed. The initial pulse may last 1 µs (1 to 10 cycles of vibration), and the pulses repeat every millisecond.

The presence of a flaw gives rise to a signal earlier than that from the rear surface. In the

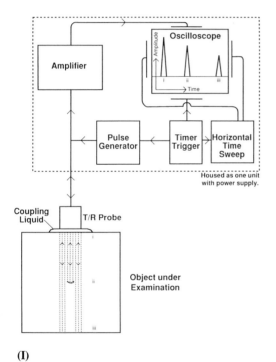

(I)

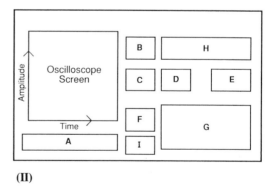

(II)

Fig. 3.4 Pulse echo (PE) A-scan. I, A transducer probe (T/R probe) acts as both transmitter and receiver and is placed against the surface of the object under examination, using a couplant. The pulser circuit and oscilloscope time-base are usually synchronized by a pulse from a timer-trigger circuit. The horizontal sweep depends on time; the vertical amplitude depends on the signal from the probe. The emitted pulse i, known as the initial pulse, is noted on the oscilloscope, and later reflections ii, iii are also displayed. In the present illustration, if signals ii and iii are observed after 30 µs and 50 µs, respectively, in steel, the specimen thickness would be 150 mm and the flaw at 90 mm from the upper surface. II, A review of the controls available. A, oscilloscope adjustments: vertical, horizontal, intensity, focus, astigmatism, scale illumination, power; B, Gate: delay, width, alarm, level, or sensitivity. Permits automatic alarm if a signal is received; C, Damping: pulse duration (improved resolution with higher damping); D, Reject: Adjust baseline to remove noise level; E, Gain dB: Increase amplitude echo signal on oscilloscope; F, Marker: Square waves below sweep line for time/distance calibration; G, Sweep: controls pulse rate to suit material and specimen thickness, delay position of initial pulse on left side or off-screen, and controls which part of the test pattern is on screen; H, Frequency/Selection: broad bands 1-15 MHz, tuned bands 1, 2, 5, 10, 15, MHz; I, DAC: distance amplitude correction to compensate for drop in amplitude with depth in specimen.

illustration in Fig. 3.4, suppose the rear surface gives rise to a signal at 50 μs on the oscilloscope. Because the velocity in the steel specimen is 6 km/s, the specimen thickness is 50 μs × 6 km/s × 0.5 (time × velocity × ½ because the signal travels back and forth in the specimen), that is 150 mm. Hence, a flaw corresponding to the signal at 30 μs will be located at 90 mm.

Interpretation of Oscilloscope Displays

The display observed on the oscilloscope needs to be interpreted by the observer, and this can only be accomplished when details of the specimen under investigation are known and fully used. It is essential to consider the nature of the specimen, to determine the most likely positions and types of defects, and to arrange the experimental examination accordingly. Only in this way can significant and useful information be extracted from the observations on the oscilloscope of the presence or absence of echo pulses.

Some examples of oscilloscope displays (amplitude pulses versus time) are given in Fig. 3.5

for a variety of defects. As shown in Fig. 3.5(a), a discontinuity perpendicular to the beam will reflect a pulse to the transducer. A discontinuity parallel to the beam will reflect at best a very weak pulse. The signal will be stronger from a defect with rough sides; smooth-sided defects have polar radiation plots (see Fig. 3.14II). A defect of large curvature with a range of slopes to the beam will give rise to a series of weak reflections. The reflection from the rear wall will decrease as flaws reflect the beam pulse. The absence of all reflections may be because the flaws are inclined and reflect the beam away from the detector and because the rear wall is not parallel to the front wall.

Flaws not parallel to the surface can be observed by using one transducer as transmitter and receiver at angular incidence. A typical indication is shown from a nonmetallic inclusion (Fig. 3.5b).

The curved specimen (Fig. 3.5c) is examined by a single transducer probe at angular incidence with the aid of a curved shoe of plexiglass to make good contact with the curved surface. There is no signal from the rear wall, unless a discontinuity is present. This method of examination is chosen if surface cracks are sus-

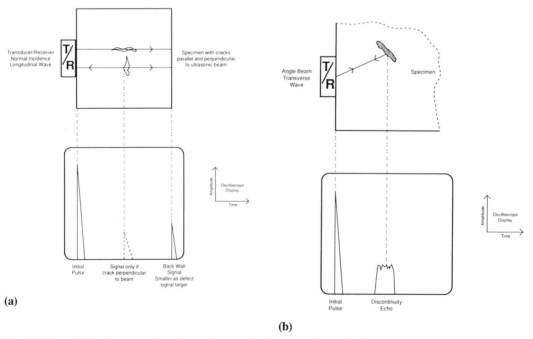

(a)

(b)

Fig. 3.5(a) and (b) Interpretation of oscilloscope displays. (a) Discontinuity parallel and perpendicular to the beam. (b) Flaws not parallel to any surface; nonmetallic inclusion. *(Continued)*

pected. In the case of a fine-grain steel, it is reasonable to use frequencies of 2.5 to 5 MHz.

For a specimen with a curved back surface (Fig. 3.5d), one transducer probe is used at normal incidence (angular incidence could also be chosen). There is no signal from the back wall, and the only signal observed will be from a

discontinuity, as shown. If the specimen were of brass, which has a relatively large grain size, lower frequencies of 0.5 to 1 MHz would be appropriate using a transducer of $1/2$ in. diameter.

In the case of multiple flaws (Fig. 3.5e), one large flaw can screen a more distant flaw. If two small defects are present, as shown in the left

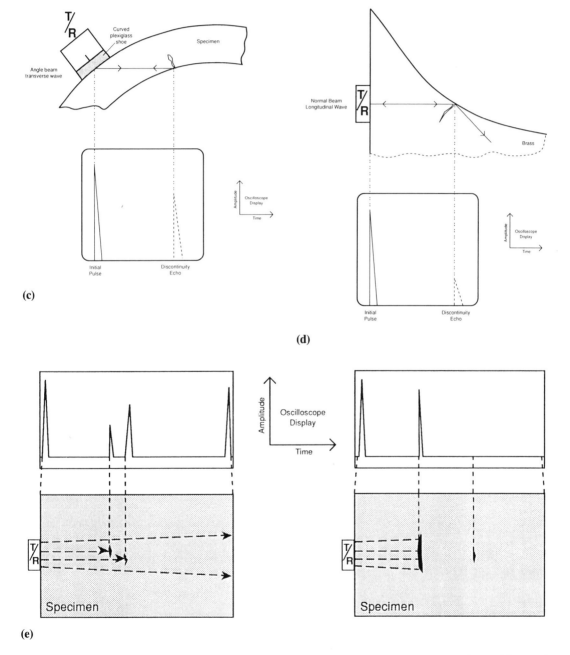

Fig. 3.5(c), (d) and (e) Interpretation of oscilloscope displays. (c) Curved specimen. (d) Specimen with curved back surface. (e) Multiple flaws.

side of Fig. 3.5(e), echo pulses may be seen from them and from the rear wall. However, if the defect is large, as shown in the right side of Fig. 3.5(e), no signal may be received from more distant defects or the rear wall. "Flaking" tends to give multiple signals.

Other oscilloscope examples related to thickness measurements, the examination of welds, immersion in water, and metal components, are described in Part III "Applications" in this chapter.

When viewing oscilloscope displays, the following general notes apply:

- The oscilloscope provides a plot of amplitude (ordinate) versus time (abscissa). The ultrasonic pulses are often shown with positive amplitude only, by rectifying the pulses. (RF waves can also be plotted.)
- The transducer probe is labeled T as transmitter, R as receiver, or T/R as transmitter and receiver.
- The normal beam (or 0° beam) is a longitudinal (compression) wave.
- The angle beam is generally a transverse (shear) wave, though a "refracted longitudinal" beam is sometimes used with an entry angle within 10° of the normal.
- The initial pulse is recorded on the oscilloscope at time $t = 0$. This can be displaced off the screen to the left using the "sweep delay" control.
- The ultrasonic beam is diverging (see Fig. 3.32e).

Practical Tips

1. Castings have a coarse grain structure that gives rise to background noise or "hash" and beam distortions. This can be reduced by increasing the wavelength compared to the average grain size (using lower frequencies or longitudinal waves) and changing the direction of the sound beam using an angle-beam transducer. Very coarse castings are unsuitable for ultrasonic testing. Most castings are probably unsuitable for ultrasonic testing, partly because of the grain structure but also because castings tend to be used for complex shapes where ultrasonic testing does not work so well.
2. Steel forgings have fine grain structures and thus have low damping. Forgings are suitable for ultrasonic testing. Inspections

should be made perpendicular to the direction of working.
3. Rolled sheet and plate are suitable for ultrasonic testing. Normal beam is used to detect laminations. Angle beam permits fast inspection of plate material for most discontinuities except smooth laminations parallel to the surface.
4. Shear waves do not travel in liquids, gases, or some plastics. Shear wavelengths are about half longitudinal wavelengths, so that shear waves can detect smaller discontinuities.
5. Ultrasonic equipment should be tested before use with a calibration block.
6. Transducer application to a rough surface results in undesirable reduction of signal amplitude in the test piece due to general scattering of the beam.
7. Discontinuities with rough surfaces scatter the sound beams more than discontinuities with smooth surfaces.
8. Nonmetallic inclusions are generally rough sided and thus scatter more than cracks do.
9. Long specimens examined by a diverging beam are likely to give nonrelevant signals; waves reflected from the sides undergo mode conversion giving spurious signals that are received after that from the rear wall. A larger diameter transducer should be employed.
10. Larger diameter transducers have sharper less-diverging beams for a given frequency.
11. Small diameter high-frequency transducers are used to detect small discontinuities.
12. Large diameter low-frequency transducers have better penetration properties.
13. The higher the frequency, the thinner the crystal transducer. Above 10 MHz, the transducers are too thin for contact and are used for immersion testing.
14. Nonrelevant indications can result from mode conversions. Shear wave reflections arrive after the rear-wall signal. This is not usually a problem unless using "full skip" for weld testing.
15. The shape of the specimen can give rise to nonrelevant indications. The transducer should be moved to a different position to confirm a discontinuity signal.
16. The velocity of longitudinal waves in water is approximately $\frac{1}{4}$ the velocity in metals such as steel and aluminum. A signal through a 1 in. path of water will appear on

the oscilloscope near that of 4 in. of steel or aluminum. For immersion testing, it is convenient to use water path greater than 1 in. for every 4 in. path of metal.

17. During immersion testing, the location of the beam entering the immersed specimen can be determined by moving a metal spoon over the specimen surface until a signal appears on the oscilloscope.

18. Tubular shape testing is carried out by shear waves. The sound zig-zags around and there is less interference.

19. False or nonrelevant indications can result from:

- Interference due to the electrical circuitry
- A broken transducer that gives rise to prolonged "ringing" or "tail" on the initial pulse
- An air bubble in the couplant between the transducer and the surface
- Transducer reflections within the wedge; these signals remain when the transducer is removed from the test specimen (contact angle beam only)
- Grain boundaries in the material
- Mode conversions

20. Nonrelevant indications from welds can be due to reflections from the crown, root, or heat-treated zone. If the reflections are consistent along the length of the weld, this would suggest nonrelevance. Reflections from the crown are modified when the crown is touched by a wet finger.

21. Couplants must be kept as thin as possible. A thick layer will probably affect the direction of the ultrasonic beam in the test piece and produce spurious reflections in the layer.

22. Cylindrical specimens can give rise to spurious reflections after the rear-wall echo, particularly if a correctly curved foot is not used with the transducer.

23. Surface waves can reflect from a specimen edge and give nonrelevant indications. Movement of the transducer will result in these indications moving with the transducer. Placing a finger on the surface in front of the transducer can remove the spurious signal.

24. Cylindrically shaped wedges focus the sound into a line, and can increase sensitivity and resolution. Spherically shaped wedges focus the sound to a point, increas-

ing the sensitivity, but decreasing the useful range.

25. In the examination of welds, as the plate thickness increases, the beam angle used for the testing should be smaller.

A glossary of some terms used in ultrasonic testing is given in Appendix 3.1.

Part II Technical Discussions

Specific Acoustic Impedance

Sounds travel through materials under the influence of a local pressure, or sound pressure, P, which is the excess pressure above atmospheric. Because molecules or atoms of a solid are bound elastically to one another, the excess pressure results in a wave propagating through the solid. An excess pressure, P, causes the particles to be displaced with a velocity, Q, then

$$\frac{P}{Q} = \frac{\text{Local excess pressure on the particles}}{\text{Particle displacement velocity}} = Z$$

where Z is the specific acoustic impedance of the medium, which characterizes the behavior of sound waves in the solid. It can be shown that

$$Z = \rho V$$

where ρ is the density of the material and V is the wave velocity (Ref 161, 221). The sound velocities of the various wave types are reviewed in Table 3.2.

Oscillations of molecules, atoms, or clusters of atoms or molecules are coupled, leading to traveling waves, and an expression for these is given by:

$$Y = Y_0 \sin 2\pi \left(\frac{t}{\tau} - \frac{x}{\lambda}\right) = Y_0 \sin (\omega t - kx) \quad \text{(Eq 3.1)}$$

where Y_0 is the maximum amplitude (of excess pressure), t is time, λ is wavelength, $k = 2\pi/\lambda$ = the wave number or propagation constant, $\omega = 2\pi/\tau$ = rotation frequency in radians/s, and τ = cycles/s; see Fig. 3.6. In Fig. 3.6(b), the amplitude variation of pressure at a distance x and time t for a traveling wave is also given by:

$$Y = Y_0 \sin 2\pi \left(v\tau - \frac{x}{\lambda}\right) \quad \text{(Eq 3.2)}$$

where $\nu = 1/\tau$ is the frequency of the wave in cycles per seconds, Hz. The wave is a function of the two variables, x and t. The period in time is τ, the period in distance (wavelength) is λ. The wave is generally described in terms of its frequency ν and wavelength λ. The velocity of the wave V is given by:

$$V = \lambda\nu = \lambda/\tau \qquad \text{(Eq 3.2a)}$$

The particle velocity, $Q = dY/dt = \omega Y_0 \cos(\omega t - kx)$, and the energy E transmitted by a wave is given by $E = P^2/(2\rho V)$ (Ref 161, 221).

Wave Types (Modes). In solids, the particles can oscillate along the direction of sound propagation as longitudinal waves, or the oscillations can be perpendicular to the direction of sound waves, as transverse waves. At surfaces and interfaces, various types of elliptical or complex vibrations of the particles occur. These are shown in Fig. 3.7. The different wave motions are known by several synonyms, and these are listed in Table 3.2 with a summary of their descriptions and properties. For example, longitudinal waves are also known as compression waves, dilatational waves, or pressure waves.

Nondestructive testing is carried out in general using longitudinal waves, or transverse waves. Limited use is also made of Rayleigh, Lamb, and other wave modes.

Velocities, V, and also wavelengths, λ, of ultrasonic waves used in NDT are listed in Table 3.3 for several materials. The waves are nondispersive, that is the velocities are virtually independent of frequency, ν, for the range 0.1 to 15 MHz used in NDT for water, oil, and most metals. Defect lengths as small as $\lambda/4$ can be detected by ultrasonic testing, although this depends on the orientation of the crack to the beam. (Lengths that can be measured depend on the ultrasonic beam diameter.)

Reflection and Transmission at Boundaries

In nondestructive testing, boundaries where there are discontinuities in Z are detected because they cause the ultrasonic waves to be reflected. The fraction of the incident intensity in the reflected waves can be derived because the particle velocity Q and local particle pressure P

Table 3.2 Wave types in solids

Wave type	Particle vibrations	Medium	Wave velocity	Comments	Ref
Longitudinal, dilatational, compression, pressure	Parallel to wave direction	All solids, liquids, gases	Plane waves V_L $V_L = \sqrt{E/\rho}$ for object size $<\lambda$ Bulk waves velocities = $V_{BL} = \sqrt{\dfrac{E(1-\sigma)}{\rho\,[1 + \sigma(1-2\sigma)]}}$ for object size $>>\lambda$	Widely used, nondestructive testing $V_{BL} \sim 1.3 V_L$ See Fig. 3.2, 3.7	195, 222, 69, 221
Transverse shear, distortion	Perpendicular to wave direction	Solids	$V_T = \sqrt{G/\rho}$	$V_T \sim 0.5 V_L$ for metals	
Surface, Rayleigh	Elliptical orbit, symmetrical mode	Any surface, major axis—perpendicular to surface, minor axis—parallel to direction of propagation		Compare to ocean waves	38, 179, 195, 223
Surface	Bleustein-Gulyaev	Piezoelectric solids		Electro-acoustic	179
Plate waves					
Love	Parallel to plane layer and perpendicular to direction waves	Surface layer bonded to solid		See Fig. 3.7	179, 222
Lamb	Component of vibrations perpendicular to surface				
Stoneley (Leaky Raleigh waves)	Wave guided along interface	Boundary of dissimilar solids		Welded boundary	69
Sezawa	Antisymmetric mode	Layered solids			

E, Young's modulus; G, shear rigidity modulus; ρ, density; σ, Poisson's ratio; V, velocity are required to be continuous across the boundary. That is,

$$Q_i + Q_r = Q_t \qquad \text{(Eq 3.3)}$$

where Q_i, Q_r, and Q_t are the particle velocities at the boundary of the incident, reflected, and transmitted waves (Ref 221, 222).

Again,

$$P_i + P_r = P_t \qquad \text{(Eq 3.4)}$$

where P_i, P_r, and P_t are the local acoustic pressure at the boundary associated with the incident, reflected, and transmitted waves.

This leads to the derivation, by Poisson in the early 19th century (Ref 50, 224), of the fraction of the energy reflected R_E normal to the boundary between regions of specific acoustic impedance Z_1 and Z_2 to be

$$R_E = \left(\frac{Z_2 - Z_1}{Z_2 + Z_1} \right)^2 \qquad \text{(Eq 3.5)}$$

and the fraction of the energy transmitted T_E at the boundary (see Fig. 3.8):

$$T_E = 1 - R_E = \frac{4Z_1 Z_2}{(Z_1 + Z_2)^2} \qquad \text{(Eq 3.6)}$$

The boundary length is taken to be much greater than the sound wavelength. For waves incident obliquely at an angle α_1 (less than the first critical angle), as in Fig. 3.9, it was shown by Green in 1838 (Ref 50) that the fraction of the energy reflected R'_E is given by:

$$R'_E = \left(\frac{Z_2 \cos \alpha_1 - Z_1 \cos \alpha_2}{Z_2 \cos \alpha_1 + Z_1 \cos \alpha_2} \right)^2 \qquad \text{(Eq 3.7)}$$

(R'_E is a real number if $\alpha_1 <$ the first critical angle; R'_E is a complex number if the first critical angle low, $<$ second critical angle; R'_E is an imaginary

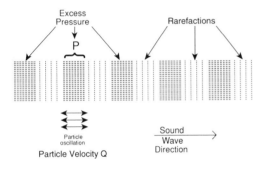

(a)

(b)

Fig. 3.6 Characteristics of a longitudinal wave with particle oscillations parallel to the wave direction. (a) The particle oscillations give rise to alternate compressed and rarified regions. An instantaneous view is presented of the particles present. The local pressure P results in a particle velocity Q, where the impedance $Z = P/Q$. (b) The wave is a function of the two variables, x and t. The period in time is τ, the period in distance (wavelength) is λ. See text for details.

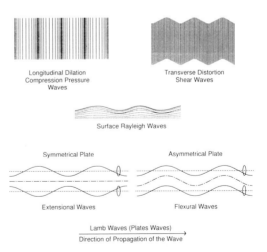

Fig. 3.7 The motion of the particles is illustrated for the various modes of waves in solids. The particles can be atoms, molecules, or clusters of atoms or molecules. Sound waves do not propagate in a vacuum.

number if α_1 > the second critical angle) and the fraction transmitted $T_É'$ is given by:

$$T_É' = \frac{4Z_2Z_1 \cos \alpha \cos \alpha_2}{(Z_2 \cos \alpha_1 + Z_1 \cos \alpha_2)^2} \qquad \text{(Eq 3.8)}$$

Similar to $R_É'$, $T_É'$ is a real, complex, or imaginary number depending on whether α_1 is less than the first critical angle, between the first and second critical angles, or greater than the second critical angle.

Figure 3.9 shows a simplifed case of the reflection of an acoustic wave incident on a planar boundary between two solid media of different specific acoustic impedances Z_1 and Z_2. The incident and reflected waves are inclined at the same angle α_1, different from α_2, the angle of the transmitted beam, as shown. The angles α_1 and α_2 are given by Snell's Law as:

$$\frac{\sin \alpha_1}{\sin \alpha_2} = \frac{V_1}{V_2}$$

where V_1 and V_2 are the wave velocities in medium I and II, respectively. The wave velocity of incident and reflected waves V_1 are the same in medium I, but different from the velocity V_2 in medium II. It is assumed that the sound wavelength λ is much smaller than the size of the boundary. For density ρ_1 in medium I, and ρ_2 in medium II, $Z_1 = \rho_1 V_1$

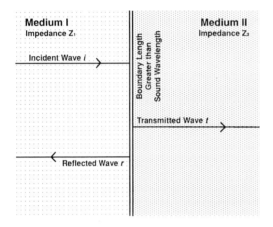

Fig. 3.8 Normal incidence of a sound wave on a boundary between two media of specific acoustic impedance Z_2 and Z_1 results in the fraction of the energy reflected R_E to be $[(Z_2 - Z_1)/(Z_2 + Z_1)]^2$ and the fraction of the energy transmitted across the boundary T_E is such that $R_E + T_E = 1$. As an example, at a steel/air interface almost all the wave energy is reflected (see Table 3.4).

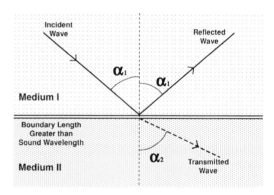

Fig. 3.9 Simplified case of no-wave-mode conversion (compare to Fig. 3.11).

Table 3.3(a)   Wavelengths of ultrasonic waves used in NDT

Material	Longitudinal wave velocities, km/s	Wavelength (λ), mm (1 MHz)	Wavelength (λ), mm (10 MHz)
Air	0.33	0.33	0.033
Water (20°)	1.49	1.49	0.149
Oil (transformer)	1.38	1.38	0.138
Aluminum	6.35	6.35	0.635
Copper	4.66	4.66	0.466
Magnesium	5.79	5.79	0.579
Steel (mild)	5.85	5.85	0.585
Lucite/plexiglass, perspex/polymethylmethacrylate	2.67	2.67	0.267
Polyethylene	1.95-2.40	1.95-2.40	0.195-0.240

Sound velocities in air, water, oil, and metals are virtually constant over the frequency v range 0.1 to15 MHz. Defects perpendicular to the sound beam as small as $\lambda/4$ can be detected by ultrasonic testing. $(V = \lambda v)$ *Note:* Shear wavelengths are about one-half of the longitudinal wavelengths of the same frequency.

and $Z_2 = \rho_2 V_2$. (Wave conversion is described in Fig. 3.11).

The ratio of the incident energy normal to a boundary to the reflected energy is plotted in Fig. 3.10 for different values of Z_1/Z_2, assuming that the boundary is perfect and of a size much greater than λ. Some typical values of the percentage of ultrasonic energy reflected at a perfect boundary for normal incidence are given in Table 3.4.

Wave Mode Conversion at Boundaries. When a longitudinal wave arrives at the boundary, as shown in Fig. 3.11, a longitudinal wave is reflected, and also there is a component of the longitudinal wave that is transmitted across the boundary, traveling with velocity V_{L2} in the second region. The angles to the normal α_1, α_2 as shown in Fig. 3.9 and 3.11 are given by Snell's Law to be:

$$\frac{\sin \alpha_2}{\sin \alpha_1} = \frac{V_{L2}}{V_{L1}} \qquad \text{(Eq 3.9a)}$$

What also happens at the boundary is that a component of the longitudinal sound wave is converted into a shear wave, with a shear wave being reflected in region 1 and a shear wave refracted in region 2. A shear wave is generated at the boundary in reflection and in transmission as shown in Fig. 3.11, with different shear velocities in the two regions V_{S1} and V_{S2}. The angles β_1 and β_2 are given by Snell's Law to be:

$$\frac{\sin \beta_1}{\sin \beta_2} = \frac{V_{S1}}{V_{S2}} \qquad \text{(Eq 3.9b)}$$

Table 3.3(b) Frequency ranges and applications commonly used

Frequency range	Applications
200 kHz-1 MHz	Coarse-grain castings: gray iron, nodular iron, copper, and stainless steels
400 kHz-5 MHz	Fine-grain castings: steel, aluminum, brass
200 kHz-2.25 MHz	Plastics and plasticlike materials
1-5 MHz	Rolled products: metallic sheet, plate, bars, and billets
2.25-10 MHz	Drawn and extruded products: bars, tubes, and shapes
1-10 MHz	Forgings
2.25-10 MHz	Glass and ceramics
1-2.25 MHz	Welds
1-10 MHz	Fatigue cracks (Ref 65)

Table 3.4 Typical values of the percentage of ultrasonic energy reflected at a perfect boundary for normal incidence (see Eq 3.5)

First medium	Impedance (Z) kgm^{-2} s^{-1} × 10^6	Steel	Nickel	Copper	Brass	Lead	Mercury	Glass	Quartz	Polystyrene	Bakelite	Water	Oil (transformer)	Air
Aluminum	17	21	24	18	13	3	0.7	2	0.4	49	38	70	74	100
Steel	46		0.2	0.1	1	10	16	31	26	77	71	88	89	100
Nickel	50			0.8	3	12	18	34	29	79	73	89	89	100
Copper	42				0.6	7	13	28	22	75	68	87	88	100
Brass	36					4	8	22	17	72	64	85	86	100
Lead	24						0.8	9	5	60	51	78	79	100
Mercury	20	(These values are 100 × R$_E$)						4	2	55	44	74	76	100
Glass	13								0.5	39	28	63	65	100
Quartz	15	(see Eq 3.5)								44	34	67	69	100
Polystyrene	3										2	11	13	100
Bakelite	4											21	23	100
Water	1.5												0.1	100
Oil (transformer)	1.4													100
Air	4×10^2 kgm^{-2} s^{-1}													

Example: At a steel/polystyrene boundary, 77% energy is reflected, and 23% transmitted. At almost all air solid boundaries, 100% energy is reflected. Note: Impedance values are given in Table 3.6. Solids and liquids have impedances of ~10^6 Rayl = 10^6 kgm^{-2}s^{-1}. Gases have impedances of ~10^2 Rayl = 10^2 kgm^{-2} s^{-1}.

Longitudinal wave velocities are greater than shear wave velocities, so that their relative angular positions are as shown. Figure 3.11 should be compared to Fig. 3.9, in which mode conversion was excluded for the sake of simplicity.

There is an angle of critical incidence where the longitudinal refraction angle α_2 is 90°, so that the longitudinal wave does not penetrate into region 2 across the boundary (Fig. 3.12).

This condition is given by:

$$\frac{\sin \alpha_1}{\sin 90°} = \sin \alpha_1 = \frac{V_{L1}}{V_{L2}} \qquad \text{(Eq 3.10a)}$$

when the longitudinal wave is totally internally reflected so that only the shear wave will penetrate the boundary. This condition is used in nondestructive testing where the ultrasonic transducer generating longitudinal waves is set at an angle greater than the critical angle, and so only the shear wave will penetrate into the specimen (Fig. 3.13). As

shown in Fig. 3.13, plexiglass is often used on a steel specimen; the critical angle of total reflection of the longitudinal waves is 27.5°. This is referred to as the first critical angle.

There is a second critical angle when the shear wave is also totally internally reflected, and no sound wave penetrates into the specimen. This occurs when

$$\frac{\sin \alpha_1}{\sin 90°} = \sin \alpha_1 = \frac{V_{L1}}{V_{S2}} \qquad \text{(Eq 3.10b)}$$

Critical incident angles are listed in Table 3.5 for total internal reflection for longitudinal and for shear waves in several materials using a plexiglass wedge and also for immersion testing. The first critical angle is for longitudinal waves, and the second critical angle is for transverse waves. When a pure transverse wave is required in the test piece, the angle of incidence used should be between the first and second critical angles. To obtain only a transverse wave in steel using a plexiglass wedge, the incident angle in the transducer should be about 40° (Ref 65, p 2-28).

The critical angles for steel are:
First critical angle for steel by water immersion test:

$$\sin^{-1}\left[\frac{V_L H_2O}{V_L steel}\right] = \sin^{-1}\frac{1.49 \text{ km/s}}{5.95 \text{ km/s}} = 14.5°$$

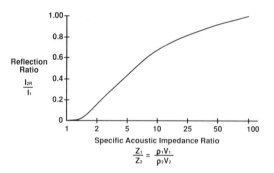

Fig. 3.10 Ratio of incident to reflected energy at normal incidence on a boundary between two media of different specific acoustic impedance Z_1 and Z_2.

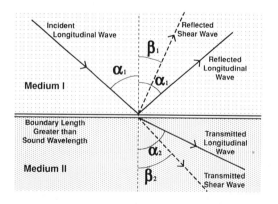

Fig. 3.11 Wave mode conversion at a boundary. Fig. 3.11 should be compared to Fig. 3.9 in which mode conversion was excluded for the sake of simplicity.

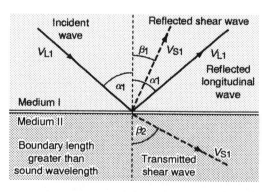

Fig. 3.12 Angle of critical incidence. There is an angle of incidence α_1 of the incoming longitudinal wave of velocity V_{L1}, such that the angle of the transmitted longitudinal wave α_2 becomes 90° (see Eq 3.10a). At angles of incidence greater than α_1 the longitudinal wave of velocity V_{L2} does not penetrate into medium II, and only the shear wave with velocity V_{S2} will be transmitted. This is used to separate longitudinal and shear waves to have only a single wave velocity traveling in medium II.

Second critical angle for steel by water immersion test:

$$\sin^{-1}\left[\frac{V_L H_2O}{V_s steel}\right] = \sin^{-1}\frac{1.49 \text{ km/s}}{3.2 \text{ km/s}} = 27.8°$$

First critical angle for steel using plexiglass wedge:

$$\sin^{-1}\left[\frac{V_L wedge}{V_L steel}\right] = \sin^{-1}\frac{2.67 \text{ km/s}}{5.95 \text{ km/s}} = 26.7°$$

Second critical angle for steel using plexiglass wedge:

$$\sin^{-1}\left[\frac{V_L wedge}{V_s steel}\right] = \sin^{-1}\frac{2.67 \text{ km/s}}{3.2 \text{ km/s}} = 56.6°$$

Surface (Rayleigh) waves are induced at solid-gas or solid-liquid interfaces, when the probe is at the second critical angle setting; the surface wave profile is mainly transverse and so the surface wave is approximately a shear wave at its critical angle. The variation of sound velocities with frequency is very small (nondisper-

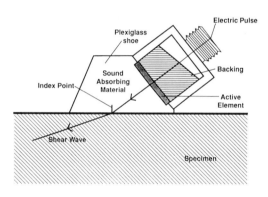

Fig. 3.13   Ultrasonic transducer arranged so that only shear waves penetrate into the specimen. The longitudinal wave is totally internally reflected within the shoe of the transducer. In a steel specimen, an angle of incidence about 40° is chosen so that only a transverse wave is produced in the steel.

sive) over the frequency range used in NDT. Some general acoustic data for a wide range of materials are listed in Tables 3.6(a to e). The data in Table 3.6 (a to e) have been selected from Ref 65, p 4-27; Ref 70, p 155, 157, 158, 162; Ref 129, p 235; Ref 161, p 14, 36; Ref 208 and 209; Ref 221, Appendix B; Ref 222, p 72, 86; Ref 225; Ref 226, p 381 to 394; and Ref 227, p 7 to 9. Rayl, the unit of specific acoustic impedance, also known as the characteristic impedance, has units of $kgm^{-2} s^{-1}$.

Beam Attenuation

In addition to the reduction in intensity by beam spreading, a wave weakens as it travels in a solid by scattering and reflection, and also by the frictional motion of the particles of the solid. This reduction in beam strength can be expressed by:

$$I_x = I_o e^{-\mu x} \tag{Eq 3.11a}$$

or

$$A_x = A_o e^{-\mu' x} \tag{Eq 3.11b}$$

where I_x, I_o, A_x, A_o are the intensities and amplitudes at distances x and o respectively; μ and μ' are the intensity and amplitude coefficients of absorption, respectively, so that:

$$\mu = 2\mu' \tag{Eq 3.12}$$

These are isotropic coefficients, although anisotropic absorption effects occur in solids. The total absorption, μ, is given by:

$$\mu = \mu_\tau + \mu_s \tag{Eq 3.13}$$

Table 3.5   Critical angles for ultrasonic waves in stainless steel calculated when using a plexiglass wedge and for water immersion testing (see Eq 3.10b)

Metal	Plexiglass wedge		Immersion testing	
	1st critical angle	2nd critical angle	1st critical angle	2nd critical angle
Steel	27	56	15	27
Type 302 stainless steel	28	59	15	29
Aluminum alloy 2117-T4	25	59	14	29
Beryllium	12	18	7	10
Magnesium alloy MIA	27	59	15	29
Titanium	26	59	14	29
Tungsten	31	68	17	31

Table 3.6(a)   Acoustic properties of metals

Material	Density (ρ), g/cm^3	Longitudinal wave Velocity, km/s	Longitudinal wave Impedance (Z), 10^6 Rayl	Shear wave Velocity, km/s	Shear wave Impedance (Z), 10^6 Rayl	Surface (Rayleigh) wave Velocity, km/s	Surface (Rayleigh) wave Impedance (Z), 10^6 Rayl
Aluminum	2.7	6.35	17.1	3.1	8.4	2.8	7.6
Beryllium	1.85	12.5	23.1	8.71	16.0	7.87	14.6
Brass	8.1	4.2-4.7	34-38	2.12	17.2	1.95	15.8
Bronze	8.86	3.53	31.2	2.23	19.8	2.01	17.8
Bismuth	9.8	2.2	21.5	1.1	10.8	...	...
Cadmium	8.6	2.8	24	1.5	12.9	...	...
Copper	8.93	4.66	42	2.26	20.1	1.93	17.2
Gold	19.3	3.2	62	1.2	23	...	...
Indium	7.31	2.5	18.3	...	...	...	...
Lead	11.4	1.96	22	0.7	8	0.63	7.2
Magnesium	1.74	5.79	10	3.1	5.4	2.87	5.0
Molybdenum	10.1	6.29	63.5	3.35	33.8	3.11	31.4
Nickel	8.84	5.7	50	2.96	26.1	2.64	23
Niobium	8.57	4.92	42.2	2.1	18.0	...	...
Platinum	21.4	4.15	89	1.73	37	...	...
Silver	10.5	3.44	36	1.59	16.7	...	...
Mild steel	7.83	5.95	46.6	3.2	25.0	2.79	21.8
Stainless steel 347	7.89	5.79	45.7	3.1	24.5	...	...
Cast iron	7.7	3.5-5.9	27-45	2.4	18.5	...	...
Tantalum	16.6	4.1	68	2.9	48	...	...
Tin	7.3	3.32	24.2	1.67	12.2	...	...
Titanium	4.54	6.1	27.7	3.12	14.2	2.8	12.7
Tungsten	19.25	5.18	100	2.87	55.2	2.65	51
Vanadium	6.03	6.0	36	2.78	17	...	...
Zinc	7.1	4.2	30	2.41	17	...	...
Zirconium	6.48	4.65	30	2.25	15	...	...
Chromium	7.0	6.65	46.6	4.03	...	...	...

Table 3.6(b)   Acoustic properties of ceramics

Material	Density (ρ), g/cm^3	Longitudinal wave Velocity, km/s	Longitudinal wave Impedance (Z), 10^6 Rayl	Shear wave Velocity, km/s	Shear wave Impedance (Z), 10^6 Rayl	Surface (Rayleigh) wave Velocity, km/s	Surface (Rayleigh) wave Impedance (Z), 10^6 Rayl
Alumina sapphire (c-axis)	3.99	11.1	44.3	6.04	24	...	...
Barium titanate	5.6	5.5	31	...	...	...	...
Beryllia	...	...	32	...	...	...	...
Boron carbide	2.4	11.0	26.4	...	...	...	...
Germanium	5.47	5.41	30	...	...	...	...
Graphite	2.17	4.21	9.1	...	...	...	...
Glass crown	2.24	5.1	11.4	2.8	6.3	...	...
Glass (pyrex)	2.24	5.64	12.6	3.28	7.3	...	...
Lithium niobate (c-axis)	4.64	7.33	34	...	...	...	...
Tourmaline	3.1	2.23	6.9	...	...	...	...
Tungsten carbide	10-15	6.7	67-100	4	40-60	...	...
Porcelain	2.4	5.6	13.4	...	...	...	...
Titanium carbide	5.15	8.27	43	5.16	...	...	...
Silicon carbide	13.8	6.66	92	...	...	...	...
Silicon	2.34	8.43	20	5.84	14	...	...
Silicon nitride	3.27	11.0	36	6.25	20	...	...
Silica fused	2.20	5.96	13	3.75	8.25	3.4	7.5
Quartz crystal (c-axis)	2.53	6.32	16	...	...	...	...
Rutile (c-axis)	4.26	7.9	34	...	...	...	...
Zinc oxide	5.68	6.33	36	2.95	17	...	...

Table 3.6(c)   Acoustic properties of plastics

Material	Density (ρ), g/cm^3	Longitudinal wave Velocity, km/s	Longitudinal wave Impedance (Z), 10^6 Rayl	Shear wave Velocity, km/s	Shear wave Impedance (Z), 10^6 Rayl
Bakelite	1.4	1.59	2.2	...	...
Lucite/plexiglass	1.15	2.67	3.1	1.1	1.3
Mylar	1.18	2.54	3	...	...
Nylon (polyamide)	1.12	2.6	2.9	1.1	1.2
Paraffin wax	1.5	1.5	2.3	...	...
Polyethylene					
Low density	0.92	1.95	1.8	0.54	0.5
High density	0.96	2.43	2.3	...	...
Polystyrene	1.05	2.4	2.5	1.15	1.2
Polyvinyl chloride (PVC)	1.38	2.38	3.3	...	...
Teflon	2.14	1.39	3	...	...
Silicon rubber	1.05	1.03	1.1	...	...
Wood (oak)	0.72	4.0	2.9	...	...
Cork	0.24	0.5	0.12	...	...

Table 3.6(d)   Acoustic properties of liquids and gases

Material	Density (ρ), g/cm^3	Longitudinal wave Velocity, km/s	Longitudinal wave Impedance (Z)
Liquids			10^6 Rayl
Benzene (25 °C)	0.87	1.295	1.1
Ethyl alcohol (25 °C)	0.79	1.207	0.95
Ethylene glycol	1.113	1.66	1.8
Oil (transformer)	0.92	1.38	1.27
Oil (silicone) (20 °C)	1.11	1.352	1.5
Mercury (23 °C)	13.53	1.45	20
Water			
20 °C	1.00	1.49	1.5
25 °C	1.00	1.40	1.4
60 °C	1.00	1.55	1.55
Seawater (25 °C)	1.03	1.53	1.58
Gases			10^2 Rayl
Air (dry atmosphere)			
0 °C	1.29×10^{-3}	0.331	4.27
20 °C	1.24×10^{-3}	0.334	4.14
100 °C	$1.11 \times 10^{0-3}$	0.386	4.28
Carbon dioxide (0 °C)	1.977×10^{-3}	0.259	5.1
Oxygen			
0 °C	1.429×10^{-3}	0.316	4.51
20 °C	1.429×10^{-3}	0.328	4.69

Table 3.6(e)   Acoustic properties of biological materials

Material	Density (ρ), g/cm^3	Longitudinal wave Velocity, km/s	Longitudinal wave Impedance (Z), 10^6 Rayl
Brain	1.03	1.51-1.61	1.55-1.66
Kidney	1.04	1.56	1.62
Aqueous humour	1.01	1.50	1.51
Eye lens	1.14	1.62	1.84
Lungs	0.4	0.65	0.26
Tooth enamel	2.95	5.8	17.1
Muscle	1.07	1.63	1.74
Bone (skull)	1.38-1.81	3.05	4.2-5.52
Fat	0.92	1.46	1.35
Liver	1.06	1.57	...
Blood	1.06	1.53	1.62

where μ_τ arises from true absorption such as internal friction between the particles of the solid, and μ_S arises from scattering of the sound waves (Ref 70, 191). Absorption mechanisms in solids depend on the wave frequency ν, and scattering mechanisms depend on the average particle size, $\overline{D}$, relative to the sound wavelength, λ:

- When $\lambda >> \overline{D}$, μ_S is proportional to $\overline{D}^3 \nu^4$
- When $\lambda \propto \overline{D}$, μ_S is proportional to $\overline{D} \nu^2$
- When $\lambda \leq \overline{D}$, μ_S is proportional to $1/\overline{D}$

Decibel and Neper. The reduction in the strength of a sound wave is usually discussed in terms of decibel or neper. An absorption coefficient in neper is defined by:

$$\mu_{NP} = \frac{-1}{x} \log_e \frac{A_x}{A_0} = \frac{-1}{2x} \log_e \frac{I_x}{I_0} \qquad \text{(Eq 3.14)}$$

so that

$$A_x = A_0 \exp - [\mu_{NP} x] \qquad \text{(Eq 3.15)}$$

and

$$I_x = I_0 \exp - [2\,\mu_{NP} x] \qquad \text{(Eq 3.16)}$$

where I_x and I_0 are wave intensities at x and o, respectively, from the sound source.

An absorption coefficient in decibel is defined by:

$$\mu_{dB} = \frac{-10}{x} \log_{10} \frac{I_x}{I_0} = \frac{-10 \times 0.4343}{x} \log_e \frac{I_x}{I_0}$$

$$= \frac{-20}{x} \log_{10} \frac{A_x}{A_0} = \frac{-20 \times 0.4343}{x} \log_e \frac{A_x}{A_0}$$

$$\text{(Eq 3.17)}$$

Therefore:

$$\mu_{dB} = 8.686\,\mu_{NP} \qquad \text{(Eq 3.18)}$$

In Table 3.7, some examples of sound wave attenuation are expressed in amplitude, intensity or power decrease, as well as by decibel and neper values.

For example, if the amplitude of a wave at $x = 10$ cm has decreased to $1/10$th of the value at the source where the distance $x = 0$, then

$$\mu_{NP} \times 10 \text{ cm} = -\ln \frac{A_x}{A_0} = -\ln 0.1 = -2.30$$

The attenuation of 2.3 per 10 cm, is expressed as 0.23 Np cm^{-1}.

The same calculation in decibel is as follows:

$$\mu_{dB} \times 10 \text{ cm} = -20 \log_{10} \frac{A_x}{A_0} = -20 \log_0 0.1 = 20 \text{ dB}$$

The attenuation is expressed as 2 dB cm^{-1}, which is therefore the same as 0.23Np cm^{-1}. It should be noted that:

- The negative sign is frequently omitted.
- Attenuations in Np (or dB) are linear with distance, and thus are additive.
- Longitudinal waves and shear waves have different attenuation characteristics.
- The attenuation is heavily dependent on the wave frequency.

Some typical values of attenuation coefficients are given in Table 3.8(a). The depths of inspection of materials and their attenuation coefficients are reviewed in Table 3.8(b). It can be seen that materials of attenuation coefficients above 1 dB/cm can only be examined in thicknesses of less than 10 cm at 2 MHz. Materials of attenuation coefficients less than 0.1 dB/cm can be examined usefully in thicknesses approaching 10 m at 2 MHz. The upper limit for attenuation in a specimen is probably at 20 dB/cm.

Gain. When one signal is increased relative to the first, the comparison is called the gain. Gain is a function of instrumentation, whereas attenuation depends on the material being examined. The units involved are dB (or Np).

$$\text{Gain} = 20 \log \frac{A_2}{A_1} \text{ dB}$$

where A_2 and A_1 are signal amplitudes, for example, if $A_2 = 2A_1$, then the gain is 6 dB.

Table 3.7 Attenuation with distance of amplitudes, intensity, power, in decibels and nepers

Amplitude ratio at 1 cm	Intensity ratio or power ratio at 1 cm	Decibel $\mu(dB)$, dB cm^{-1}	Neper $\mu(Np)$, Np cm^{-1}
1	1	0	0
1/2	1/4	−6	−0.69
1/4	1/16	−12	−1.38
1/8	1/64	−18	−2.07
1/16	1/4096	−24	−2.76

The decibel values are additive. The neper values are additive. Power, intensity, or amplitude changes must be multiplied. A change of 1 Np corresponds to a change of 8.686 dB.

98

Couplants

When a transducer probe is placed against a test piece, a coupling liquid is used to provide a suitable sound path between transducer and test surface to increase the transmission of the ultrasonic pulse energy into the test piece. This is done to avoid a layer of air between the probe and solid surface, which would transmit only a very weak beam (see Table 3.4). As an example, the percentage reflected between steel and air is approximately 100%, and between steel and water 88%, so that the presence of water will permit some of the energy, although still a small fraction, to enter a steel specimen.

The couplant must:

- Wet both the transducer and the test surface
- Exclude all air
- Fill all irregularities to provide a smooth surface
- Allow free movement of transducer probe
- Be easy to apply, easy to remove, harmless to surfaces

Special couplants are available for difficult circumstances

- *Vertical surfaces*: Use a heavy oil or grease
- *Rough surfaces*: Use a heavy oil or grease
- *Liquids not feasible*: Use rubberlike materials

The couplant must be as thin as possible, otherwise it may alter the direction of the ultrasonic beam.

The amount of energy transmitted across a layer can be increased by choosing an intermediary layer of specific acoustic impedance equal to the geometric mean of the surrounding media and of thickness equal to $[(2n - 1)/4]\lambda$, where n is an integer and λ the sound wavelength in the intermediate layer (Ref 161).

Gases have extremely low acoustic impedance; liquids have values closer to those of solids, as can be seen from Table 3.6. Some couplants are listed below:

Couplants	Specific acoustic impedance, 10^6 Rayl
Water	1.5
Silicone oil	1.58
Glycerine (2 parts water + 1 part glycerine)	2.4
Thick oil, grease, petroleum jelly (for vertical surfaces)	3
Perspex, Plexiglass, Lucite, thin rubber or rubberlike materials	2-3.5
Toothpaste	...
Wallpaper paste	...

Table 3.8(a) Typical values of attenuation coefficients (longitudinal waves)

Material	Np cm^{-1}	dB cm^{-1}	Frequency range, MHz	Ref
H_2O	0.025	~0.1	10	70
Silicone oil	~0.1	~1	...	195
Biological materials				
Soft tissue	~0.03-0.2	~0.2-2	~1	161
Bone	...	2-20	~2	...
Ferritic mild steel	0.05	0.434	10	222
Stainless steel	0.13	1.1	2.25	
Acrylic plastics	0.044	0.38	2.25	208
Aluminum	0.1	0.9	2.25	...

These values are heavily dependent on the frequency of the waves.

Table 3.8(b) Depths of inspection and attenuation coefficients for various materials

Attenuation coefficients, dB/cm	Useful depth inspection, cm	Typical materials
0.01-0.1	100-1000	Al, Mg; wrought metals: steel, Al, Mg, Ni, Ti, W; worked metals
0.1-1	10-100	Steel, high-strength cast Fe; Al alloys, Mg alloys; Wrought metals: Cu, Pb, Zn; Nonmetals: sintered carbides, plastics, rubber
>1	<10	Low-strength cast Fe, Cu, Zn, Pb, Sn (softer metals) white cast iron, brass, bronze, gray cast iron, cast Al; Nonmetals: porous ceramics, filled plastics and rubbers

The specific acoustic impedances of coupling media can be compared to that of air 3.3×10^2 Rayl, and of steel 44.7×10^6 Rayl (see Table 3.6). High viscous media are required for vertical surfaces. The couplant must be as thin as possible to avoid affecting the ultrasonic beam direction. Various commercially made coupling agents are available, for example, Sonotech (Ref 228), Panametrics (Ref 229), and Magnaflux Ultrasonic Couplants (Ref 230).

Pulse Shape and Beam Shape

Pulses of ultrasonic waves are used in the echo and pulse transmission methods. These pulses are generated for approximately 1 μs (1 to 10 cycles of vibration) and consist of waves of different frequencies as illustrated in Fig. 3.14I. The pulse may be rectified and smoothed as shown. The pulse shape changes with time and spreads out so that the pulse will distort as it travels through the medium. The beam shape varies with direction depending on the relative size of the transducer to sound wavelength. Assuming a circular transducer of diameter Δ the longitudinal, transverse, and surface waves spread out as shown in Fig. 3.14II for different λ/Δ ratios. The beam of ultrasonic waves resembles a "searchlight" when the diameter of the transducer Δ is very much greater than the wavelength of the sound wave.

These diffraction effects apply to a continuous train of waves and are less pronounced in the case of a short pulse when the side lobes in Fig. 3.14II become less important. This is an added advantage to the use of wave pulses rather than continuous wave trains.

Near Field and Far Field (Ref 179). The intensity along the beam is not uniform, but varies due to the limited size of the source that gives rise to diffraction effects. The conventional way of presenting the ultrasonic beam is given in Fig. 3.1, as if the beam were a slightly diverging

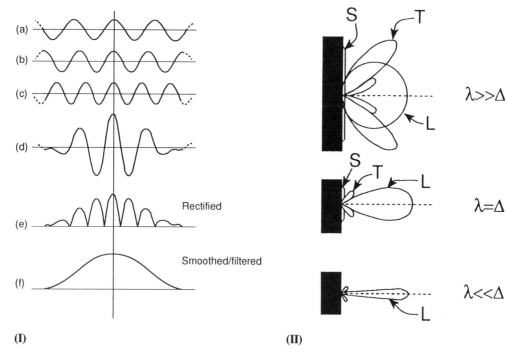

(I)   **(II)**

Fig. 3.14 Pulse shape and beam shape. I, Pulse shape. A pulse of ultrasonic waves consists of waves of different frequencies which travel with different attenuations through the solid. The synthesis is shown of a 1 MHz pulse (d) composed of sinusoidal waves of (a) 0.8, (b) 1, and (c) 1.2 MHz. The length of the pulse (d) for NDT in time is usually of the order of 1 μs (1 to 10 cycles of vibration). The pulse may be rectified (e) and smoothed (f). II, Beam shape. The directional character of the sound wave increases with the relative size of sound source Δ to wavelength λ due to diffraction effects and, because $V = \lambda v$, the directional character increases with frequency. These diffraction effects apply to continuous wave trains and are less applicable to short pulses. L, longitudinal wave; T, transverse; and S, surface wave.

searchlight. In fact, there are extensive fluctuations near the source, known as the near field or near zone or Fresnel zone. Further away, the beam is more uniform as portrayed in the far field or far zone (Fig. 3.15). The length of the near field N is given by:

$$N = \frac{\Delta^2}{4\lambda} = \frac{\Delta^2 v}{4V} \qquad \text{(Eq 3.19)}$$

so that the near field is shorter for longer wavelengths (Ref 65). The ultrasonic beam is more uniform in the far field, or Fraunhofer Zone, where the beam spreads out as if originating from the center of the transducer. The angular spread is given by (Ref 65):

$$\sin \alpha = \frac{1.22\,\lambda}{\Delta} \qquad \text{(Eq 3.20)}$$

Some typical values of α and N for steel for transducers of diameter $\frac{3}{8}$ to 1 in. are given in Table 3.9.

Ultrasonic Transducers

Ultrasonic waves are generated by suitable transducers, in which a single electrical "spike" of short rise-time, <10 ns, is converted into high frequency mechanical vibrations of the solid. Several processes can accomplish this and these are listed in Table 3.10.

Ultrasonic transducers contain an active element made of piezoelectric materials such as single crystal α-quartz or of a ferroelectric polycrystalline ceramic such as PZT5, and these are listed in Table 3.11.

The electro-acoustic transducer is characterized by the strain ε' produced by an applied electric field F (in volts/m) and also by the electric field F produced by an applied stress σ'. These can be expressed as:

$$\varepsilon' = \alpha F \qquad \text{(Eq 3.21a)}$$

$$F = \beta \sigma' \qquad \text{(Eq 3.21b)}$$

Table 3.9 Typical values of frequencies, v, wavelengths, λ, angular spread, α, transducer diameters, Δ, and the near field lengths N for steel specimens

Frequency (v), MHz	Wavelength (λ), cm	Transducer diameters, Δ							
		at $\frac{3}{8}$ in. (0.95 cm)		at $\frac{1}{2}$ in. (1.27 cm)		at $\frac{3}{4}$ in. (1.9 cm)		at 1.0 in. (2.54 cm)	
		α	N, in.	α	N, in.	α	N, in.	α	N, in.
1	0.581	48° 10′	0.15	34°	0.27	21° 52′	0.61	16° 10′	1.1
2.25	0.259	19° 23′	0.34	14° 25′	0.61	9° 25′	1.37	7° 33′	2.45
5	0.116	8° 34′	0.77	6° 25′	1.4	4° 16′	3.06	3° 10′	5.5
10	0.058	4° 16′	1.53	3° 11′	2.73	2° 8′	6.13	1° 36′	10.95

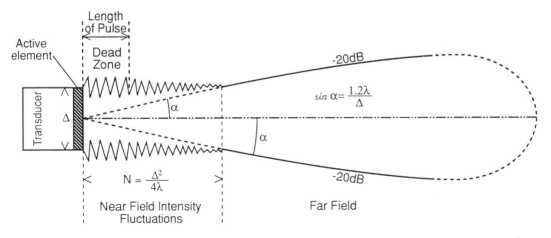

Fig. 3.15 An ultrasonic beam from a transducer diameter Δ has intensity fluctuations out to a distance $N = \Delta^2/4\lambda$, known as the near field or Fresnel zone, beyond which the beam is more uniform in the far field or Fraunhofer zone. The leading and trailing edges of an ultrasonic pulse are indicated, corresponding to the extent of the dead zone.

where α mV^{-1} and β Vm^{-1}Pa^{-1} are the piezoelectric constants (Ref 234). Strain ε' and stress σ' are related by Young's modulus, so that:

$$\text{Young's Modulus} = \frac{1}{\alpha\beta} \text{ (from Hooke's Law)}$$
$$\text{(Eq 3.22)}$$

To have a "good" transmitter, β should be large. To have a "good" receiver, α should be large. Unfortunately, it is not possible for both α and β to be large at the same time, because they are related to Young's modulus by Eq 3.22. The use of dual crystal transducers can overcome this. In practice, PZT5 is found to be very satisfactory for nondestructive testing purposes. Values of α and β are given in Table 3.11. For example, it can be seen that quartz has a large value of β and is a good transmitter of mechanical waves, but it has a small value of α and thus is a poor receiver.

Another factor in the choice of transducer is the damping of the crystal once a pulse of mechanical waves has been produced. In quartz, the inherent damping is poor, as indicated by a very sharp resonance response of the crystal with applied electric field. A material with less sharp resonance would produce a more damped pulse of waves. Damping is important because the transmitter-receiver transducer cannot receive until the transmitted pulse has completely died down, although receiver paralysis is also a severe limitation; this is considered the "dead zone" of the system as shown in Fig. 3.15. PZT5 possesses good damping characteristics, although the damping is probably more influenced by the "backing material" applied to the active element.

The construction of a typical transducer probe is shown in Fig. 3.16. The active element (most likely PZT or PMN), backing, wear plate, and electrical connections are shown. The backing is of high density, with impedance matched to the transducer, and a high absorption of the energy radiated from the back face of the transducer active element. The wear plate (shoe) protects the transducer and can also enhance the sound energy penetrating into the specimen. Alu-

Table 3.10 Physical bases of ultrasonic transducers

	Materials	Frequencies	Ref
Piezoelectric and ferroelectric effect			
When crystals with asymmetric structures are subject to pressure, electrical charges develop on opposing crystal faces, and the crystal changes size. The inverse also occurs where crystals change size when subjected to an electric potential.	See Table 3.11	0.1-25 MHz	69, 191
Magnetostriction effect			
Change in dimension of ferromagnetic materials on magnetization. (Very limited in NDT application.)	Ferromagnetic materials, Ni	<200 kHz (upper limit determined by heating of specimen)	231
Electromagnetic acoustic EMA, EMAT			
Induced eddy currents at radio frequencies in an electrical conductor in a magnetic field cause the surface to vibrate.	Electrical conductors	~2 MHz	222
Laser generation			
Localized heating by laser pulses causes material to expand and contract very rapidly.	All solids and liquids	...	69, 222
Others			
Several other physical effects are used in transducers to produce ultrasonic waves.	...	...	209, 232

Note: Ferroelectric and EMAT methods are the predominant methods.

mina or titanium oxide are used as the "common contact"; for water immersion, epoxy resin is used. On rough surfaces, a soft wear plate is used to provide maximum contact with the specimen and minimum use of couplant. This is found to reduce the noise level.

Compression waves can be generated using a quartz crystal having an X-cut and shear waves can be generated using a quartz crystal having a Y-cut (Fig. 3.17) (Ref 195, 198, 234, 235). A crystal of α-quartz is shown in Fig. 3.17(a) with the crystallographic axes X, Y, and Z drawn in. The optic axis is parallel to Z, and electric axis is parallel to X. Plates are usually cut perpendicular to the X-direction, "X-cut", or perpendicular to the Y-direction, the "Y-cut". Upon applying an alternating electric field across the X and Y plates, vibrations take place as shown in Fig. 3.17(b) so that the X-cut can be used to generate longitudinal compression waves, and the Y-cut can generate shear transverse waves. Other cuts are also used. Quartz crystals are operated most

efficiently at their natural frequencies where their resistance to vibration, that is their acoustic sound impedance, is minimum. This requires that the thickness of the quartz slice be equal to half the wavelength, so that the relationship between crystal thickness Δ and frequency ν is given by $\Delta = V/2\nu$. The thickness of an α-quartz crystal for 0.5 MHz is 6 mm, and for 10 MHz is 0.3 mm. Surface waves can be generated if the critical angle is chosen precisely, as shown in Fig. 3.12. It is now very common to use the ferroelectric PZT5 or PMN for the transducer material. Other ultrasonic transducers are discussed and illustrated in Ref 70.

Dead Zone. A single transducer probe consists of one active element which emits an ultrasonic pulse, lasting about 1 μs, and then detects the reflected ultrasonic pulses, perhaps 0.1 ms later. This means that in the pulse echo technique there is a dead zone, that is a certain time, or distance into the specimen, which cannot be examined (see Fig. 3.15). Reflections from discon-

Table 3.11 Piezoelectric and ferroelectric materials used for ultrasonic transducers

Material	Chemical formula	Material type	Piezoelectric constants		Specific acoustic impedance ($Z = \rho V$), 10^6 kg/m² · s	Comments
			α mV^{-1} × 10^{-12}	β mV^{-1} Pa^{-1} × 10^{-3}		
α-quartz	SiO$_2$	Piezoelectric single crystal	2.3	58	15.2	X-cut longitudinal, Y-cut shear, poorly damped, good transmitter, poor receiver
Barium titanate	BaTiO$_3$	Ferroelectric polycrystalline ceramic	149	14	29-31	Perovskite structure
Lead zirconate titanate(a)	PbZrO$_3$-PbTiO$_3$ solid solution PZT (PZT5)	Ferroelectric polycrystalline ceramic	374	15	28-30	Perovskite structure, PZT5 most commonly used for NDT
Lead metanio-bate(a)	PbNb$_2$O$_6$ PMN	Ferroelectric polycrystalline ceramic	85	43	16-21	Highly damped sharp pulses
Lithium sulfate (hydrated)	Li$_2$SO$_4$ LSH	Piezoelectric single crystal	16	175	11.2	Decomposes at 75 °C
Polyvinylidene fluoride	[CH$_2$-CF$_2$]$_n$ PVDF	Ferroelectric plastic	...	...	4.1	Sheet form highly damped; single pulse 0.1-25 MHz
Rochelle salt	KNaC$_4$ H$_4$O$_6$·4H$_2$O	Piezoelectric single crystal	...	...	...	Soluble in H$_2$O
Backing materials (resins)	...	...	...	...	3-5	...

Other crystals: Tourmaline (H$_6$Na$_2$Fe$_5$Al$_{14}$B$_6$S$_{12}$O$_{63}$); ammonium dihydrogen phosphate (NH$_4$H$_2$PO$_4$); potassium dihydrogen phosphate (KH$_2$PO$_4$). (a) PZT and PMN are used most generally. Source: Ref 162, 191, 208, 209, 222, 227, 233

tinuities are confused with the "tail" of the initial pulse. To overcome this:

- A shorter pulse may help.
- Immersion testing: the initial pulse is separated from the reflection from the specimen by the water path. However, there is still a dead zone near the liquid-solid interface because of the high-amplitude reflection from the interface. Immersion testing permits the elimination of near-field effects.
- Delay versions of pulse echo. A length of soft material (polymer) is added between the active element and the specimen. The sound velocity in the soft material is slower than that in the specimen, so that the echo is delayed. The oscilloscope display can be adjusted so that the specimen echoes arrive between two echoes from the specimen-

polymer interface. Frequencies of 10 to 25 MHz are typically used.

- Use separate transducers as transmitters and receivers. The dual crystal ("combined double") pitch-catch) probe consists of two crystals, one acting as the transmitter, the other detecting the reflected pulses (Fig. 3.18). The two crystals are slightly inclined, especially for use with thin specimens (Ref 191, 236). Delay versions of dual crystals can also be prepared by adding a length of polymer between each transducer and the specimen.

Focused Transducers. Ultrasonic waves can be focused and simple geometric optics apply particularly at the higher frequencies. A spherical or a cylindrical transducer surface shape can be used to focus the ultrasonic wave towards a point or a line focus (Fig. 3.19). The focal spot is replaced by a focal zone; the beam diameter and length of the focal region are indicated in Fig. 3.19(b), where Δ is the transducer diameter.

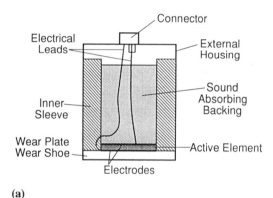

(a)

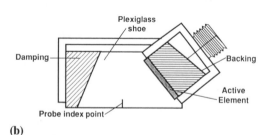

(b)

Fig. 3.16 (a) Construction of a transducer probe. (b) Angular probe for shear waves, showing the Plexiglass shoe and the probe index point from where the shear waves are considered to be emitted. The diameters of commercially available transducers range generally from ¼ to 2 in. Normal beam transducer frequencies are from 0.5 to 15 MHz. Shear beam transducer frequencies used in the United States are 1, 2.25, 5, and 10 MHz; frequencies used in Europe are 1, 2, 4, and 10 MHz.

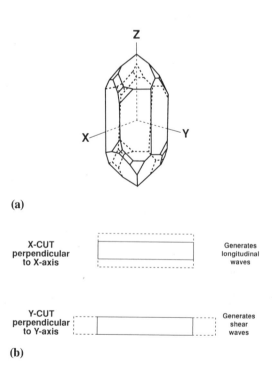

Fig. 3.17 Quartz piezoelectric transducers. (a) Crystal of α-quartz with axes X, Y, and Z. (b) X-cut generates longitudinal compression waves; Y-cut generates shear transverse waves. *Note*: The use of quartz has been almost completely replaced by the ferroelectrics PZT and PMN.

Focusing arrangements are used to advantage in the water immersion technique (see the section "Immersion Inspection" in this chapter). Flat transducers are used in general with thick specimens. The properties of focused transducers are summarized as follows (Ref 69, 208, 209, 227, 229):

• The beam diameter for a 6 dB drop B_D at the focal zone is:

$$B_D \approx \frac{fV}{v\Delta} \qquad \text{(Eq. 3.23)}$$

where v is the frequency, f is the focal length, Δ is the diameter transducer, and V is the sound wave velocity in the specimen (see Fig. 3.19).

• The length of the focal zone F_z (depth of field) is:

$$F_z = NS_F^2\left[\frac{2}{1 + 0.5\,S_F}\right] \qquad \text{(Eq 3.24a)}$$

• The near-field distance N_W for a flat transducer in water is:

$$N_W = \frac{\Delta^2}{4\,\lambda}\left[1 - \left(\frac{\lambda}{\Delta}\right)^2\right] \qquad \text{(Eq 3.24b)}$$

Selection of a Transducer. The selection of a transducer depends very much on the properties of the test specimen, particularly its sound attenuation. Ultrasonics of high frequency can give rise to a good resolution, which is the ability to separate echoes from closely spaced defects. Ultrasonics of low frequency penetrate deeper into materials, because attenuation is generally lower. However, backscattering "noise" from grain boundaries is usually more important than attenuation, although the net result is the same because the signal-to-noise ratio also usually decreases with frequency. This is shown in Fig. 3.20. This means that the two requirements of high penetration and high resolution are mutually exclusive. For example, a specimen such as steel of high attenuation is examined by

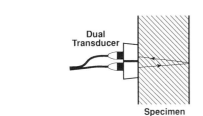

(a)

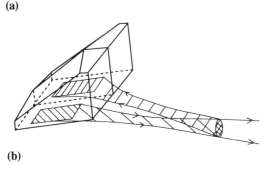

(b)

Fig. 3.18 Dual or pitch-catch transducers consist of separate transmitting or receiving transducers set at a slight angle to each other; see (a). Dual transducers are not as accurate as single transmitting-receiving transducers but are needed for thin specimens to overcome dead-zone effects. Dual transducers are used, therefore, for near-surface resolution and for corroded or rough surfaces. The crossed-beam configuration gives rise to an effective sensitive focusing at the "cross-over" point; see (b).

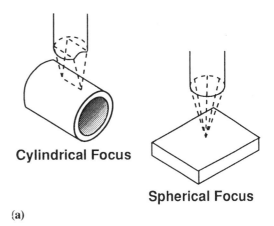

(a)

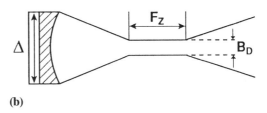

(b)

Fig. 3.19 Focused transducers. Ultrasonic waves can be focused using cylindrical or spherical active elements (a). A more realistic description of the focusing is shown in (b), where the focal zone is indicated (see Eq 3.23 and 3.24a).

a low frequency beam of about 0.5 MHz and a large transducer diameter about 2 in., which will provide a high penetration, but a relatively low lateral resolution of about 6 mm. Improved resolution can be obtained by using shear waves, because these have shorter wavelengths than compression waves of the same frequency in the solid. (The velocity of a longitudinal wave in a solid is greater than that of the shear wave of the same frequency.)

The large diameter transducers are chosen to produce a narrow focused beam; see Eq 3.24, where the beam diameter at the focal zone $B_D \sim fV/v\Delta$; this will enhance the lateral resolution.

Immersion Inspection

A convenient way of maintaining probe coupling to the specimen surface is by using water as the couplant. Indeed, total immersion is a general and useful procedure that offers the advantages of:

- Uniform coupling giving uniform sensitivity
- Easier to deal with complex shapes
- Suitable for automated scanning
- The focusing of immersion transducers increases the sensitivity to small reflecting surfaces and the detectability of mis-oriented defects.

For longitudinal waves at 0° incidence, the distance W_p of the specimen surface from the transducer active element is given by (Fig. 3.21):

$$W_p = f - M_d \left(\frac{V_1}{V_2}\right) \qquad \text{(Eq 3.25)}$$

where f is the ideal focal length in water, M_d is the depth of the focal point within the specimen, V_1 is the sound velocity in the specimen, and V_2 is the sound velocity in water.

Water-bubbler technologies are also available and are used to provide a flow of water around the transducer for access to curved or restricted areas; see Ref 65.

Pulse Echo Display Systems

There are several procedures in examining a test specimen, and these are listed in Table 3.12.

The A-scan or A-scope method is the pulse echo, pulse transit time, system shown in Fig. 3.4.

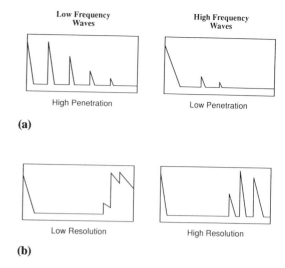

(a)

(b)

Fig. 3.20 Oscilloscope displays using ultrasonic transducers of: (a) high or low penetration (ability to detect defects at distances within the solid) and (b) high or low resolution (ability to separate echoes from closely-spaced defects).

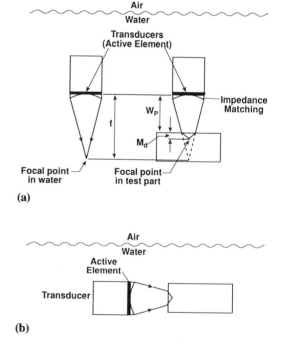

(a)

(b)

Fig. 3.21 Focused immersion transducer probes are used in water tanks to permit the propagation of compression waves into the specimen, maintaining good coupling over the whole surface. The ultrasonic beam is modified in the specimen as shown. Examination can be from (a) above or from (b) alongside.

A one-dimensional view of the defects is given from the ultrasonic pulse echoes (Ref 70, 237).

The B-scan or B-scope method involves a series of parallel A-scans and leads to a two-dimensional view of the defects present in the specimen. This is shown in Fig. 3.22.

The C-scan or C-scope method involves a series of parallel A-scans, which are carried out over a surface, and the pulse echoes are restricted to those returning during a fixed time interval. This is the equivalent of examining a particular

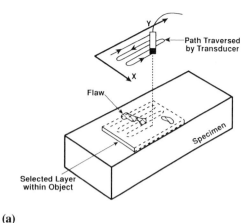

(a)

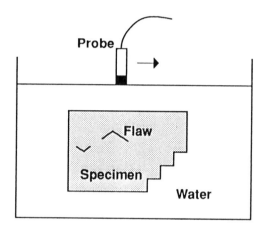

Experimental Arrangement

(a)

Oscilloscope Display

(b)

Fig. 3.23 C-scan (C-scope). The ultrasonic instrument is gated so that only signals received at a fixed time are displayed on the oscilloscope. The probe is moved over the X-Y plane as shown in (a). The result is a two-dimensional image from one particular level of the object under examination (b). The echo pattern will be from discontinuities at that level of the object.

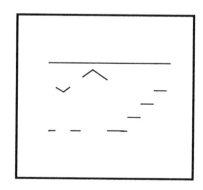

Oscilloscope Display

(b)

Fig. 3.22 B-scan (B-scope). The transducer probe moves across the surface of the water tank, which acts to couple the ultrasonic waves into the specimen. Reflections are recorded on the oscilloscope as the probe is displaced. In this way a two-dimension scan of the object is achieved.

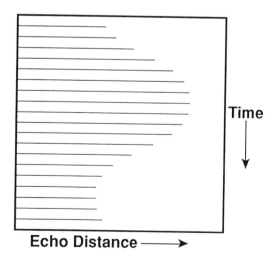

Fig. 3.24 M-mode. Sequence of parallel A-scans displaced with time.

two-dimension level within the specimen. By changing the fixed time interval, it is possible to examine the specimen in all three dimensions: see Fig. 3.23.

The M-mode, listed in Table 3.12, is known by several synonyms; a sequence of parallel A-scans is carried out over a period of time and displaced with time as shown in Fig. 3.24.

Automated Systems

Specimens can be examined automatically with the transducer probe moving mechanically across the surface and echoes presented by a series of A-scan displays, gated if necessary to reveal a particular feature. The probe-surface coupling can be maintained constant and ensured using total immersion in water, with a compression wave-emitting probe traversing the surface a few centimeters above the specimen. The initial pulse can be discarded as well as the echo from the specimen surface.

There are many automated computer controlled systems available of this type. A few of these include Hyscan (Panametrics) (Ref 229), NDT Systems (Ref 239), and Zipscan (Harwell) (Ref 69). Other automated systems are described in Ref 69. Several computer-based systems are available that can acquire data and reconstruct and display A, B, and C scans in different presentations.

Calibration and Reference Standards

The signal on the oscilloscope of a reflection from a flaw in the test piece or from the far side of the specimen cannot be interpreted directly and easily in terms of flaw size. The most convenient way to explain the signal position and amplitude is by comparison with the signal from a calibrated reference test object. Several reference blocks are in use, and some of these are discussed in Fig. 3.25 to 3.27.

Table 3.12 Nondestructive testing display systems

Display system	Description	Fig.	Ref

A-scan
  A-scope
  A-mode
	One-dimensional data.	3.4	161, 237, 238

B-scan
  B-scope
  B-mode
  Section scan
	Two-dimensional data. Parallel set of A-scans.	3.22	129, 161

C-scan C-scope	Third dimension added to two-dimensional scan by restricting the echoes to a particular time, which corresponds to a constant depth in the specimen.	3.23	129, 161, 237
D-scan	Two-dimensional cross section of specimen, normal to the test surface and along the direction of the scanning beam.	...	69
P-scan	Simultaneous display of B and C scans (proprietary name).	...	69

M-mode
  Time-motion scan
  T-M
  Time-position scan
  Motion scan
  UCG
  Ultrasonic
  Cardiogram
	Sequence of parallel A-scans displayed against time. The time of the scan must be much less than the time of the event.	3.24	237, 238

Stop-action scan
  ECG-gated B-scan
  Echo cardiogram
	In periodic motion, a particular phase is examined by ultrasonic pulse. Thus, a particular phase of the heart action can be "frozen" on the display screen.	...	237

Real-time scan
  Rapid B-scan
  Linear arrays of A-scan
	Linear arrays of A-scans.	...	139, 237

Doppler scan	Shift in frequency of ultrasound when reflected by a moving object. Continuous wave: Beat frequency between outgoing and reflected wave is related to the velocity of the object. Pulsed wave: Range and velocity of object.	...	237

The ultrasonic echoes can be displayed in several different modes, providing one-, two-, or three-dimensional data. There are also several time-motion methods of examination.

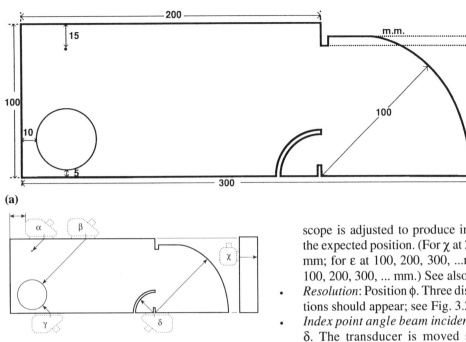

(a)

(b)

Fig. 3.25 The calibration block type recommended by the International Institute of Welding. (a) Dimensions in mm. (b) Calibration procedures (see text for details).

Test block IIW-A2 (International Institute of Welding) is used extensively because this provides for the calibration of distance, sensitivity, resolution, index point of sound emission and entry, and angle of sound propagation for angular incidence (shear, transverse waves) for contact inspection (Fig. 3.25) (Ref 65).The calibration procedures (Fig. 3.25b) are as follows:

- *Sensitivity setting* (gain): Position χ is for a straight beam; position α is for an angle beam. The gain is adusted until the first back reflection nearly fills the oscilloscope screen.
- *Distance calibration* (time base): Position χ or ε is for a straight beam; position δ is for an angle beam. The time base of the oscillo-

scope is adjusted to produce indications at the expected position. (For χ at 25, 50, 75, ... mm; for ε at 100, 200, 300, ...mm; for δ at 100, 200, 300, ... mm.) See also Fig. 3.26.

- *Resolution*: Position ϕ. Three distinct indications should appear; see Fig. 3.20(b).
- *Index point angle beam incidence*: Position δ. The transducer is moved sideways to obtain maximum signal. This occurs when the point of sound emission from the transducer is opposite the mark on the calibration block.
- *Beam angle*: Position β is 45 to 60°. Position γ is 60 to 70°. The transducer is moved sideways to obtain maximum reflected signal from the curved surface (see Fig. 3.27).

Thickness calibration, for normal incidence compression waves, can be achieved using a series of steps of known height (Fig. 3.26). The time base of the oscilloscope is adjusted to permit the direct conversion of the horizontal axis to distance in inches (or cm). The transducer (normal incidence) is moved from one thickness to the next, with adjustments to the oscilloscope controls to calibrate the time base in terms of distances—from 0.75 to 0.05 in. in Fig. 3.26(b). In the case of nondispersive media, it is preferable to use a higher order reflection pulse, such as the fourth reflection (Fig. 3.26c). See also the sections "Resolution" and "Thickness Measurements" in this chapter.

American Welding Society (AWS) block (Fig. 3.27) permits the resolution (echo pulses clearly resolved) of an angle beam transducer to be determined at 45, 60, and 70°. The transducer is placed at A for 70° angle, B for 60°, and C for 45° angle (Fig. 3.27b). In each case, three peaks should be resolved in the oscilloscope display, as shown in Fig. 3.27(c). The beam profile of the angle beams can be determined using the AWS reference block. The procedure is outlined in Ref 69. This block is also known as the British Standard Institute BSI-A5 test block (Ref 240).

ASTM E 127 block is described in *Nondestructive Testing, Annual Book of Standards* (Ref 61).

Other reference and calibration test pieces include distance and sensitivity block type A3 (Ref 69), and type V2. Calibration test pieces are also available from ASTM and from the Aluminium Industries Council (U.K.) (Ref 240).

Both the IIW and ASTM blocks are widely used in the United States.

Resolution

Resolution is the ability of the system to separate signals from neighboring defects. The reso-

(a)

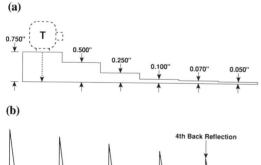

(b)

(c)

Fig. 3.26 Thickness calibration steps. (a) Steel six step calibration block. (b) Transducer (normal incidence) is moved from 0.75 to 0.05 in. (c) Higher order reflection pulse—the fourth reflection. The thickness is then a quarter of this measured distance. An estimate of the dead zone of the transducers is provided by the width at the base of the initial pulse of ultrasonic waves. *Note*: The baseline is adjusted using either the baseline break (the commencement of the echo) or the peak position.

lution of an angular beam from a shear wave transducer can be determined using the AWS reference block; the procedure is outlined above and shown in Fig. 3.27. The profile of the angle beam spread can also be determined using the AWS block and is determined by finding the position of maximum signal from a target hole, then displacing the transducer until the amplitude signal decreases to 10% of its maximum value (20 dB) or 50% of its maximum value (6 dB). The displacement is carried out in both directions from the angle beam to lower and higher values of the beam angle. In this way, the beam spread is determined. Further details of the procedure are given in Ref 69.

An estimate of the dead zone of the transducer probe is obtained by measuring the base width of the initial pulse of ultrasonic waves (Fig. 3.26c). The material of the reference block should be the same as that of the test object. Most commercial reference blocks are made of carbon steel or aluminum, and thus allowance must be made for the different acoustic properties of the test object and of the reference block.

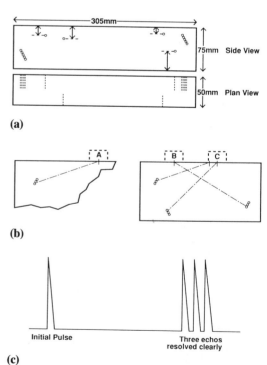

(a)

(b)

(c)

Fig. 3.27 Calibration of angle beam transducer with AWS reference block. (a) Dimensions of the AWS block. (b) Positions of the transducer for the different resolution tests. (c) Typical display on the oscilloscope, with the echo pulses resolved clearly.

Size of the Defect

There is no simple relationship between echo and defect size; signal amplitude depends on a wide range of quite different factors; these include metal microstructure, grain size, defect distance, shape, orientation, as well as impedance difference. The depth of the defect can generally be determined from the time-of-flight, so that some empirical methods have been set up to allow the estimation of "equivalent defect size."

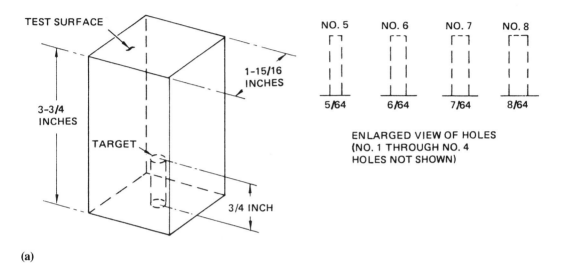

(a)

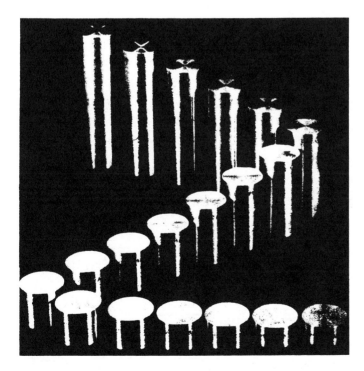

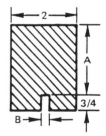

DIMENSION A	
1/16	1 3/4
1/8	2 1/4
1/4	2 3/4
3/8	3 1/4
1/2	3 3/4
5/8	4 1/4
3/4	4 3/4
7/8	5 1/4
1	5 3/4
1 1/4	

DIMENSION B
3/64
5/64
8/64

(b)

Fig. 3.28 Standard reference blocks. (a) Area/amplitude (Alcoa-Series A). (b) Distance/amplitude (Alcoa-Series B; Hitt). Reprinted with permission of The American Society for Nondestructive Testing.

Echo amplitudes are calibrated by comparison with signals from standard sizes of ideal-shaped holes over a range of distances. A calibration curve is drawn, providing a distance correction to the amplitude to give the minimum size of defect giving a comparable signal at the same distance.

For area-amplitude blocks (Fig. 3.28a), flat-bottomed holes are drilled at a depth of ¾ in. and in diameters from ¹⁄₆₄ to ⁸⁄₆₄ in. For distance-amplitude blocks (Fig. 3.28b), cylinders of different heights have a flat-bottomed hole drilled in the base as shown. A distance amplitude correction DAC calibration curve can be prepared.

Part III Applications

Oscilloscope Interpretations

Immersion in Water (Fig. 3.29). A plate immersed in water is examined by a single transducer probe, at normal incidence because only longitudinal waves can propagate in a liquid (Fig. 3.29a). The initial pulse can be displaced to the left of the oscilloscope screen, with the reflected pulse from the interface water/solid arranged at the origin of the oscilloscope screen,

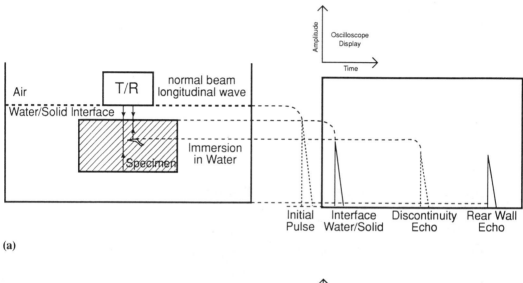

(a)

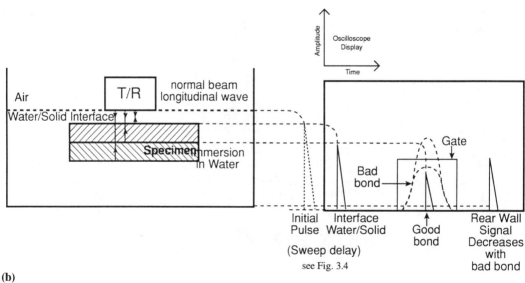

(b)

Fig. 3.29(a) and (b)   Oscilloscope interpretations: immersion in water. (a) A plate immersed in water examined by a single transducer probe at normal incidence. (b) Examination of laminates and brazed joints.   *(Continued)*

using the "sweep delay" control. A discontinuity perpendicular to the beam is likely to reflect a pulse, and the larger the pulse reflected by the defect, the smaller the pulse reflected by the rear wall. The water path should be more than $\frac{1}{4}$ of the metal path. For examination of laminates and brazed joints (Fig. 3.29b), the initial pulse can be displaced, with the pulse from the specimen upper surface at the origin of the display on the oscilloscope, using the "sweep delay" control. The signal from the bond in the laminate is sharp for a good bond, but broader and more intense for a bad bond. A gate can be used to differentiate between satisfactory and unsatisfactory bonded laminates. Where the signal from the bond region increases in intensity due to poor bonding, the signal from the rear wall will decrease in intensity and it is this signal that can also be gated. A curved specimen immersed in water can be examined by a single transducer probe at normal incidence (Fig. 3.29c). There is virtually no sig-

nal from the interface or rear wall. Only a discontinuity at a suitable angle will reflect a signal as shown in Fig. 3.29(c). The angle of the beam entering the specimen must be less than the critical angle.

Examination of Metal Components (Fig. 3.30). For the examination of a casting (Ref 129), indications are shown in Fig. 3.30(a) from shrinkage cracks and solid inclusions or blow holes. The shape of the echo from the flaw can be used to identify the nature of the discontinuity. To detect shrink fit (Ref 219) (Fig. 3.30b), a single transducer probe is used at normal incidence, so that the shrink fit and the back wall (bore wall) reflect pulses. For a very good tight fit, the shrink fit pulse will be very small compared to the back wall signal, whereas a bad looser fit will give a very large signal compared to the back wall. For examination of a thin plate (Fig. 3.30c), a normal incidence probe will give rise to multiple ripple echoes. Changes in thickness, or in the lamina-

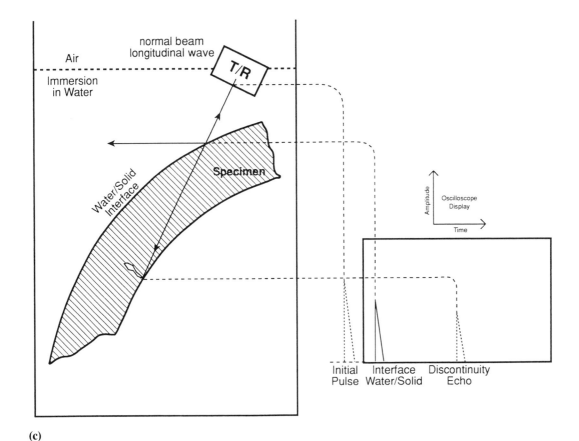

(c)

Fig. 3.29(c) Oscilloscope interpretations: immersion in water. (c) Curved specimen immersed in water examined by a single transducer probe at normal incidence.

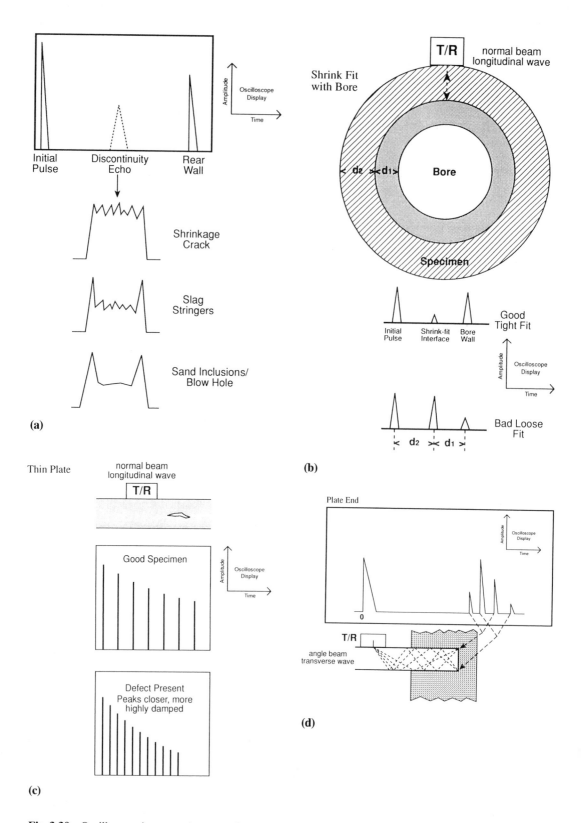

Fig. 3.30 Oscilloscope interpretations: metal components. (a) Castings (Ref 129). (b) Shrink fit (Ref 219). (c) Thin plate. (d) Plate end. (Ref 65)

114

tion of a thin plate will cause the multiple ripples to be closer together and more highly damped (Ref 65). During angle beam examination of the end of an embedded plate (Fig. 3.30d), the echoes take on a complex pattern. This plate can be embedded, free-standing, or soldered to another part. If a normal beam were used, mode conversion to a shear wave would occur. These reflections occur after the signal from the rear wall and thus can be distinguished.

Examination of Welds; Skip Distance

The examination of welds by ultrasonics has been described and discussed very extensively, in particular in Ref 3, by the Institute of Metallurgists, U.K, as well as in Ref 69, 129, and 219. The American Welding Society (AWS) has issued a series of specifications used extensively throughout NDT. Some defects that can occur in welds are shown in Fig. 3.32.

A shear wave angle beam transducer is used to examine the weld, and this results in the ultrasonic angular beam arriving at the back surface at a halfskip distance. This is shown in Fig. 3.31 where the halfskip, fullskip, and the node distances are indicated in (a), (b), and (c). The angular beam enables the underside of the weld to be examined and is a very useful approach. In Fig. 3.31(c), the method of examining the root of the weld and the interior of the weld is shown, where the transducer probe is displaced from the halfskip to the fullskip position. The ultrasonic beam is not as parallel as the straight line indicated in most figures. The beam is diverging, and this is shown in Fig. 3.31(d), where the 6 dB ($\frac{1}{2}$ maximum amplitude) on either side of the beam direction is given. However, this is a simplified representation, and the beam has a wider spread than shown. The skip distance S_K can be calculated:

$$S_K = 2\,\Delta \tan \phi \qquad \text{(Eq 3.26)}$$

where Δ is the thickness of the specimen, and ϕ is the angle of the beam; see Fig. 3.31(a). For example, if an angular beam of 70° is being used to examine a 1 in. plate, the skip distance will be (2 tan 70) 5.5 in. Thicker plates require the use of

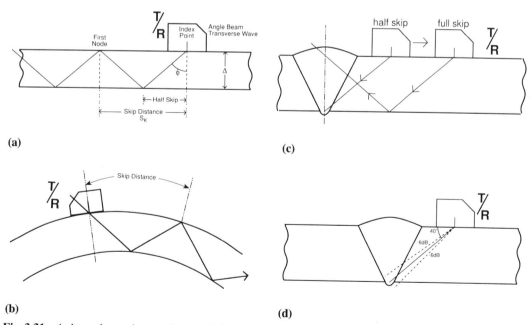

Fig. 3.31 A shear ultrasonic wave is an angle beam that gives rise to reflections at a distance down the specimen. The halfskip, fullskip, and node positions are as shown in (a) or in (b) for straight or curved objects. The positions of the first node and second node can be determined using a second angular transducer as a detector, enabling the skip distances to be measured. (c) Welds can be examined by displacing the transducer from the halfskip to the fullskip position. (d) The ultrasonic beam spreads out, and the –6 dB spread ($\frac{1}{2}$ maximum amplitude) is shown. The skip distance is also known as node or leg. The skip distance $S_K = 2\Delta \tan \phi$.

smaller beam angles. For example, steel plates of 1 in. thickness examined using a lucite wedge would use a 70° angular beam, resulting in a skip distance of approximately 5 in.; a 3 in. thick plate would be examined by a 45° angular beam with a skip distance of about 6 in.

To accomplish useful measurements of a weld, the following should be determined or known in advance:

- The thickness of the material, of known ultrasonic wave velocity.

- The index position, angle of incidence, and spread of the transducer beam.
- The skip distance; this can be determined by using a second transducer and detecting the position of the nodes.
- The nature of the weld, its ideal shape and dimensions.
- The original material must be free of defects and laminates in the vicinity of the weld.

With this information, it is now possible to proceed to examine a weld, and the following

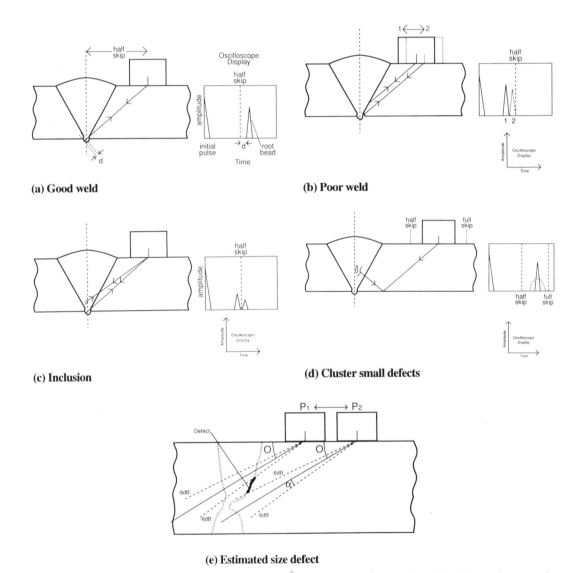

(a) Good weld

(b) Poor weld

(c) Inclusion

(d) Cluster small defects

(e) Estimated size defect

Fig. 3.32 Examination of a weld by an angle beam shear wave ultrasonic transducer. Useful procedures require that the skip distance, node positions, beam angle, beam spread, index position of the transducer are all known, as well as the sound velocity in the material, the exact position of the weld and its ideal shape. The thickness of the wall needs to be known, and other defects, laminations, and porosity should be absent; see text.

observations can be made (Ref 240). A good weld will give a sharp signal pulse a little after the half-skip position (Fig. 3.32a). If there is a poor weld with a lack of complete fusion on the side nearest the transducer, a signal will be received before the halfskip position; and on moving the transducer the signal will move toward the halfskip position, as shown in Fig. 3.32(b). The presence of an inclusion in the root bead, as in Fig. 3.32(c), will give rise to a signal that becomes a maximum when the probe is at less than the halfskip distance.

There is information to be obtained by considering the shape of the signal pulse from the crack or inclusion. A smooth, sharp crack will give rise to a sharp pulse. In the case of a jagged-edge crack, roughly shaped inclusion, or cluster of small defects, a broad, poorly defined signal will be given as shown by the dashed line in Fig. 3.32(d). The position of the defect, crack, or inclusion can often be decided by observing the signal change on displacing the transducer. The size of a defect can be estimated as shown in Fig. 3.32(e); a transducer of known beam spread is required, where the angle is known between the central beam maximum and the direction of 6 dB, that is $\frac{1}{2}$ maximum amplitude. The transducer is then displaced from P_1 to P_2, when the signal amplitude will go from $\frac{1}{2}$ maximum, through the maximum value, and return to $\frac{1}{2}$ maximum amplitude. The distance P_1-P_2 moved

by the transducer is related to the size of the defect as shown in Fig. 3.33. The crack length based on an assumed beam spread, C can be determined from measurements of L_1, L_2, P_1, P_2, and the angles θ and d; see Fig. 3.33.

It is often essential to examine a weld in its entirety, and this involves the complete scanning of the weld length and breadth. The procedure is shown in Fig. 3.34(a), where the angle beam transducer is moved from the halfskip to the fullskip position in a zig-zag path along the centerline of the weld. Another procedure is shown in Fig. 3.34(b) and (c) where a guide bar is set up at the correct halfskip distance.

Further information about the defect can be obtained by observing the shape of the echo pulse and how this changes with the positioning of the transducer probe. The signal is different from gas pores, cracks with smooth faces, inclu-

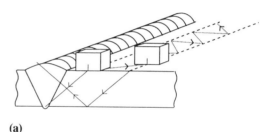

(a)

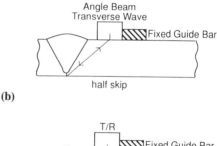

(b)

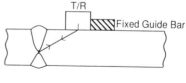

(c)

Fig. 3.34 Scanning procedure to examine a complete weld line. A zig-zag path parallel to the weld direction is followed, from halfskip to fullskip positions. (a) The angle beam probe zig-zag movements from halfskip to fullskip positions. Another procedure is illustrated in (b) and (c), where a guide bar can be attached to the specimen so that the probe is set up to examine a particular feature of the weld region; in (b), it is the weld root, and in (c) it is the central portion of the double-V weld.

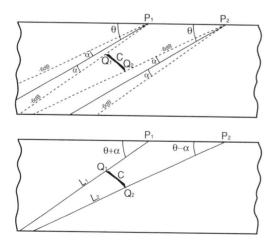

Fig. 3.33 Calculation of crack length C using the beam spread at –6 dB and displacing the transducer from P_1 to P_2. This is only an estimate based on an assumed spread angle.

sions with ragged edges. The examples given of weld examination involve angle beam transverse waves. Examination can be undertaken using normal beam longitudinal waves; see for example Ref 219.

Thickness Measurements

The principle of thickness measurements is illustrated in Fig. 3.35. A normal beam, longitudinal transducer probe is placed in good contact with the surface, and the pulse echo of multiple reflections observed. Alternatively, a dual crystal transducer can be used (see the section "Ultrasonic Transducers" in this chapter).

Small hand-held devices are available commercially that simplify the procedure to measure specimen thickness. However, care must be taken to correct for the velocity of sound in the material, and the transducer should be pressed closely to the surface to avoid an air gap. Accuracies of the order of ±0.2 mm are attainable typically for a smooth-sided, fine-grain steel specimen about 10 mm long. The accuracy of the measurement depends on the calibration, and a calibration curve should be set up rather than just at one distance. Using a normal A-scan type measurement, accuracies are 50×10^{-3} in. in a specimen length of 10 in., and resolution of about 10^{-3} in.

Resonance Methods. The thickness of an object can be determined by measuring resonance frequencies. Resonance occurs when the reflected wave is exactly in phase with the incident wave, so that a standing wave forms. This is illustrated in Fig. 3.36 when thickness Δ is a multiple of $\lambda/2$. In general:

$$\Delta = \frac{n\lambda_n}{2} = \frac{nV}{2v_n} \qquad \text{(Eq 3.27)}$$

because $\lambda_n v_n = V$ the sound velocity, and λ_n and v_n are the wavelength and frequency of the nth standing wave. The value of n is not known in general, but can be removed from consideration by measuring adjacent resonance frequencies, v_n and v_{n+1}. Then

$$\Delta = \frac{nV}{2v_n} = \frac{(n+1)V}{2v_{n+1}} = \frac{nV}{2v_{n+1}} + \frac{V}{2v_{n+1}} \qquad \text{(Eq 3.28)}$$

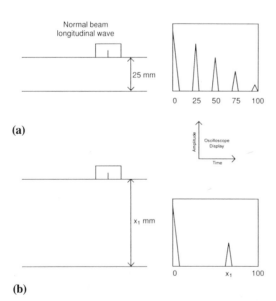

(a)

(b)

Fig. 3.35 Thickness measurements from multiple echoes. When a parallel-sided plate is examined by a single transducer probe at normal incidence, multiple echo pulses are displayed on the oscilloscope screen, and their positions will be as shown in (a). Calibration plates can be used to enable the measurement of thickness, as in (b). A calibration curve should be set up, rather than a single calibration thickness.

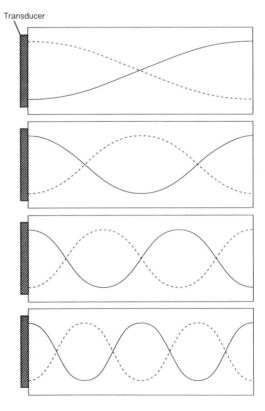

Fig. 3.36 Resonance thickness measurements using standing waves (Ref 65, 139) (see Eq 3.27 to 3.29).

118

Using $n = (2\Delta v_n)/V$ from Eq 3.27, then rearranging

$$\Delta = \frac{V}{2[v_{n+1} - v_n]} \qquad \text{(Eq 3.29)}$$

As an example, a copper plate is observed to have neighboring resonance frequencies at 1.213 MHz and at 1.111 MHz, using longitudinal wave of velocity in copper of 4.66 km/s. Then the thickness of the copper plate is given by:

$$\frac{V}{2[v_{n+1} - v_n]} = \frac{4.66 \text{ km/s}}{2[1.213 - 1.111]\, 10^6/\text{s}} = 2.28 \text{ cm}$$

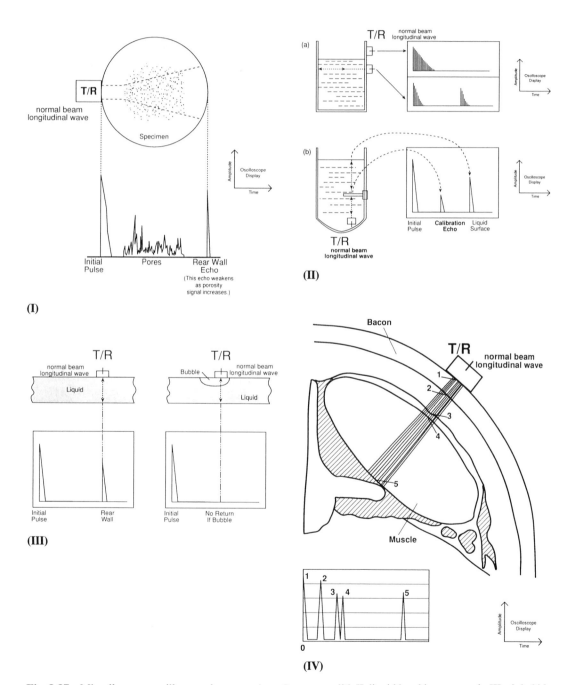

Fig. 3.37 Miscellaneous oscilloscope interpretations. I, porous solid. II, liquid level in a reservoir. III, air bubble in a pipe (Ref 219). IV, bacon thickness (Ref 219).

General Applications

Oscilloscope displays for a range of specimens are shown in Fig. 3.37, covering porous materials, liquid in a reservoir, bubble in a pipe, and bacon thickness.

Porous Solid (Fig. 3.37, part I). The indications from pores, inclusions, or fine cracks are very similar. A series of small pulses are observed. Pores give rise to stronger signals compared to inclusions, because the strength of the pulse depends on the impedance difference, which is usually greater at an empty pore than at an inclusion. Small pores may tend to scatter while larger inclusions may reflect more. However, some inclusions may have an air gap, which will reflect strongly. The signal from the rear wall will be reduced and may not be visible due to the scattering of the pores; interpretations will generally be based on the rear wall echo, as well as the baseline oscillations.

Liquid Level in a Reservoir (Fig. 3.37, part II). In the absence of liquid, one transducer probe at normal incidence will receive strong reflected pulses due to multiple reflections in the first wall, with little pulse energy transmitted through the air inside the reservoir. However, where there is liquid present, some pulse energy will be transmitted to the far wall of the region, resulting in diminished intensity of the first wall multiple reflections, followed later by a reflected pulse from the far wall if this is perpendicular to the beam. This is illustrated in Fig. 3.37, part II(a). Alternatively, a transducer probe can be located inside the reservoir, giving rise to a reflected pulse from the liquid surface. A simple internal calibration is possible by inserting an obstacle at a known distance in the liquid; see Fig. 3.37, part II(b).

Air Bubble in Pipe (Fig. 3.37, part III) (Ref 219). A single normal incidence (longitudinal) transducer probe will detect pulses returning from the far wall of the pipe. If an air bubble is present, no echo will be received due to lack of transmission through the air bubbles.

Bacon Thickness (Fig. 3.37, part IV) (Ref 219). The thickness layers of a live pig can be determined using a single normal incidence (longitudinal wave) transducer providing the various velocities involved are known: 1, surface; 2 and 3, fascia; and 4 and 5, muscle.

Lists of applications with references are given in Tables 3.13 to Tables 3.16. The inspection of metals is covered in Table 3.13 and includes ferrous and nonferrous metals, different

Table 3.13 Inspection of metals: a bibliography

Application	Ref
Welded joints, aluminum welds	69, 81, 82, 129, 211, 208 (Chapter 26), 222
Castings	69, 129, 208 (Chapter 25)
Wires, thin plates	69, 241
Tubes and bars	69, 129
Machine parts	208
Rods, billets, wires	208 (Chapter 33)
Heavy and medium plate and sheet	208 (Chapter 22)
Soldering, brazing, gluing joints	208 (Chapter 27)
Rivets and holes	208
Pipes and cylinders	208 (Chapter 24), 222
Boilers	129
Heavy forgings	129, 208
Railways	208 (Chapter 21)
Nuclear reactors	208 (Chapter 28)
Intergranular cracks	222
Crack growth	243
Fatigue cracks	211, 222, 244-246
Steel and cast iron	208
Metals, nonferrous	208
Stress, stress corrosion, residual stress	129, 211, 222 (Chapter 9), 247, 248, 249
Spot welding	195
Extrusion	129
Grain refinement dispersion	195
Porosity	208, 243
Microstructure	243
Phase transformation	243
Corrosion	129
Powder metallurgy	129, 243
Defect sizing	69, 198
Honeycomb structures	211, 250-253

Table 3.14 Inspection of nonmetals: a bibliography

Application	Ref
Ceramic materials	208 (Chapter 30), 243
Rocks	208
Concrete	69, 208
Composite materials	69, 243
Glass-reinforced polymers	129, 254
Carbon and graphite	208
Wood, leather	208
Viscoelastic materials, plastics, rubber	208
Thin fibers	242
Piezoelectric solids	223
Abrasives	208
Adhesives	211, 243
Anisotropic solids	179
Preferred orientation	211
Interfaces and surfaces, boundaries, layered solids, coatings, platings, thin films	81, 139, 179 (Chapter 10), 195

structures from boilers, tubes, rods, pipes, cylinders, powder metallurgy, corrosion problems, porosity, and a range of mechanical parts. Nonmetallic objects are listed in Table 3.14, ranging from wood, ceramics, thin fibers, coatings, interfaces, surfaces, and abrasives. Some general applications are listed in Table 3.15, such as thickness measurements, medical studies, viscosity and flow measurements, surface hardness, locating plugs or air bubbles in pipes, some chemical and precipitation effects. In Table 3.16, several other ultrasonic techniques are listed with appropriate references.

Table 3.15 General applications: a bibliography

Application	Ref
Air bubbles in pipes	208
Plug in waterline	208
Level liquid in reservoir	195 (Chapter 11), 208
Precipitation effects	195
Dispersion and coagulation	195
Electrochemistry	195
Chemical effects	195
Cavitation	195
Absorption sound	208
Cutting tool surface	208
Surface hardness	208
Outer space	211
Burglar device, blind guidance flowmeters	195 (Chapter 11)
Seismology	38 (Section 50)
Medical	
EKG	195, 237
Blood flow	161, 238
Doppler measurement of velocity	161 (Chapter 7), 237 (Chapter 15), 238 (Chapter 8)
Sea depth	198, 255
Wall thickness	208
Thickness gage	69
Ranging	195, 255
Bats	255
Underwater detection, submarines	194 (Chapter 2), 195, 220, 255
Fish detection	194 (Chapter 4)
Navigation	194 (Chapter 3)

Table 3.16 Other ultrasonic techniques: a bibliography

Application	Ref
Schlieren technique	39, 41, 242
Acoustic microscopy	39, 254, 256
Scanning acoustic microscopy	69, 254
Ultrasonic spectroscopy	69
Holography	69
Noncontact laser	243
Time-of-flight	30, 211, 218
Critical reflection	222
Ring-wheel surface wave	211, 257

Worked Examples

Example 1. A composite plate is 10 cm thick and consists of a layer of lead bonded to nickel. The time required for an ultrasonic pulse to traverse the composite is 4×10^{-5} s. How thick is the layer of lead? The ultrasonic longitudinal velocity in lead is 2.16 km/s and in nickel 5.63 km/s. The time to traverse the composite equals the sum of the times to traverse the lead (t_{Pb}) and the nickel (t_{Ni}). Let d cm be the thickness of lead. Then:

$$4 \times 10^{-5} \text{ s} = t_{Pb} + t_{Ni} = \frac{d}{2.16 \times 10^5} + \frac{(10-d)}{5.63 \times 10^5}$$

Therefore:

$$d = 7.79 \text{ cm}$$

The thickness of lead is 7.79 cm and nickel is 2.21 cm.

Example 2. What is the time required for an ultrasonic pulse to traverse a composite plate consisting of 1 in. thick copper between two 4 in. lead plates? [Ultrasonic longitudinal velocities: copper 4.66 km/s, lead 2.16 km/s]

Time required equals time to traverse copper t_{Cu} plus time to traverse lead t_{Pb}.

$$\therefore \text{Time} = t_{Cu} + t_{Pb} = \frac{1 \times 2.54}{4.66 \times 10^5} + \frac{8 \times 2.54}{2.16 \times 10^5}$$

$$= 99.52 \text{ } \mu s$$

Example 3. In an ultrasonic test, a pulse of longitudinal waves requires 74.9 μs to pass through a thick rod of polymer of length 20 cm. Determine:

- The ultrasonic velocity in the material

$$V = \frac{20 \text{ cm}}{7.49 \times 10^{-6} \text{ s}} = 2.67 \text{ km/s}$$

- The probable identity of the material—probably plexiglass (see Table 3.6c)

Example 4. A longitudinal ultrasonic pulse is introduced into a 10 in. long rod of a metal. After 87.7 μs, a pulse returns to the transducer. Determine:

- The velocity of the ultrasonic pulse

$$V = \frac{2 \times 10 \times 2.54}{87.7 \times 10^{-6}} = 5.792 \text{ km/s}$$

- The probable identify of the metal—probably magnesium (see Table 3.6a)

Example 5. An ultrasonic pulse in a 4 in. copper plate returns a signal to the transducer after 2×10^{-5} s. Is there a flaw in the copper plate? The ultrasonic velocity in copper is 4.66 km/s (see Table 3.6a). The time t required for the ultrasonic pulse to traverse the copper bar and return, a total of 8 in., is:

$$t = \frac{8 \times 2.54}{4.66 \times 10^5} = 4.36 \times 10^{-5} \text{ s}$$

Because the echo returns after 2×10^{-5} s, a flaw must be present at a distance from the surface of:

$$\frac{2 \times 10^{-5} \text{ s} \times 4.66 \times 10^5 \text{ cm/s}}{2} = 4.66 \text{ cm}$$

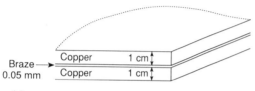

(a)

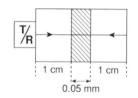

(b)

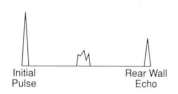

(c)

Fig. 3.38 For Example 6. (a) Lap joint to be made by brazing. (b) Testing schematic. (c) Oscilloscope display.

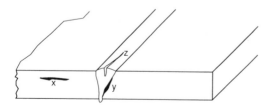

Fig. 3.39 For Example 7.

Example 6. A lap joint between two sheets of copper of thickness 1 cm is to be made by brazing. Describe the ultrasonic nondestructive testing technique that might be used to determine if the joint is completely filled. Assume a braze thickness of 0.05 mm. The ultrasonic longitudinal velocity in copper is 4.66 km/s. The echo from the rear wall occurs at:

$$\frac{4 \text{ cm}}{4.66 \times 10^5 \text{ cm/s}} = 8.58 \text{ μs}$$

The echo from the braze will occur at 4.29 μs. The ultrasonic testing technique is shown in Fig. 3.38. The expected oscilloscope display is given in Fig. 3.38(c), with the rear wall echo occurring at 8.58 μs. The braze echo is expected at 4.29 μs, although this signal will vary, becoming multiple, broad, or diffuse depending on the state of the braze.

Example 7. Three types of cracks in a weld are portrayed in Fig. 3.39. Give some general comments about nondestructive testing methods that might be applied including x-radiography (Xrad), ultrasonics (UT), magnetic particle inspection (MPI), liquid penetrant inspection (LPI).

Method	Defects		
	Z	Y	X
Xrad	Orientation required	Suitable	Suitable
UT	Orientation required	Probably most convenient method; use transverse beam with known half skip distance	Can estimate depth
MPI	Orientation required if ferromagnetic; depends on surface finish	Only if close to surface	Not suitable
LPI	Cheapest; simplest; depends on surface finish	Not suitable	Not suitable

Conclusion: Ultrasonic is probably the best overall.

Example 8. What angle of incidence should be used on a tungsten specimen using a wedge of rubber so that only a shear wave is transmitted in the metal?

- *Tungsten*: longitudinal velocity = 4.44 km/s
- *Tungsten*: shear velocity = 2.87 km/s
- *Rubber*: longitudinal velocity = 2.3 km/s

First critical angle = $\sin^{-1}\dfrac{2.3}{4.44} = 31.2°$

Second critical angle = $\sin^{-1}\dfrac{2.3}{2.87} = 53.3°$

A suitable angle would be ~45°.

Example 9. What angular beam should be used to examine a plate 2 in. thick to have a skip distance of at least 7.5 in.?

$$S_K = 2\,\Delta\tan\phi \qquad \text{[see Fig. 3.31 and Eq 3.26]}$$

then

$$\phi = \tan^{-1}\dfrac{7.5}{2\times 2} = 62°$$

A 70° beam should be used.

Example 10. A plate 2 in. thick is examined by a 70° angular beam, ϕ, when a discontinuity is detected after a sound path, P_s, of 8 in. What is the distance, D, along the plate from the probe exit point? Now

$$D = P_s\sin\phi = 8\sin 70 = 7.52 \text{ in.}$$

Example 11. The thickness of a steel plate is to be determined using the resonance method with longitudinal sound wave of velocity, V, 5.79 km/s.

- Resonance frequencies were observed at 2.103 MHz and subsequently at 1.978 MHz. What is the thickness, Δ, of the steel block?

$$\Delta_{cm} = \dfrac{V}{2(v_{n+1}-v_n)} = \dfrac{5.79\times 10^5 \text{ cm/s}}{2(2.103-1.978)\,10^6/\text{s}}$$

$$= 1.93 \text{ cm} \qquad \text{[see Eq 3.29]}$$

Neighboring resonance frequencies were observed in another steel block at 2.103 MHz and at 2.081 MHz:

$$\Delta_{cm} = \dfrac{5.79\times 10^5 \text{ cm/s}}{2(2.103-2.081)\,10^6/\text{s}} = 13.16 \text{ cm}$$

ASTM Standards

Various documents and codes are published by ASTM (Ref 61) on an annual basis and are listed in Table 3.17. The topics covered include:

- *Reference blocks*: ASTM E 1158, E 127, E 428
- *Flat-bottom hole sizes*: ASTM E 804
- *Immersion procedures*: ASTM E 664, E 1001, E 214
- *Examination of welds*: ASTM E 164
- *Thickness measurements*: ASTM E 797

Table 3.17 ASTM standards for ultrasonic testing

Standard No.	Description
E 114	Ultrasonic Pulse-Echo Straight-Beam Examination by the Contact Method (1990)
E 127	Aluminum Alloy Ultrasonic Standard Reference Blocks (1992)
E 164	Ultrasonic Examination of Weldments (1990)
E 213	Ultrasonic Examination of Metal Pipe and Tubing (1990)
E 214	Immersed Ultrasonic Examination by the Reflection Method Using Pulsed Longitudinal Waves (1991)
E 273	Ultrasonic Examination of Longitudinal Welded Pipe and Tubing (1989)
E 317	Evaluating Performance Characteristics of Ultrasonic Pulse-Echo Testing Systems (1990)
E 428	Steel Reference Blocks Used in Ultrasonic Inspection (1991)
E 494	Ultrasonic Velocity in Materials (1992)
E 587	Ultrasonic Angle-Beam Examination by the Contact Method (1988)
E 588	Large Inclusions in Bearing Quality Steel by the Ultrasonic Method (1988)
E 664	Apparent Attenuation of Longitudinal Ultrasonic Waves by Immersion Method (1989)
E 797	Ultrasonic Pulse-Echo Contact Method Measuring Thickness (1987)
E 804	Ultrasonic Test System Flat-Bottom Hole Sizes (1988)
E 1001	Immersed Pulse-Echo Ultrasonic Method Using Longitudinal Waves (1990)
E 1065	Evaluating Characteristics of Ultrasonic Search Units (1982)
E 1158	Reference Blocks for the Pulsed Longitudinal Wave Ultrasonic Examination of Metal (1990)
E 1315	Ultrasonic Examination of Steel with Convex Cylindrically Curved Entry Surfaces (1989)
E 1324	Ultrasonic Examination Instruments (1990)
E 1454	Computerized Transfer of Digital Ultrasonic Testing Data (1982)

Note: These standards are frequently revised and published annually in Ref 61.

Appendix 3.1: Glossary of Some Terms used in Ultrasonic Testing (Ref 61, 190)

acoustic impedance (Z). The propagation of waves in solids depends on the resistance of the atoms (or particles) to vibrate; that is Z = (Local excess pressure on the particles)/(Particle displacement velocity). It can be shown that Z is given by the product of sound velocity and density of a material.

angle transducer. Designed to project ultrasonic waves at an angle to the surface to generate shear, surface, or other types of wave mode conversion; see Fig. 3.16(b).

area amplitude response curve. Shows changes in the amplitude from planar reflectors at normal incidence of a range of areas located at equal distances from the transducer probe; see Fig. 3.28.

attenuation (absorption). The decrease in ultrasonic intensity with distance expressed in decibel (dB) per unit length or nepers (Np) per unit length. 1 dB cm^{-1} = 8.686 Np cm^{-1}.

A scan. A plot of signal amplitude against time that can be related to distance in a specimen; see Fig. 3.4.

B scan. A plot of signal amplitude displaying a cross section of the specimen perpendicular to the upper surface; see Fig. 3.22.

C scan. A series of B scans carried out over the surface of the object with echoes restricted to a fix time delay, so that a plane perpendicular to the beam and at a particular range of depths is examined; see Fig. 3.23.

characteristic impedance. See *specific acoustic impedance.*

compression wave (longitudinal). Composed of a series of alternate surfaces of compressions and rarefactions traveling perpendicular to these surfaces. Particle motion is in the direction of travel; see Fig. 3.6(a).

contact inspection. The transducer probe makes direct contact with the material, with a minimum of couplant film.

couplant. A substance used between the transducer probe and the surface to improve transmission of ultrasonic energy.

critical angle. The angle of incidence beyond which a particular refracted wave is not transmitted; see Fig. 3.12.

dead zone. Directly in front of an ultrasonic transducer where no reflections can be detected due to the length of the initial pulse, or to amplifier overload resulting from the initial pulse; see Fig. 3.15.

decibel (dB). 10 log$_{10}$ (intensity/reference intensity). The reference intensity must always be stated; see *attenuation.*

distance amplitude correction (DAC) curve. The curve of amplitude responses obtained from reflectors of equal area situated at different depths within an object; see Fig. 3.28.

edge effect. A false indication due to reflections from the edge of a specimen.

electromagnetic acoustic transducer (EMAT). Ultrasonic signals generated in conducting materials by electromagnetic coupling between a coil and the surface layers of the material; see Table 3.10.

far field (Fraunhofer). Extends beyond the limit of the near field where the ultrasonic beam amplitude on the beam axis decreases exponentially with distance (see Fig. 3.15).

Fraunhofer zone. See *far field.*

Fresnel zone. See *near field.*

fundamental frequency. At which the wavelength is twice the thickness of the examined material; see Fig. 3.36.

gate. An electronic process of selecting a segment of time for further processing; see Fig. 3.29(b).

grain effect. When grain size in a material is comparable to wavelength, an ultrasonic echo occurs from each grain and these give rise to "noise" on the display.

immersion testing. Transducers and specimen are immersed in a water bath, which acts as an acoustic coupling; see Fig. 3.21.

impedance ratio. The ratio of the acoustic impedance of solids at an interface; see Fig. 3.10.

Lamb wave. A mode of propagation in which the two parallel boundary surfaces of the plate or the wall of a tube establish the mode of propagation; see Fig. 3.7.

longitudinal wave. See *compression wave.*

mode. The different modes of ultrasonic wave are characterized by different particle motions; see Fig. 3.7.

mode conversion. The conversion of the mode of ultrasonic wave propagation, which can occur at an interface between materials of different acoustic impedances; see Fig. 3.11.

near field (Fresnel zone). The region of the ultrasonic beam adjacent to the transducer which is dominated by interference effects; see Fig. 3.15.

piezoelectric effect. Observed in certain crystals with low symmetry atomic structures where a potential difference develops across opposite faces when the solid is subjected to a stress; see Fig. 3.17.

pitch-catch. Separate ultrasonic transmitter and receiver transducers are employed; see Fig. 3.18.

pulse echo method. The presence and position of a discontinuity are given by the echo amplitude and time; see Fig. 3.4.

Rayl. The unit of specific acoustic impedance and has units of $kg\ m^{-2}s^{-1}$.

Rayleigh wave (surface wave). Surface wave in which the particle motion is elliptical and the effective penetration is of the order of one wavelength; see Fig. 3.7.

refraction. The change in the direction of a beam of radiation on passing through an interface; see Fig. 3.9.

resolution. The ability to give simultaneous, separate indications from discontinuities having nearly the same range and lateral positions; see Fig. 3.20.

resonance method. Continuous ultrasonic waves are varied in frequency to identify resonant frequencies; see Fig. 3.36.

search unit. An electro-acoustic transducer used to transmit or receive ultrasonic energy, or both.

shear wave. Waves in which the particle motion is perpendicular to the direction of propagation; see Fig. 3.2.

skip distance. Using shear wave examination, the distance along the test surface from the entry point of ultrasound to the point at which the beam returns to the same surface (see Fig. 3.31).

specific acoustic impedance. See *acoustic impedance,* in units of Rayl.

standing wave (stationary wave). The particle vibrations do not change with time at any point in the medium; see Fig. 3.36.

stand-off. Separation of transducer from specimen by a water path, or low-loss plastic block to separate the initial signal from the returning signals.

through transmission technique. The ultrasonic vibrations are emitted by one transducer and received by another at the opposite surface of the specimen; see Fig. 3.3(b).

transducer. A device that converts energy from one form to another. An electro-acoustical transducer converts electrical energy into acoustic energy and vice versa.

transverse wave. The particle displacement at each point in a material is perpendicular to the direction of wave propagation; see *shear wave.*

ultrasound (ultrasonic radiation). A mechanical wave at a frequency above about 20 kHz; see Table 3.1.

Chapter IV

Liquid Penetrant Inspection

Part I General Discussion

Historical Background

Liquid penetrant inspection (LPI) is a simple yet effective method of examining surface areas for cracks, defects, or discontinuities, and has its origin almost certainly in the observation by blacksmiths that quenching liquids could be seen to seep out of cracks and stain the surface after quenching a hot piece of ironware. A very early surface inspection technique involved the rubbing of carbon black on glazed pottery, whereby the carbon black would settle in surface cracks rendering them visible (Ref 236). Later it became the practice in railway workshops to examine iron and steel components by the "oil and whiting" method. In this method, a heavy oil commonly available in railway workshops was diluted with kerosene in large tanks so that locomotive parts such as wheels could be submerged. After removal and careful cleaning, the surface was then coated with a fine suspension of chalk in alcohol so that a white surface layer was formed once the alcohol had evaporated. The object was then vibrated by striking with a hammer, causing the residual oil in any surface cracks to seep out and stain the white coating. This method was in use from the latter part of the 19th century through to approximately 1940, when the magnetic particle method was introduced and found to be more sensitive for the ferromagnetic iron and steels: see Table 4.1.

A different (though related) method was introduced in the 1940s, where the surface under examination is coated with a lacquer, and after drying the surface is vibrated by hitting with a hammer, for example. This causes the brittle lacquer layer to crack generally around surface defects. The brittle lacquer (stress coat) has been used primarily to show the distribution of stresses in a part and not finding defects.

Many of these early developments were carried out by Magnaflux, Chicago, IL USA (Ref 49) in association with the Switzer Bros., Cleveland, OH, USA. More effective penetrating oils containing highly visible (usually red) dyes were developed by Magnaflux to enhance flaw detection capability. This method, known as the visible or color contrast dye penetrant method, is still used quite extensively today. In 1942, Magnaflux introduced the Zyglo system of penetrant inspection where fluorescent dyes were added to the liquid penetrant. These dyes would then fluoresce when exposed to ultraviolet light (sometimes referred to as "black light") rendering indications from cracks and other surface flaws more readily visible to the inspectors' eyes. The historical development of liquid penetrant inspection is outlined in Table 4.1.

Outline of Penetrant Inspection

The liquid penetrant inspection method is used very extensively, perhaps more so than any other method. It is relatively simple to carry out, and there are few limitations due to specimen material or geometry and it is inexpensive. The equipment is very simple and the inspection can be performed at many stages in the production of the article as well as after the article has been placed in service. Relatively little specialized training is required to perform the inspection, however experience is very helpful.

The method is restricted to defects that are open to the surface. The inspection depends heavily on the visual acuity and abilities of the operator. There are several delicate stages of preparation involving precleaning of the inspection surface, application of the penetrant, observance of a dwell time to allow the penetrant to seep into flaws, removal of excess penetrant, application of a developer again with a dwell time to allow the penetrant to seep out of any surface flaws to form visible indications. The surface is then ready for inspection under a well-lit area. Finally, the surface may need to be completely cleaned before further use. A permanent record can be made of the liquid penetrant test results by photographic methods.

The steps in liquid penetrant inspection are as follows (experience is helpful).

1. *Surface preparation.* The surface must be free of oil, grease, water, or other contaminants that may prevent penetrant from entering flaws. Penetrant inspection should be scheduled prior to the mechanical operations such as grit or vapor blasting, which have a tendency to smear the surface of parts, thus closing the defects.

2. *Penetrant application.* Penetrant may be applied by spraying or immersing parts in a penetrant bath.

3. *Penetrant dwell.* The penetrant is left on the surface for a sufficient time to allow penetration into flaws. The time involved is based on experience.

4. *Excess penetrant removal.* This is a most delicate procedure because it is essential not to remove penetrant from fine surface cracks. Penetrants may be washed off directly with water, or first treated with an emulsifier and then rinsed with water, or they may be removed with a solvent.

5. *Developer application.* Developer may be applied by dusting (dry powdered) or immersion or spray (water developers) applications. Nonaqueous wet developers may be applied by spray only. The developer should then be allowed to dwell on the part surface for sufficient time (usually 10 min minimum) to permit it to draw penetrant out of any surface flaws to form visible indications of such flaws. Longer times may be necessary for tight cracks.

6. *Inspection.* Inspection is then performed under appropriate lighting to detect any flaws which may be present.

7. *Clean surface.*

Table 4.1 Historical development of liquid penetrant systems

Period	Description
Prior to 19th century (blacksmiths)	Stains on ironware as quenching liquids seep out of surface cracks
Mid-19th century (railway industry)	"Oil and whiting." The penetrant used is heavy oil in kerosene, and the developer is white chalk in alcohol. Tap metal to force oil out of cracks, discoloring the whiting. Examination of locomotive parts
1940s	Magnetic particles for ferromagnetic materials replaces oil and whiting
	Brittle lacquer coating, cracks on vibrating object by tapping
	Color-contrast colored dyes added to penetrant. Avoids use of ultraviolet light for large structures. Developed by Tabor de Forest and others at Magnaflux
~1940s	Ultraviolet (black) light and fluorescence of oil. Hot oil method. Weak fluorescence and low sensitivity
Late 1940s	Water washable self-emulsifying fluorescent penetrant liquids introduced
1941	R.C. Switzer: U.S. Patent 2,259,400. Fluorescent dye additives in the penetrant liquid
1942	Zyglo system of fluorescent penetrant inspection introduced by Magnaflux
1950s	Postemulsifying penetrant liquids introduced. Improved penetration into surface cracks

General references: Ref 49, Chapter 1; Ref 211, Chapter 2

The actions of the penetrant and the developer are shown in Fig.4.1 and in Fig. 4.2.

The penetrant methods complement the magnetic-particle method, which can only be applied to ferromagnetic surfaces because liquid penetrants can be applied to all surfaces. The penetrant method is generally found to be more sensitive than x-radiography or ultrasonics for fine surface cracks. When examining large surface areas, the liquid penetrant method is much more effective as it is much less time consuming and much less expensive (Ref 211).

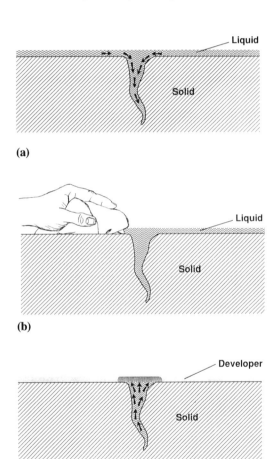

(a)

(b)

(c)

Fig. 4.1 Action of penetrants and developers. (a) Penetrant liquid is drawn down into the open crack by capillary action. (b) Excess surface penetrant is removed. This can be carried out by wiping with a cloth. (c) Developer powder on the surface draws out penetrant liquid that seeps into and discolors the powder. A colored dye or a fluorescence compound is usually added to the penetrant liquid. The liquid should not seep too far away from the crack. (However, the crack width appears increased by a factor of 100×.)

Liquid Penetrant Processes

The liquid penetrant used in NDT can contain colored dye (usually red); fluorescent dye, visible using a UV lamp; or dual sensitivity dye with both visible coloring and a fluorescent compo-

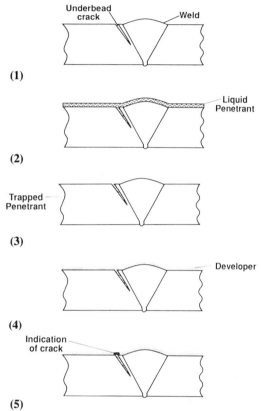

(1)

(2)

(3)

(4)

(5)

Fig. 4.2 Examination of a surface crack. *Steps in the Procedure.* 1: Clean surface. Any contamination may hinder the wetting of the surface cracks. The surface must be dried. However, the surface may be dry, yet the crack may be full of water. 2: Apply liquid penetrant. A visible (usually red) or fluorescent dye is present in the liquid. Sufficient dwell time must be given for the liquid to penetrate fine cracks. 3: Wipe surface to remove excess surface penetrant, leaving some penetrant trapped in the surface crack. Do not flood the part with cleaner to remove excess penetrant. The cleaner should be applied by wiping with a cloth. 4: Apply developer to the surface and allow sufficient time for the developer to draw penetrant out of any surface cracks. 5: The penetrant liquid seeps out into the developer forming a visible indication of the surface crack. The crack size may appear increased by a factor of 100×. 6: Subsequent to examination, the surface must be cleaned once again to prevent penetrant inspection residues from adversely affecting the reliability of the part.

nent. The removal of liquid penetrant can be carried out as follows:

- The penetrant may be washable with plain water.
- An emulsifier has to be added subsequent to testing to remove excess penetrant by water washing.
- The penetrant may require a special cleaning solvent; this is typical of penetrants from spray cans.

The process sequences are presented schematically in Fig. 4.3.

The sensitivity of liquid penetrants depends on additives that improve the penetrant's resistance to over-removal, as well as heat and UV fading of fluorescent indications.

Commercial Systems

There are several commercial systems of penetrant testing available, and lists of these can be found in *Materials Evaluation* (Ref 258), in *Thomas Register* (Ref 259), and are described in Ref 211. Other data concerning nondestructive test-

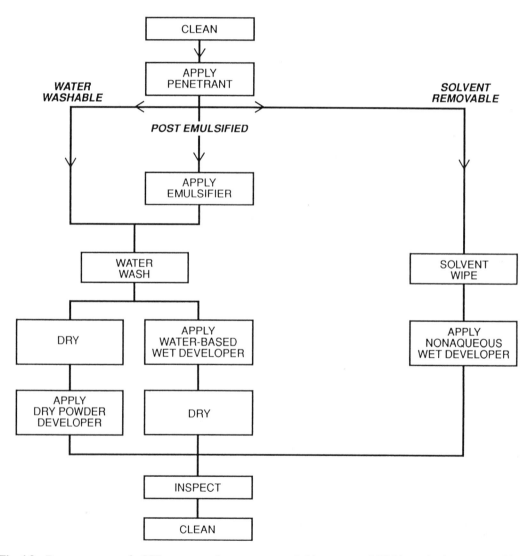

Fig. 4.3 Process sequence for LPI penetrants that are water washable, post-emulsifiable, and solvent removable.

ing commercial systems are given in Ref 70 and 196.

Many of the commercial systems have been developed by the companies concerned so that the chemical details are proprietary and not available. Thus, the company Magnaflux (Ref 230) has been connected with the developments of Switzer and others in the 1940s (see the section "History" in this chapter).

The selection of a commercial penetrant system will depend on a series of local and individual factors (Ref 211), such as:

- Suitability of penetrant inspection compared to other methods.
- Cost, time involved, skill of observers, fatigue of observers, environment of test object.
- Possible interference of penetrant liquid with future use of the part.

- Nature of object—size and shape, porosity, surface roughness, importance of those defects not open to the surface.
- Fineness of surface defects and sensitivity required.

Penetrant dwell times are governed by the alloy and type of defects. For example, fatigue discontinuities in titanium are extremely tight and therefore require much longer penetration times. There are two distinct postemulsification methods in use:

- *Lipophilic—oil based.* Penetrated parts are dipped in the emulsifier and dwelled for a designated time.
- *Hydrophilic—water based.* Parts are prerinsed with plain water spray to remove excess penetrant (physical force) and then immersed in a dilution of the emulsifier in

Table 4.2(a)   Properties and method of employment of a penetrant/developer system (Zyglo liquid penetrant system from Magnaflux) (Ref 230)

Tradename	Spotcheck	Zyglo
Manufacturer	Magnaflux	Magnaflux
Penetrant type	Liquid red dye	Fluorescent liquid
Applied by	Dip, brush, or spray can	Dip, brush, or spray can
Temperature range	Above 50 °F	Above 50 °F
Removed by	Solvent-based cleaner (some by water)	Water or solvent-based
Developer	Volatile liquid suspension of white powder	Powder or liquid
Indication visible by	White light	Near ultraviolet light
		Black light
Auxiliary apparatus	None	Sources of water and electric power

Table 4.2(b)   Typical penetration time for a dye penetrant (Zyglo from Magnaflux)

Surface under examination	Type of defect	Penetration time, min
All	Heat-treatment cracks	2
	Grinding	10
	Cracks	10
Ceramics	Cracks	2-5
	Porosity	2-5
Aluminum welds	Cracks and pores	10-20
Steel welds	Cracks and pores	10-20
Forgings	Cracks	20
	Laps	20
Metal rollings	Seams	10-20
Die castings	Surface porosity	3-10
	Cold shuts	10-20
Metal-permanent mold casting	Shrinkage porosity	3-10
Carbide-tipped cutting tools	Poor braze	2-10
Cutting tools	Cracks in steel	2-10
	Cracks in tip	2-10

All penetrant times are by experiment and should be determined in each case.

water (usually one part of emulsifier concentrate to four parts of water) for 60 to 120 s.

Examples of some commercial liquid penetrant systems are given in Table 4.2(a) and (b). A typical processing system for liquid penetrant (fluorescence) inspection is shown in Fig. 4.4. The system consists of seven work stations including a penetrant tank, rinse tank, dryer station, and inspection booth, as required for the water-wash process. Penetrant kits are available for the examination of small areas. A general-purpose kit consists of aerosol penetrant cans, aerosol developer cans, aerosol cleaner cans, UV (black) light source, and cleaner materials. The procedure is shown in Fig. 4.5 in steps 1 to 5.

Interpretation of Penetrant Indications

The observation of indications on the developed surface after a penetrant-developer treatment is very dependent on the individual inspector and therefore quite subjective. Some general descriptions of defects are given in Tables 5.6 to 5.11.

True LPI indications can be considered to be of three general types (Fig. 4.6):

- *Continuous Lines* (Fig. 4.6a): Cracks are observed generally as jagged lines; cold shuts appear as smooth, narrow straight lines; forging laps appear as smooth wavy lines; scratches tend to be shallow.
- *Broken Lines* (Fig. 4.6b): Continuous lines become partially closed by working, such as grinding, peening, forging, or machining, resulting in a discontinuous line.
- *Small Round Holes* (Fig. 4.6c): These are typical of general porosity, gas holes, pinholes, or very large grains.

Examples of fluorescent LPI indications are shown in Fig. 4.7; examples of visible dye LPI indications are shown in Fig. 4.8. A range of

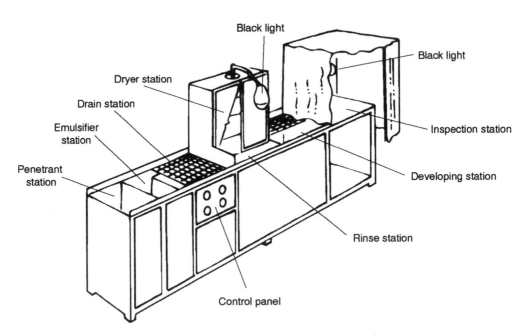

Fig. 4.4 Multistation processing unit for liquid penetrant inspection (Ref 63). *Preliminary precleaning station* (not shown), remove grease and oil to expose surface cracks. *Station 1 Penetrant tank*, dye application; allow dwell time. *Station 2 Drain station*, remove excess dye solution. *Station 3 Emulsifier tank*. *Station 4 Rinse tank*. remove residual surface dye so that only dye in flaws remain; involves emulsifier dwell and rinse times. *Station 5 Developer tank*, apply suspension of developer particles; involves developer dwell time. *Station 6 Dryer oven*, dry developer coating. *Station 7 Inspection booth*, illuminate by UV lamp $1000\,\mu W/cm^2$ (100 watts at a distance of 15 in.) or visual observation (150 watts at 3 ft). *Post-cleaning station* (not shown). Source: American Society for Nondestructive Testing.

acceptable and rejectable indications are given in Ref 211 for the fluorescent penetrant inspection of tubing. A darkened visible light environment should be used, less than 2 ftc white light for the fluorescent method; by comparison, the comfortable reading of a book requires about 3 ftc of illumination (Ref 260). The fluorescent lighting should be at least 1000 μw/cm^2 intensity, given by 100 watt lamp at 15 in. Penetrant inspection has a very wide range of use in many industries of penetrant inspection, and a list of these uses is presented in Table 4.3.

Practical Tips

The following are practical hints, "rules of thumb" comments, and suggestions for performing liquid penetrant inspection.

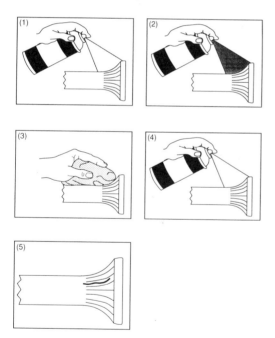

Fig. 4.5 LPI using spray can. A penetrant kit for the examination of small areas consists of aerosol cans of cleaner, penetrant, developer, UV lamp, and cleaner materials. The surface area to be examined is cleaned by spraying the part (1), then wiping with a clean rag after a short delay. The penetrant is sprayed, left for a suitable dwell time (2), about 10 min for new clean castings, welds, and most defects. The surface is now wiped gently with a clean cloth (3), sprayed with developer, just sufficient to whiten the surface (4), and then examined by UV (black) light after delay of approximately one-half penetrant dwell time (5).

1. *False, Nonrelevant, and True Indications.* False indications are not related to the presence of defects and are shown in Fig. 4.9 (Ref 63). False indications can arise from

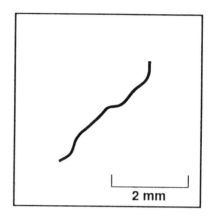

(a)

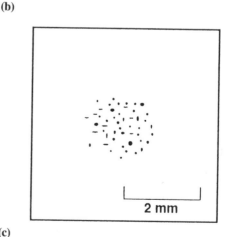

(b)

(c)

Fig. 4.6 LPI indication. True LPI indications can be divided into three general types: (a) continuous lines, (b) broken lines, and (c) small round holes.

Table 4.3   Examples of use of penetrant inspection

Penetrant method	Specimen
Fluorescent	Anodizing cracks on aluminum and magnesium
	Cracks in nickel plate; aluminum block; sand casts; aluminum, magnesium steel shrinkage
	Tungsten wires
	Turbine blades
	Light alloy aircraft castings (porosity and shrinkage)
	Shrinkage cracks in aircraft parts
	Welds
	Ceramic (dental porcelain)
Oil and whiting	Locomotive parts
Post-emulsion fluorescence	Turbine blades (cracks)
Fluorescent water wash	Billet control. Shrinkage cracks in rolled copper bar seams
	Age cracks in brass tube
Color contrast	Seam welded tank
Dye penetration	Stress-corrosion cracking of metallic sheet
Filtered particles	Porous material
	Aerospace industry
	Liquid oxygen system
	Composite and bonded structure
Various	Nickel-chrome surface cracks
	Primary metal production aluminum, steel
	Light alloy foundries
	Metal cutting tools

Source: Ref 49, 261

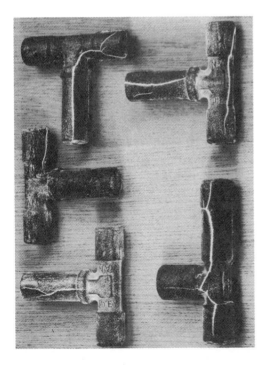

Fig. 4.7 Fluorescent LPI indications of shrinkage cracks in coupling castings (Ref 63). Reprinted with permission of The American Society for Nondestructive Testing.

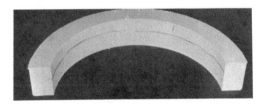

(a)

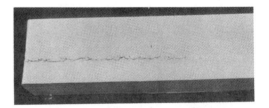

(b)

(c)

Fig. 4.8 Visible dye LPI indications. (a) Laminations. (b) Weldment shrink cracks. (c) Cracks in aluminum forging (Ref 63). Reprinted with permission of The American Society for Nondestructive Testing.

poor cleaning, the presence of lint, dirt, operator handling, contamination of the developer, or by contact with another surface covered with penetrant. Nonrelevant indications could be, for example, due to a fluorescent background on a rough surface. It may be useful to repeat the LPI in such a case.

2. *Dwell Times.* The dwell times for penetrant and for the developer should be similar. Dwell times need to be determined in each case. These can be sometimes very long, sometimes very short; minimum time is 10 min, although for titanium and nickel alloys at least 30 min is required.

3. A "rule of thumb" is to use a developer time equal to half the time of dwell penetration. For water-based developers, the time to dry in the drying oven is sufficient. For dry or nonaqueous developers, the normal time is between 7 and 30 min.

4. *Cleaning.* Even after careful cleaning and drying, surface cracks may remain full of water.

5. *"Open Tank" Penetrant Systems.* The penetrant solution needs to be checked regularly for effects of evaporation, UV light, humidity, or external solid contamination.

6. Subsurface defects may be opened to the surface at a later stage after grinding or machining.

7. Screening tests by other NDT methods is sometimes useful. For example, ultrasonic testing may be used, followed by LPI on selected parts (Ref 262).

8. Sandblasting to prepare a surface is not recommended because open surface discontinuities may be closed by the shot or sandblasting; etching may be needed.

9. Surface contaminants such as dirt, grease, rust, scale, acids, and cleaning solvents must be removed.

10. LPI can be carried out successfully on metal surfaces of aluminum, magnesium, brass, copper, iron, stainless steel, titanium, and most alloys and powder metallurgy parts.

11. LPI can be carried out successfully on nonmetallic surfaces of glass. Among plastics, only Teflon and nylon can be inspected with penetrants; rubber only after compatibility testing.

12. LPI is somewhat limited when examining surfaces of very porous objects.

13. LPI is not applicable to subsurface cracks.

Part II Technical Discussions

Penetration of a Liquid into a Crack

Liquid penetrant testing can only be carried out with a liquid that freely wets the surface under examination. The surface-liquid interface, where the liquid molecules may prefer to approach the molecules of the wall, is shown in Fig. 4.10a; in this case, wetting of the surface occurs. The contrary applies when the liquid molecules prefer to be closer to other liquid molecules and no wetting occurs (Fig. 4.10b). Liquids wet surfaces when the contact angle, θ, is less than 90°.

The relationship of contact angle θ to surface energies γ is shown in Fig. 4.11. The surface energies between solid and liquid (γ_{SL}), between liquid and vapor (γ_{LV}), and between vapor and solid (γ_{VS}) are all different.

The penetration of liquids into surface cracks is shown in Fig. 4.12 for both upper surface cracks (a) and for lower surface cracks (b). The capillary force driving the liquid into the crack of radius, r, is due to the surface tension γ_{LV} acting at contact angle θ and is equal to $(2\pi r \gamma_{LV} \cos \theta)$.

As shown in Fig. 4.12(a), for a crack on the upper surface of the specimen, the forces due to capillary action and weight are additive, resulting in the compression of the residual gas at the foot of the crack. Depending on the geometry of the crack, the residual gases may or may not be

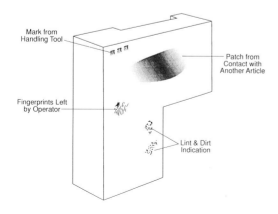

Fig. 4.9 False LPI indications (Ref 63).

able to escape. As shown in Fig. 4.12(b), for a crack on the lower surface of the specimen, the forces due to capillary action and weight are opposed.

Then, the capillary force balances the weight of the liquid so that

$$h = \frac{2\gamma_{LV} \cos \theta}{rg \rho} \qquad \text{(Eq 4.1)}$$

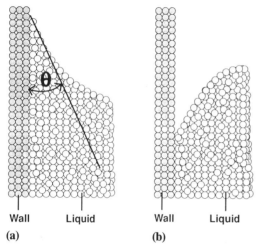

Fig. 4.10 Molecular arrangements at a liquid-wall interface. (a) The contact angle, θ, as shown, between the liquid and the wall is less than 90°, and the surface of the wall is wetted. (b) Cohesive forces between like molecules are greater than the adhesive forces between unlike molecules. In this case, the surface of the wall is not wetted and the liquid is in the form of separate droplets.

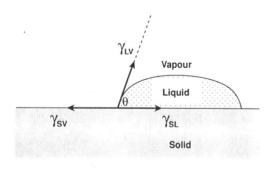

Fig. 4.11 Surface tension contact angle and surface wetting. The contact angle θ and the surface tension between solid and vapor (γ_{SV}), solid and liquid (γ_{SL}) and liquid and vapor (γ_{LV}) are related as shown. The surface tensions γ_{VS}, γ_{VL}, and γ_{LS} are not the same. Good surface wetting requires θ to be much less than 90°. Otherwise, if $\theta > 90°$, the liquid will form droplets and no surface wetting will occur.

where h is the height that the liquid penetrates, g is the acceleration due to gravity, and ρ is the density of the liquid.

The height h that liquid penetrates into a crack:

- Depends on the capillary force due to the surface tension
- Requires a contact angle less than 90°
- Increases as the crack width decreases
- Does not depend on the viscosity of the liquid

However, the time taken for the liquid to penetrate into the crack does depend on viscosity; therefore, adequate dwell time is required to enable the penetrant to enter very fine cracks. (It can be advantageous to leave the penetrant overnight on the test piece.) *Note*: Surface tension is affected markedly by contaminants.

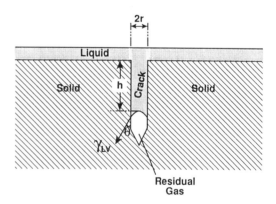

(a)

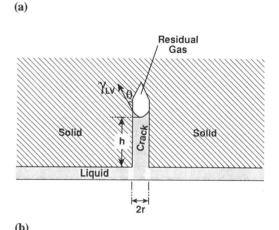

(b)

Fig. 4.12 Penetration of liquids by capillary action into surface cracks for (a) upper surface and (b) lower surface cracks.

Limits of Penetration Inspection (Ref 49, 69, 261)

The finest cracks that can be detected using liquid penetrant inspection have been estimated to be about 5 µm (0.2×10^{-3} in.) wide, by 10 µm (0.4×10^{-3} in.) deep. Several measurements have been attempted to determine this sensitivity level, which depends on a wide range of factors that vary from one inspection to the next. Factors which determine sensitivity include:

- The wetting of the solid by the penetrant
- The geometry of the crack; the width, depth, aspect ratio, length, and general shape
- Cleanliness of the crack and of the specimen surface, and the absence of contaminants that, even present only in minute concentrations, might alter the surface tension of the liquid penetrant

- The time available to perform the inspection
- The skill of the inspector, method of observation, and visibility and contrast of the indications
- The penetrant materials must be in good condition; "open tank" penetrants should be verified regularly, because they may be affected by open-tank evaporation, effects of ultraviolet light, heat, water, and external solid contamination.

Estimates have been made of the crack detection efficiency of several types of penetrant and developer systems. A tabulation of observations and efficiencies of penetrant inspection is given in Table 4.4.

Comparator and reference materials, and standardization codes are reviewed in Ref 261, Section 9. The military specification: MIL-1-25135E (1989) is widely used (Ref 265). A selection of liquid penetrant inspection specifications

Table 4.4 The probability of detecting fine cracks based on past experience

Ref	Details	Probability of observing crack detection efficiency	Fine crack dimensions
49 (Walters and McMaster)	Glass plate separations	...	0.13-0.33 µm (0.005-0.1×10^{-3} in.)
59 (McCauley and Van Winkle)	Cracks in chrome-plated brass plate	60%	0.5 µm (0.02×10^{-3} in.)
59, 263 (Fricker)	Cracks in chrome-plated steel	...	25 µm (10^{-3} in.)
59, 264 (Lord and Hollaway)	Fatigue cracks in Ti-6Al-4V	...	5 µm (0.2×10^{-3} in.)
59 (Packman et al.)	Cracks in aluminum, steel cylinders	65-80%	...
	Cracks in aluminum, steel alloys	90%	...
49 (Betz)	Chromium plating over nickel steel	...	...

Table 4.5 Dwell times for water washable penetrants

Material	Form	Penetration time, min (at room temperature)
Aluminum	Castings	15
	All other forms	30
Magnesium	Castings	15
	All other forms	30
Steel	Castings	15-30
	Extrusions, forgings, welds	30-60
	Other forms	30
Brass and bronze	Castings	10
	All other forms	30
Glass	All forms	Up to 30
Carbide-tipped tools		Up to 30
Ceramic	All forms	5

In practice, shorter penetration times are very often used. Adapted from ASME Boiler and Pressure Vessel Code (Ref 261, 267).

are given in Ref 49, 211, and 266. Penetration dwell times for water washable penetrants are listed in Table 4.5 taken from Ref 261. These are the suggested dwell times given by theAmerican Society of Mechanical Engineers (Ref 267) and depend on the material and the type of defect as well as on the temperature of the penetrant and part to be inspected.

Penetrants, Developers and Crack Visibility

The desired properties of penetrants include (Ref 170, 236):

- Ability to penetrate and remain in fine and coarse openings

- Easily cleaned off surface
- Highly colored or fluorescent in thin film form
- High flash point, low cost, stable and non-toxic

Penetrants are generally kerosene oil-based, but recently other solvents have become available (Ref 49, Chapter 4). The various types of penetrants are (Ref 222; 261, Section 4):

- *Water washable, with fluorescent or colored dye.* Emulsifiers are added to water washable oil-base penetrants so that they are easily removable with a water wash.
- *Post-emulsifiable with visible or fluorescent dye additives.* A separate emulsification step treats the excess surface penetrant so that it

Table 4.6 Penetrant properties at 20 °C (Ref 49)

Liquid penetrant	Density, g/cm^3	Surface tension, $10^{-3} J/m^2$	Flash point, °C	Dynamic viscosity, $10^{-3} Pa \cdot 5$
Water	1	7.28	...	1.002
Ethyl alcohol (C_2H_5OH)	0.789	23	14	1.20
Kerosene	0.79	23	...	1.30
Ethylene glycol	1.115	47.7	110	19.9
SAE No. 10 (lubrication oil)	0.89	31	230	100

Table 4.7 A comparison of penetrant systems (Ref 211, 268)

Water washable	Postemulsifiable	Solvent removable
Visible dye penetrants		
Lowest in sensitivity	Higher sensitivity than water washables	Where water rinse is not feasible, or desirable
Suited for large surface areas	Suited for large surface areas	For spot inspections
	Suited for large quantities of similar objects	Recommended for small areas and simple geometries
Fluorescent penetrants		
Lowest in sensitivity of fluorescent penetrants	Higher sensitivity than water washable fluorescent penetrants	Higher sensitivity than solvent-removable visible penetrant
Suited for large surface areas	Suited for large quantities of similar articles	Where water rinse is not feasible or desirable
Suited for large quantities of similar objects	Suited for wide, shallow discontinuities and tight cracks	For spot inspections
Suited for deep, narrow discontinuities	Contaminants must be removed prior to inspection	Recommended for small areas and simple geometries
Recommended for rough surfaces (i.e., sand castings)	Suited for stress, intergranular, or grinding cracks	

Note: Recently for environmental reasons, water washable penetrant systems have been used even though the solvent system would be preferable.

can be removed with a water rinse. These are among the most sensitive penetrants because of their reduced susceptibility to inadvertent removal from wide open or shallow flaws during rinsing.

• *Reverse fluorescent dye penetrant.* A dark dye is added to the penetrant with a fluorescent agent added to the developer.

• *Solvent removable visible dye penetrant.* These are employed where water is not available or harmful to the part being inspected, or where only a small area is to be inspected in situ.

The properties of some liquid penetrants are given in Table 4.6. The extent of the liquid penetration increases with the surface tension. The viscosity affects only the time required for the liquid to penetrate. The flashpoint is of importance for safety considerations. A comparison of water washable, post emulsifiable, and solvent removable liquid penetrant systems are given in Table 4.7.

The length of time that the penetrant is in contact with the surface (penetrant time) is important, and is generally from 10 to 60 min. This is to ensure that the finest surface flaws are completely penetrated.

The emulsifying agents are based on non-ionic surfactants. Using sodium compounds can affect a high-temperature corrosion of Inconel or titanium alloys.

White contrasting coating of developers is essential for visible penetrants. Fluorescent penetrants do not require contrasting background. The required properties of a developer are:

• Visible white coating for contrast
• The developer must be wetted easily by the penetrant
• High absorption of the penetrant seeping from the surface flaw
• Capable of forming a thin uniform coating on the surface
• Easily cleaned away after inspection
• Avoid excessive bleeding of the penetrant liquid from the surface defect

Penetrant/developer systems that provide a recorded image of defects on surfaces have been attempted using a strippable plastic developer, and several versions are available (Ref 211). Film type plastic developers are based on PVC of low adhesion to the surface of the specimen (Ref 261).

ASTM Standards

The documents listed in Table 4.8 are the reference standards for NDT from ASTM (Ref 61). Some reference photographs are presented in ASTM standard E 433. In Chapter I, Fig. 1.1, a valve bridge forging has been examined by liquid penetrant, revealing a surface crack. Magnetic particle inspection gave comparable results. X-radiography was not able to reveal this surface crack. A glossary of some terms used in liquid penetrant inspection is given in Appendix 4.1.

Table 4.8 ASTM standards for liquid penetrant inspection

Standard No.	Description
E 165	Liquid Penetrant Examination (1991) p 55-71
E 433	Liquid Penetrant Inspection (1985) p 162-166. Reference Photographs
E 1135	Fluorescent Penetrants (1986) p 512-516
E 1208	Fluorescent Liquid Penetrant Examination Using Lipophilic Post-Emulsification (1991) p 536-541
E 1209	Fluorescent Liquid Penetrant Examination Using Water-Washable Process (1991) p 542-547
E 1210	Fluorescent Liquid Penetrant Examination Using Hydrophilic Post-Emulsification (1991) p 548-554
E 1219	Fluorescent Liquid Penetrant Examination Using Solvent-Removable Process (1991) p 566-571
E 1220	Visible Penetrant Examination Using Solvent-Removable Process (1991) p 572-576
E 1417	Liquid Penetrant Examination (1992) p 669-677
E 1418	Visible Penetrant Examination Using Water-Washable Process (1991) p 678-683

Note: These standards are frequently revised and published annually (Ref 61).

Appendix 4.1: Glossary of Some Terms used in Liquid Penetration Inspection (Ref 61, 190)

black light. Electromagnetic radiation in the near-ultraviolet range of wavelength (330 to 390 nm).

developer. A material applied to the specimen surface to accelerate bleedout and to enhance the contrast of indications.

- *Aqueous.* A suspension of developer particles in water.
- *Dry.* A fine free-flowing powder.
- *Liquid film.* A suspension of developer particles in a solvent which leaves a resinous film on the surface after drying.
- *Soluble.* A developer completely soluble in its carrier.
- *Solvent.* Developer particles suspended in a nonaqueous vehicle prior to application.

developing time. The time between the application of the developer and the examination of the part.

dwell time. The time that the penetrant or emulsifier is in contact with the specimen surface, including the time required for application and to drain (see Fig. 4.2).

emulsification time (emulsifier dwell time). The time that an emulsifier remains on the specimen to combine with the surface penetrant prior to removal.

emulsifier. A liquid that interacts with an oily substance making it water washable.

hydrophilic. Having a strong affinity for water. Substance that is attracted to a water phase rather than to air in an air-water interphase (e.g., –OH).

hydrophilic emulsifier. A water-based liquid that interacts with the penetrant oil rendering it water washable.

lipophilic. A fat-liking molecular group (e.g., CH_3) that has affinity for hydrocarbons.

lipophilic emulsifier. An oil-based liquid that interacts with the penetrant oil rendering it water washable.

lyophilic. Solid-liquid mixture with surface active molecules containing several molecular groups with both affinities and repulsions for the liquid phase; "solvent-loving" (opposite of *lyophobic*).

lyophobic. "Solvent-hating" (opposite of *lyophilic*).

penetrant. A liquid that is applied to the surface of a part so that it penetrates surface-breaking defects.

- *Dual purpose.* Produces both fluorescent and color contrast visible indications.
- *Fluorescent.* Emits visible radiation when excited by black light.
- *Solvent-removable.* Traces removable by wiping with a material lightly moistened with solvent remover.
- *Visible.* Characterized by an intense color, usually red.
- *Water-washable.* With a built-in emulsifier.

postemulsification penetrant. Requires the application of a separate emulsifier to render the excess surface penetrant water washable.

Chapter V

Magnetic Particle Inspection (MPI)*

Part I General Discussion

Historical Background

As early as 1868, Saxby used a magnetic compass to locate defects and inhomogeneities in iron gun barrels (Ref 52). Subsequently, Herring in 1879 obtained a patent in the United States for the detection of defects in railway lines using a compass needle. The use of iron particles to locate defects in magnetized articles stems from the observations in 1917 by Hoke in the United States and others in the United Kingdom (Ref 51, 211, 269). They noticed that when steel articles were machined while held in a magnetic chuck, the iron grindings formed patterns around cracks on the surface. The use of iron filings to display the magnetic field of a bar magnet has been well known for a very long time, and typical patterns around bar magnets are shown in Fig. 5.1.

A.V. de Forest in 1928 and 1929 carried out deliberate studies of the use of magnetic particles

for nondestructive testing of ferromagnetic materials. He described the use of longitudinal magnetization for the location of transverse cracks, and circular (or transverse) magnetization for detecting longitudinal cracks. Indeed, it was de Forest who demonstrated that the best way to observe cracks in a particular direction required the correct choice of direction of the applied magnetic field. In 1934, de Forest and F.B. Doane founded the Magnaflux Corporation for the commercial exploitation and development of the magnetic particle method of nondestructive testing of ferromagnetic materials. The liquid suspension technique was introduced in 1935 by de Forest and Doane and consisted of a black magnetic oxide powder suspended in kerosene. In 1936, Unger and Hilpert described in a patent in Germany a liquid suspension of magnetic particles in water using wetting agents and rust inhibitors.

The inspection of aircraft engine parts, locomotive axles, and engines for fatigue cracks developed rapidly during this period. As early as 1936, magnetic particle testing was made mandatory at the Indianapolis Motor Speedway.

The recent development of magnetic particle inspection is described in Ref 51, 211, and 269 and has remained closely linked to the Magnaflux Corporation. Many other companies have

*This chapter has been reviewed and revised with the aid of A. Lindgren (L&L Consultants, 1629 Eddy Lane, Lake Zurich, IL 80047 USA) and Dr. L.J. Swartzendruber (Office of Nondestructive Testing, NIST, Gaithersburg, MD 20899 USA).

recently introduced important innovations. The historical development of MPI is described briefly in Table 5.1.

Magnetic Leakage Field

If a bar magnet is imperfect, cracked, or broken into two parts, each part will act as a separate magnet with north-south poles so that opposite the two broken ends, or across a crack, the lines of force will be disturbed, resembling those in Fig. 5.1(d). The magnetic lines of force are highly distorted across the crack and this is known as the magnetic leakage field. This situation is shown in Fig. 5.2, where the magnetic leakage field is shown for defects parallel and perpendicular to the main magnetic field. Iron filings cluster around the magnetic leakage field and reveal the presence of a defect, particularly if the field is perpendicular to the defect.

In magnetic particle inspection (MPI), detection of a crack depends on the magnetic leakage field. This leakage field must be sufficiently strong if the crack is to be observed by magnetic particle inspection (known as flaw visibility).

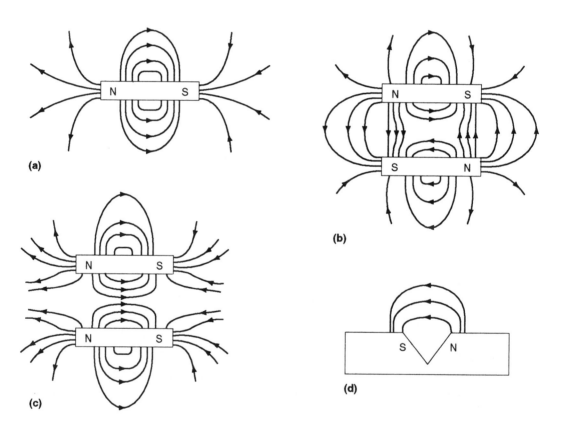

Fig. 5.1 A magnetic line of force is an imaginary line whose direction at each point gives the direction of the magnetic field at that point. The magnetic lines of force of a bar magnet can be plotted using a small compass or can be displayed using a very fine powder of iron filings. Iron filing plots can be prepared by spreading wax paper over a magnet and sprinkling iron filings over the paper. The filings become magnetized and link up to form chains along the magnetic lines of force particularly if the paper is gently vibrated to enable the filings to undergo small displacements. By mildly heating the paper to soften the wax and then cooling, a permanent record can be obtained (see, for example, Ref 260). Several magnetic patterns of bar magnets are shown in (a) to (d). (a) The lines of force around an individual bar magnet. The north pole is by convention the end of the magnet that lines up in the direction of the Earth's north pole. (b) and (c) Two bar magnets are arranged either with north poles adjacent, or with north-south poles adjacent. (d) A crack in a magnet or magnetized part will give rise to north-south poles on the two sides of the crack.

There is however, no simple method to derive the level of magnetic leakage field suitable for MPI (Ref 222). From a practical point of view, it is necessary to proceed by empirical methods and these are discussed particularly in terms of specifications for MPI inspection; see the section "Required Level of Magnification" in this chapter.

Some indication terms used in magnetic particle inspection and general magnetic property definitions are given in Appendix 5.1. Because only ferromagnetic materials can be examined by MPI, the magnetic terms employed can be simplified; the magnetic induction B, the permeability of free space μ_o, the magnetizing force H, magnetization M, and the magnetic susceptibility χ_m are defined in Appendix 5.1.

The relationships between B, μ_o, H, and M are given by (Ref 270):

$$B = \mu_o H \text{ in vacuum}$$

$$B = \mu H = \mu_o (H + M) \text{ in SI system of units}$$
$$= H + 4\pi M \text{ in cgs units}$$

Because $M \gg H$ for ferromagnetic materials:

$$B \cong \mu_o M = J$$

where J can be called the ferromagnetic magnetization in units of Tesla (T) or gauss (G).

Graphs of B_{gauss} or B_{Tesla} versus $H_{(amp/m)}$ are usually plotted for the behavior of a material in a magnetic field. Because only ferromagnetic materials are studied by MPI, J versus H is nearly equivalent.

Outline of Magnetic Particle Inspection

The workpiece surface must be cleaned, the specimen magnetized, magnetic particles spread over the surface, and the excess removed. Careful examination and evaluation of indications must be carried out, and finally the specimen must be demagnetized and cleaned. A very good introduction to MPI procedures is given in the specification ASTM E 709 and E 1444 (see Table 5.14 and Ref 271).

Surface Preparation. The surface must be sufficiently free of grease and dust to permit the magnetic powder particles to move freely over the surface and concentrate in regions where leakage fields exist. In cases where a current is to pass through the specimen, the surface areas must allow good electrical contact.

Magnetization of the Specimen. Magnetic particle inspection is best carried out when the magnetic field direction is perpendicular to the direction of any suspected flaws. This means that "A defect will be revealed best if parallel to the direction of electric current flow." The magnetization of the workpiece can be carried out by contact or noncontact methods. In contact methods, it is important to avoid electric arcing at the surface contact points. The part is placed between electrodes, known as the headstock and the tailstock. The current is passed, and this is called a "headshot" or "shot." Most wet horizontal units make use of an air-operated headstock for the clamping of parts. Air pressure is generally applied to clamp the part in position once the stocks are adjusted in position (see Fig. 5.4). Magnetization can also be

Table 5.1 Historical development of magnetic particle inspection

Year	Development	Ref
1868 (Saxby)	Magnetic compass locates defects in iron gun barrels	52
1876 (Ryder)	Carbon content of iron by magnetic studies	211
1879 (Herring)	U.S. Patent for detection of defects using a compass	211
1911 (National Bureau of Standards)	Magnetic testing of steels	211
1916 (ASTM)	Committee on magnetic testing	211
1917 (W.E. Hoke)	Iron filings accumulate around cracks in steel machined while held in a magnetic chuck	269
1928 (A.V. deForest)	Developed magnetic particle method, including circular magnetization, equipment and applications	51
1934 (A.V. deForest and F.B. Doane)	Founded Magnaflux Corp. (Chicago, IL)	51
1935 (A.V. deForest and F.B. Doane)	Liquid suspension method	51
1936 (Unger and Hilpert)	Patent in Germany on liquid suspension of iron filings in water	51
1936	Indianapolis Motor Speedway makes MPI mandatory	51

carried out using prods (see the section "Magnetization Methods" in this chapter).

In noncontact methods, the electric current flows around the object using a coil, or through the center using a copper rod. Another noncontact method uses a magnetizing yoke placed on or near the surface of the test object. (See Fig. 5.16 and 5.20 in this chapter.)

It should be recalled that the direction of magnetization must be greater than 45° to the direction of any possible defect. Direct or alternating currents can be used, although subsurface cracks require the use of direct current magnetization methods. The level of magnetization must be selected carefully. If the level is too low, important defects can be missed. If the level is too high, a heavy background forms that can mask the indications from important defects. The ideal level of magnetization must be determined empirically. (This is discussed further in the section "Required Level of Magnification" in this chapter.)

Application of Magnetic Powder Particles. The magnetic powder particles may be:

- Dry, applied using a shaker can, puff bottle, or spray blower
- Suspended in H_2O with a rust inhibitor, antifoam agent, and dispersion agents
- Coated with a fluorescent agent or a dye
- Applied by special techniques such as magnetic rubber, magnetic paint, or magnetic plastic

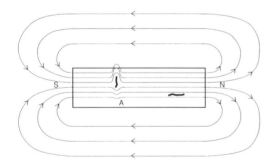

Fig. 5.2 Magnetic leakage field. When the direction of magnetization is perpendicular to the flaw as in case A, the magnetic lines of force are distorted the most so that the probability of detection of the flaw is greatest. Flaws such as B, parallel to the magnetic field direction, do not result in any magnetic leakage field and are therefore not detectable with MPI. Subsurface flaws are only visible if very near the surface (Ref 51).

- Suspended in a light petroleum distillate of low viscosity such as kerosene

The magnetizing electric current must continue during the powder application unless the object has a high magnetic retention. The application methods are discussed in the section "Magnetic Powders" in this chapter.

Surface Illumination. Good visible surface lighting is required, 1000 lux (100 ftc) as given, for example, by an 80 W fluorescent lamp, or 150 W incandescent lamp, at a distance of 1 m. Fluorescent powders require black light examination; that is ultraviolet (UV) light from a mercury vapor arc lamp in the wavelength band 320 to 400 nm, with principal output at 365 nm. The minimum recommended black UV light intensity is 1000 $\mu W/cm^2$ at the examination surface. This can be achieved over a 6 in. diam surface area using a 100 W black UV light source at a distance of 15 in.

- 10 Lux are approximately 1 ftc. The footcandle (ftc) is the English unit of 1 lumen $(lm)/ft^2$, and the lux is the SI unit of 1 lm/m^2, so that 1 lux = 0.093 ftc.
- Black lights are "flood" or "spot" type, where "flood" spreads the UV light, and "spot" focuses the UV light.
- A separate glass filter is needed for some UV lamps to remove the hard UV light (shorter wavelengths) and to remove visible light; see Ref 289. In some units, the filter is built in permanently.

Black light (UV) arc lamp is the popular name for mercury vapor arc lamps, giving light in the near ultraviolet. Some points to be noted are:

- The arc is affected by magnetic field, so that the black light should be kept away from the magnetizing or demagnetizing systems.
- The arc is struck over a limited voltage range; the precise instructions issued by the manufacturer must be followed.
- Black lights can be kept on continuously, avoiding frequent switching on and off which affects lamp life. The lamp output decreases with usage, dropping by as much as 25% after the first 100 h of use.

Surface Inspection. Inevitably, the inspection depends on the good eyesight, acuity, and experience of the inspector. Where a powder indication of uneven shape occurs, the inspector may need to repeat the process to see if the powder indication is reproducible. Some general comments are (see Ref 272):

- Surface defects tend to give rise to sharp, narrow particle patterns, held tightly together. There is a buildup of particles, and the deeper the crack the greater the buildup.
- Subsurface defects give rise to broad, fuzzy particle patterns, where the magnetic particles are less tightly adherent.

False indications are not the result of magnetic forces. Some patterns of particles are formed by surface roughness or held together mechanically. Such indications will often not reappear on reprocessing. False indications can result from the presence of lint, hair, fingerprints, or due to surface tension liquid drain patterns.

Nonrelevant Indications. The causes of various nonrelevant indications, where the magnetic leakage field is in no way related to the serviceability of the part, are described below. A nonrelevant indication is one caused by flux leakage that has no relation to a discontinuity that is considered to be a defect.

- *Accumulations* of powders occur at corners, where there is a change in the cross section of the part or change of specimen shape. Care and discretion must be exercised because cracks are also likely to occur in these regions.
- *Magnetic labeling and external magnetic fields.* The surface of a part may become partially magnetized by a neighboring electric current or by a permanent magnet. Sharp indications may result, though these will disappear after demagnetization. Residual magnetization may result from use of a magnetic chuck, and this will usually give a fuzzy indication. Demagnetization before MPI is required in these cases.
- *Sharp corners, thread roots, keyways, fillets.* Sharp edges give rise to magnetic leakage fields. Fuzzy indications can result. The magnetizing force should be reduced and MPI repeated.
- *Metal path constrictions.* Size reductions can give rise to apparent magnetic leakage fields. Fuzzy indications can result.
- *Junction of dissimilar metals.* Differences in magnetic permeabilities at the junction can give rise to a magnetic leakage field. Fuzzy or sharp indications result.
- *Boundary zones of welds.* The edges of the heat-affected zone (HAZ), or of a decarburized zone can give rise to fuzzy indications. This may indicate welds which have not been heat treated or stress relieved. The harder the HAZ, the more pronounced the indication.
- *Cold working.* Machining or cold working can cause changes in the magnetic permeability. Fuzzy indications can be seen, which are often not affected by demagnetization only by appropriate heat treatments.
- *Grain boundaries.* When the grain size is large and high magnetizing currents are being used, magnetic permeability changes at the grain boundaries may be sufficient to give rise to a fuzzy indication.
- *Pressed fits.* Forced fits, such as shaft and pinion gear, may have a very fine air gap, and this can give rise to a sharp indication. The tighter the fit, the smaller the indication. If shaft and gear are of different materials, a sharp indication can result even for tight fits.

Demagnetization of the Workpiece. It is frequently necessary to demagnetize the workpiece subsequent to testing. This can be accomplished by reversing and reducing the applied magnetic field successively, removing the object from the magnetizing coil.

Demagnetization can be performed in as little as 4 s, using about 30 cycles of demagnetization (see the section "Magnetization—Demagnetization" and Fig. 5.5 in this chapter).

Materials Suitable for MPI. MPI can be used to test some but not all ferromagnetic materials (Ref 273-275). A simple test is to determine whether a small permanent magnet is strongly attracted to the surface of the part. Some generalizations for steels are as follows:

- The magnetic induction B in a ferritic steel needs to be

$$B \geq 10,000 \text{ G in a magnetic field } H \leq 2500 \text{ A/m}$$

that is, the relative permeability μ_R

$$\mu_R = \frac{B}{\mu_0 H} = 300 \text{ for } H \leq 2500 \text{ A/m}$$

- Stainless steels with ferritic content of more than 70% are probably suitable. Austenitic stainless steels with high chromium and nickel content are *not* suitable.

Some limitations include:

A crack or linear discontinuity should have a length about three times greater than its width.

The defect must be at or very near the surface. Sensitivity for locating defects is greatest to within:

- About $\frac{1}{4}$ in. below surface using dry magnetic powder and half-wave rectified current
- About $\frac{1}{100}$ in. below surface using wet magnetic powder and ac magnetization
- About $\frac{1}{20}$ in. below surface using wet magnetic powder and dc magnetization

The applied magnetic field must be 45 to 90° to the length of the flaw, otherwise the flaw's leakage field is likely to be insufficient to permit detection; see Fig. 5.2.

The defect should have a magnetic susceptibility very different from that of the parent material; the greater the difference, the more distinct the MPI observation.

Inspection must be carried out at temperatures well below the Curie temperature, above which ferromagnetism is zero. The Curie temperature of iron is 770 °C.

Useful comments have been suggested by A. Lindgren (Ref 275):

1. Defects become visible when parallel to the electric current flow.
2. Do not lay parts (e.g., bolts) on a central conductor to magnetize. The field produced will not be suitable for all defects.
3. Never magnetize a coil spring using a coil.
4. Magnetic powder particles intended for the dry method must not be used in the wet method.
5. A crack or linear discontinuity must have a length at least three times greater than its width to be detected.
6. Subsurface defects can be detected if within $\frac{1}{4}$ in. of the surface by the dry method using half-wave magnetizing current. Using the wet method, the depth limitations are: using ac $\frac{1}{100}$ in., using dc $\frac{1}{20}$ in.
7. Surface cracks give powder patterns that are sharply defined, tightly held, usually built up heavily. The deeper the surface cracks, the heavier the buildup of powder.
8. Subsurface cracks produce less sharply defined, fuzzy patterns. The powder is also less tightly adherent.
9. Cylindrically shaped parts, such as rings, gears, tubes, and nuts, can be circularly magnetized using a central nonmagnetic conductor (for example, made of aluminum or copper). This type of magnetization reveals longitudinal cracks. Internal and external examinations are possible. Many parts can be magnetized by the same central conductor at the same time.
10. Half-wave rectified ac current provides excellent mobility for dry magnetic powder particles, and the depth of penetration into the part is quite good. Full-wave rectified ac (single or three-phase) has a penetration that is very good, but only with fair magnetic particle mobility using dry powder.
11. Surface cleaning. Prior to the inspection of a part, the surface must be very clean and dry. Cleaning can be carried out using a detergent, solvents, descaling solutions, vapor degreasing, sand blasting or grinding.
12. Surface coatings. Coatings of paint or plating less than 0.003 in. thick are unlikely to interfere with MPI. However, ferromagnetic coatings can only be tolerated if extremely thin, probably less than 0.001 in. thick.
13. The wet method, where the magnetic particles are suspended in an oil- or water-based liquid, is more suitable for the detection of very fine surface defects. The dry method is more suitable for the detection of subsurface defects.
14. Localized inspection (of welds, castings, and flat surfaces) can be carried out using contact prods. The prod spacings range from 3 in. minimum to 8 in. maximum. Currents should be at least 100 A/in. prod spacing. A current of high amperage and low voltage is passed parallel to the direction where defects are suspected.
15. Cracks in case hardened parts may not show unless the cracks go deep.
16. The residual method should only be used on steels having a high retentivity.
17. When examining cylindrical parts of different cross sections, magnetization shots should be carried out separately for each different part diameter.
18. Springs. Longitudinal crack examination is carried out by passing the current through the spring. Transverse crack examination re-

Magnetic Particle Inspection

quires a central conductor through the spring.

19. For parts with a hole, such as a connecting rod, the electric current is passed through the part and the particle bath applied to detect longitudinal defects. A central conductor through the hole is suitable to detect cracks near the hole. Coil magnetization is used to find transverse cracks.

20. Large diameter disk gears with central hole. Use should be made of a central shot to locate radial defects and induced conductor current magnetizing for circumferential defects. Pancake (spiral) coils work well on round disk-shaped parts.

21. Examination of long crankshafts. Head shot examination, passing the current through the part, should reveal lengthwise (longitudinal) defects. Coil shot examination should reveal transverse (radial) cracks. In the case of very long crankshafts, the coil magnetization is valid over a limited length only, approximately equal to the diameter of the magnetizing coil.

22. Magnetization current direction. A current of low voltage and high amperage is passed parallel to defects to be detected. The current in a coil is passed parallel to the defects to be detected.

23. Examination of welds. Prods are held very firmly on the weld, spaced 3 to 8 in. apart. The part of the weld between the prods becomes magnetized when the current flows; dry powder should be lightly dusted onto the magnetized areas. Excess powder is gently blown off. The electric current is switched off, followed by examination for indications.

24. Large areas are inspected by overlapping smaller areas. Prods 8 in. apart give adequate magnetization of areas approximately oval in shape 8×4 in. The direction between the prods is the direction of best detection, so that the whole surface needs to be re-examined with two prod angles, one at right angles to the other.

25. Transverse cracks in a pipe, or shaft. Several turns (2 to 4) of a cable are wrapped closely around the part. While the current is flowing, powder is lightly dusted onto the magnetized area, and excess powder gently blown away. The cable wrap creates a suitable magnetic field for about 8 in. on either side of the cable wrap. The process should

be repeated along the pipe, with reasonable overlapping areas.

26. Parts with holes. These can be examined using 2 to 4 turns of cable through the hole. While the current is on, powder is dusted lightly onto the surface, and then the excess blown off gently. The current is switched off, and examination carried out for indications.

27. Demagnetization is carried out using a coil of 2 to 4 turns, with an ac current as least as strong as that used for magnetization. Electric current is passed through the coil, and then the coil is removed to a distance greater

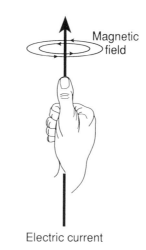

(a)

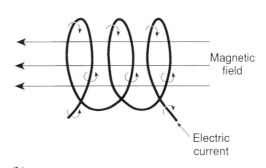

(b)

Fig. 5.3 Electromagnetism. (a) The right-hand rule for the direction of the magnetic field due to an electric current. (b) Uniform magnetic field (H) within a solenoid due to electric current (i) in a long solenoid or coil of n turns and length b when $H = ni/b$.

than one yard. Only then can the electric current be switched off.

28. A white background is helpful in making the MPI easier to observe and is also useful for photographic records.

Magnetization

Electromagnetism. Magnetic fields generated by electric currents are shown in Fig. 5.3 for the case of a straight conductor and due to a solenoid. The right-hand rule (Fig. 5.3a) gives the direction of the magnetic field due to the flow of an electric current. The rule is given by the following (Ref 276):

Imagine the conductor to be placed in the palm of the right hand and the fingers closed upon it, the thumb being outstretched; then if the thumb indicates the direction of the current, the fingers indicate the direction of the magnetic field.

An electric current i flowing along a straight wire conductor gives rise to a magnetic field H where:

$$H = \frac{i}{2\pi r} \text{ at a distance } r$$

If i is in amperes and r is in meters, then H is in amperes/meter (the SI unit of H).

Figure 5.3(b) shows that the magnetic field H due to an electric current i in a long solenoid or coil of n turns and of length b is uniform within the solenoid where

$$H = \frac{ni}{b}$$

For i in amperes and b in meters, the magnetic field H is expressed in units of amperes/meter (the SI unit of H).

Magnetization Equipment. The types of magnetization equipment available cover a range of needs and are discussed in Ref 51 and 269; see also the section "Magnetization Methods" in this chapter. A typical system is illustrated in Fig. 5.4.

Demagnetization. Problems may arise if demagnetization is not carried out to avoid residual magnetization in the part (Ref 51). The problems are likely to be:

• Residual magnetism of the part results in the attraction of metal chips that remain on the surface.
• These adhering metal chips will cause difficulties for subsequent machining and processing of the part; applying paints, plating,

or coatings; proper cleaning of the part; performance of the part.

• The residual magnetism can interfere with electric arc welding; instruments sensitive to magnetic fields; moving parts, for example, gear teeth, races of ball bearings.
• In other cases, residual magnetization of the part is less troublesome. This is probably so in the cases of (1) structural parts where the performance of the part is unaffected, for example, large castings, boilers, or circular magnetization of seamless pipes; (2) subsequent heat treatment near to or above the Curie temperature (770 °C for steel) where the part becomes demagnetized; (3) subsequent remagnetization in a magnetic chuck, or for other MPI; (4) soft steels of low retentivity.

A demagnetizer with track and carriage is shown in Fig. 5.5. An ac current through the coil gives rise to a continually reversing magnetic field. As the part is moved through and away from the coil (Fig. 5.5a), the magnetization (and magnetic induction) is reduced (Fig. 5.5b; see also Fig. 5.13). Alternatively, the part may remain in the coil and the ac current progressively reduced to zero. About 25 to 30 magnetic field reversals are required for demagnetization, and a long part such as a shaft would need to be demagnetized in short lengths of about 50 cm. Parts with a high ratio of length to diameter (L/D) are more readily demagnetized than those of low ratio. Long parts may become magnetized by the earth's magnetic field, if vibrated or cold worked. Most portable and mobile magnetization units include an ac magnetizing circuit, which may be used as a demagnetizer. Batches of small parts may be demagnetized by passing them through an ac coil, in a (nonmagnetic) container. Where a part has been magnetized by an ac generated field, demagnetization can also be accomplished by an ac generated field. Fields generated by ac can be used to demagnetize parts magnetized by dc generated fields as long as the cross section of the magnetized part is 2 in. or less.

Demagnetization techniques have been compiled in Ref 275.

Magnetic Powders (Ref 69, 269)

Powder Types. The magnetic powders used in MPI consist of finely divided iron, black, brown, or red iron oxide (magnetite Fe_3O_4), brown iron oxide (γ Fe$_2$O$_3$), ferrospinel ferrites (Ni$_x$Fe$_2$O$_4$), or some Ni alloys. The properties required of these powders concern:

- Their magnetic behavior
- Particle geometry that can determine the fluidity and free flowing of the powders
- Visibility of the patterns of the magnetic particles formed on surfaces
- Particle size distribution; a mixture of sizes for detecting both large and small defects

The particles must have a low retentivity so when not in the presence of a magnetic field, they will not cling together to form clusters or agglomerates. The particles do need to become magnetized by small magnetic fields, so that a high permeability is essential.

Particle shapes may be spherical, needlelike, or rodlike in form. Elongated needles tend to develop into little magnets with north-south poles and thus form into definite and distinct patterns which give a more clear indication of the presence of a weak magnetic field. However,

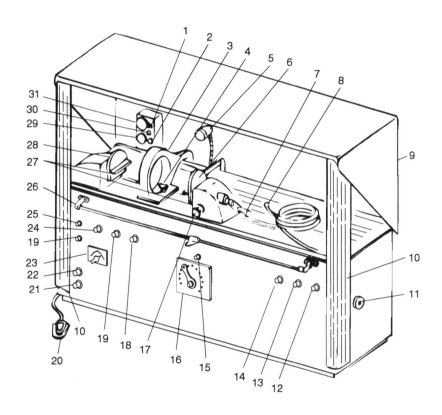

Fig. 5.4 Typical wet horizontal magnetic particle test system (Ref 64). Source: American Society for Nondestructive Testing.

1. d.c. pilot light
2. a.c. pilot light
3. Coil
4. Coil locking handle
5. Black light
6. Tailstock contact plate
7. Tailstock crank handle
8. Nozzle
9. Hood
10. Curtain
11. Flow control valve
12. Demag current pilot light
13. a.c. transfer switch
14. Magnetizing current control
15. 30-point switch start button
16. Current regulating switch
17. Tailstock locking handle
18. d.c. transfer switch
19. d.c./a.c. selector switch
20. Foot switch
21. Control cable receptacle
22. 110 V a.c. outlet
23. Current control
24. Line pilot light
25. Pump switch
26. Actuator
27. Headstock and tailstock ledges
28. Headstock contact plate
29. a.c. ammeter
30. Magnetizing current pilot light
31. d.c. ammeter

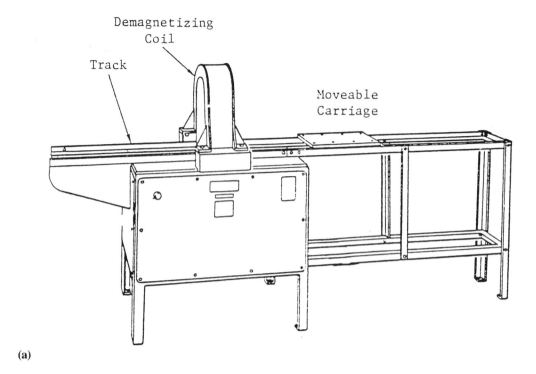

(a)

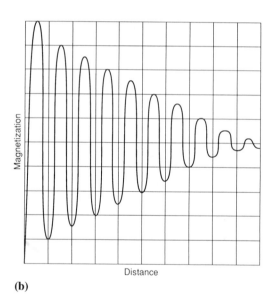

(b)

Fig. 5.5 Demagnetization of the workpiece. (a) The part is moved through and away from the continually reversing magnetic field in the coil. (b) The magnetization of the part is reduced as it is moved away. About 25 to 30 magnetic field reversals are required for demagnetization. The part should be moved at least 1 yd away from the coil. This can be carried out in 4 to 5 s (Ref 230). (a) Reprinted with permission of The American Society for Nondestructive Testing.

there is an optimum elongation aspect ratio of particles, because spherical particles provide a more freely flowing powder. The behavior of wet powder (suspensions) is less dependent on particle shape.

Particle visibility is an important consideration. Visibility and contrast increase with choice of color, so that pigments are added to provide powders that range in color from white to black, gray, red, or yellow. A fluorescent agent can be used to coat the magnetic particle for higher visibility.

The different types of powder are:

Fluorescent. These powders have been treated with a fluorescent component, usually organic, which responds to UV black light. Typical colors are green, yellow, red, or pink. The inspected surfaces should be illuminated with a minimum of $1000 \ \mu W \ cm^{-2}$ of black light, with a maximum of general visible light of 2 ftc at the inspection point.

Color Contrast (Visible Light). These powders, colored black, red, white, yellow, or gray, are viewed by ordinary light with a minimum of 100 ftc at the inspection point. The color of magnetic particle is chosen to give maximum contrast with the surface under inspection.

Dual purpose powders are available and suitable for UV black light, visible light, or examination with a combination of UV and visible.

Application Methods. The methods of powder application are by dry or wet methods. In the dry method, the powder is sprinkled, or sprayed onto the part. This method is often used for the examination of welds and castings.

In the wet method, a suspension (or slurry) of magnetic particles in water or oil is poured, sprayed, or brushed onto the surface, or the part is submerged in a bath. The suspension has to be carefully agitated at all times, with careful control of settling rates, suspension concentrations, and uniformity throughout the suspension; manufacturer instructions need to be followed very carefully.

A comparison of the wet and dry methods shows:

Dry Method

- The magnetic particles are relatively coarse, from 100 to 1000 μm
- Stable to temperatures of about 700 °F
- Maintenance is simple
- The powder is used as received
- Limited to relatively coarse defects
- Powder reclamation possible

Wet Method

- The magnetic particles are very fine, from 1 to 80 μm
- Stable only to about 110 °F
- Elaborate maintenance is required
- The liquid bath has to be prepared very carefully following the manufacturer's instructions
- Finer defects can be detected
- This method is very reproducible and reliable

The MPI Wet Method. A suspension of magnetic particles is applied by either the continuous method or the residual method. These are shown in Fig. 5.6.

Under the continuous method, flow from a nozzle over the part is carried out immediately prior to and during the electric current flow. In the residual method, when examining a steel of high magnetic retentivity, the electric current is passed prior to the flow of suspension from the nozzle, or the part can be immersed in a bath after the electric current is switched off.

The suspension of magnetic particles is either a water- or oil-based liquid carrier. The equip-

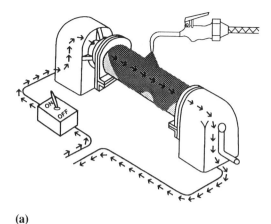

(a)

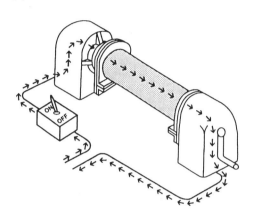

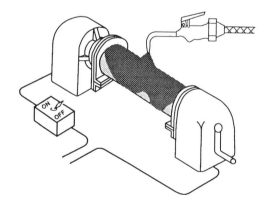

(b)

Fig. 5.6 (a) Flow of MPI suspension during electric current flow. (b) Flow of MPI suspension after electric current flow. The magnetic particle slurry or suspension can be applied to the part by (a) the continuous method using a nozzle immediately prior to and during the passage of the electric current or by (b) the residual method, after the electric current by nozzle, or by dipping in a bath. This method applies to steels of high retentivity, such as high carbon or alloy steels especially in the hardened state.

ment must be free of contaminants, the correct concentration of suspension must be maintained as well as of all required additives such as wetting agent, rust inhibitor, antifoam agent.

Some comments are listed concerning the bath preparation:

- An oil bath must be free of water.
- A water bath must be free of oil.
- The recommended level of powder or liquid concentrate to carrier must be maintained.
- A uniform concentration in the bath must be maintained by adequate circulation of the particles in the bath.
- The settling test of the suspension must be correct. This test is specified by the manufacturer, and usually involves the use of a pear-shaped tube filled with suspension, required to settle under specified time conditions.

Precautions have to be taken to avoid the breakdown of the magnetic particles with the loss of the color or fluorescent pigment. This can arise due to a shearing action of the spraying equipment, the effect of the carrier, or too high a temperature. The recommended test of the manufacturer should be regularly carried out.

A comparison of water and oil bath suspensions is given in Table 5.2.

Troubleshooting for Wet Method. The wet method requires careful adherence to a strict procedure. Various problems can arise and these are listed with probable cause and solution in Table 5.3.

Techniques. There are several special techniques of MPI that can provide permanent recorded data, or enable examinations to be carried out underwater or in areas otherwise inaccessible such as the inside of holes and bolts. These techniques include magnetic rubber inspection, magnetic printing, magnetic painting, pressure-sensitive tape transfers, alginate impression recordings, and photographic MPI. These techniques are discussed briefly in Table 5.4 with appropriate references.

Sensitivity

Size Limits for Visible Defects. The visibility of cracks by magnetic particle inspection is very difficult to analyze precisely, because it depends on a wide range of totally different factors. Some of the factors that affect the ability to detect a surface or subsurface crack with MPI are:

- *Crack geometry*, that is crack width, length, depth, crack edge shape, and profile, the variation of the crack dimensions along its length, and orientation relative to the surface
- *Specimen condition*, that is surface roughness, the frequency of cracks occurring, the presence of surface and subsurface cracks, and the shape and size of the specimen
- Magnetic field, which needs to be applied in a direction perpendicular to the length of any crack as far as possible. Direct, half-wave, or alternating current can be applied directly to the specimen, by a solenoid coil or cable wrap. The magnitude of the applied field is very important, depending also on the magnetic properties of the article. The part may be processed using a continuous magnetic field or by the residual method

Table 5.2 Comparison of water and oil baths for magnetic particle inspection

	Oil bath		Water bath	
	Advantage	Disadvantage	Advantage	Disadvantage
Cost		Higher	Lower	
Flammability		Flash point below 200 °F requiring extra protection	No problem	
Wetting	Excellent			Requires wetting agent
Fumes and odor		Exhaust required	No problem	
Foaming	No problem			Antifoam agent required
Skin irritant		Possible	None	
Evaporation	Negligible			Considerable
Suspension settling		Slower	Faster	
Rust formation	No problem			Rust inhibitor required

Source: Ref 275

- *Magnetic powder* must be carefully selected and is discussed in detail in the section "Magnetic Powders" in this chapter.
- *Observer/inspector.* Once again the detection of the flaw depends on the skill, experience, acuity of the observer, and this can vary with fatigue during a working day. The lighting around and on the specimen must be suitably arranged to help the observer.

All of these factors give rise to a vast variation in practice in the ability of detecting a crack. Indeed, some specific comparative tests have been carried out to compare the results of

Table 5.3 Troubleshooting for wet magnetic particle inspection

Possible problem	Probable explanation and solution
Oil bath	
The surface of the test part not uniformly wetted	Test part has water on surface and should be carefully dried
Clusters and agglomerations of particle, spottiness on test part surface, large particles in bath or in nozzle tube	Water contamination in bath or bath above 110 °F for long periods. Add special emulsifier or, if excessive, change bath
Excessive background on test part	Loose pigment from breakdown of particles. This requires change of bath. Alternatively, excessive concentration particle that requires additional oil carrier
Water bath	
Surface of test part not uniformly wetted	Presence of oil on surface, or insufficient wetting agent in bath. The test part needs to be carefully cleaned, or additional wetting agent added to the bath
Clusters and agglomerations of particles, with a spotty test surface, large particles in nozzle or in bath	Oil contamination in the bath gives a cloudy liquid appearance or scum on top of liquid layer. Excessive concentration of antifoamer, or temperatures above 110 °C give similar effects. An emulsifier can be tried, otherwise it is necessary to change the bath
Excessive background on test surface	Probably breakdown of pigment on particles, requiring change of bath
Excessive foam in bath or on test surface	Probably need to add more foam inhibitor or add more liquid to increase bath liquid level

Source: Ref 275

Table 5.4 Special techniques of magnetic particle inspection

Technique	Special techniques of magnetic particle inspection	Ref
Magnetic rubber	Homogeneous suspension of magnetic particles in silicone rubber is applied to workpiece, which is then magnetized, and the rubber left to cure at room temperature. It may be necessary to construct a dam around the area to contain the liquid rubber suspension.	129, 269
Magnetic printing	Spray surface workpiece with white plastic coating and add magnetic powder. Apply vibrating magnetic field that causes magnetic particles to stain the plastic coating in regions of magnetic field changes. Remove excess particles, spray clear plastic coating to fix particles. Strip off dried film.	129
Magnetic painting	A slurry containing magnetic particles is brushed onto the surface of the workpiece and the magnetic field applied. Flaws become visible as dark lines on a light gray background. The painting can be considered semipermanent, and by rebrushing the slurry on the surface, the test can be repeated. This test can be carried out underwater.	129
Pressure-sensitive tape transfer	The MPI is carried out as usual, then a piece of pressure-sensitive tape is pressed firmly over the surface of the workpiece. The tapes provide permanent records if properly stored.	269
Alginate impression records	A rubbery solid is formed by mixtures of potassium alginate, calcium sulfate, sodium phosphate, and a filler. This is used to replicate accurately the workpiece surface lifting off the magnetic particles. Areas of difficult access can be examined and a permanent record obtained.	269
Photography	This is discussed extensively in many references. It is useful to apply a white spray to the background before use of dry powder MPI. (The white spray used for LPI is suitable.)	269

MPI inspection; this is shown in Fig. 5.7. The probability of detection of cracks of different length in jet turbine blades has been studied (Ref 269, 277) using specimens with independently identified and known crack lengths, and it was found that the probability of detection dropped off rapidly for cracks less than 0.5 mm in length.

As a consequence of the widely varying degrees of detection, empirical rules have been introduced based on general experience and experimental data. Industrial standards are introduced based on calibration slots or holes of known width and size drilled and cut into metal testpieces. Sets of precise specifications for testing procedures have been introduced and are described in the section "Magnetic Particle Inspection Specifications" in this chapter.

The minimum size defect that can be detected varies widely with the conditions of the exami-

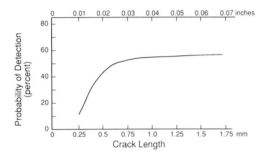

Fig. 5.7 The probability of detection of a defect in jet turbine blades has been studied using specimens with independently identified cracks of known length (Ref 269, 277). The probability of detection approached 60% for cracks of lengths greater than 0.5 mm (0.02 in.).

nation. As a consequence, various estimates of the minimum detectable flaw size have been made depending on experience in examining defects. Several such estimates were derived from the experience of Magnaflux Corporation (Ref 230) and are reviewed in their literature.

In Table 5.5, estimates of surface crack visibility are given for a series of cracks of known depth in steel billets (size 60 × 60 mm cross section). Ultraviolet light examination was performed using wet fluorescent magnetic particles, of 25 to 50 µm diameter, using direct current magnetization of 200 and 800 A/in.; crack widths are not specified. Crack lengths were of the order of 0.5 in. The visibility of the cracks was estimated in arbitrary units from 10 for a clear indication versus 0 for no indication. It can be seen that crack visibility improves as crack depth and magnetizing electric current increase.

The finest crack widths observed on carburized steel are reported (Ref 269) to be ~1 µm using direct current magnetization methods with wet fluorescent magnetic particles. In the case of most steels, the finest crack widths were reported to be ~10 µm using direct current magnetization methods with wet fluorescent magnetic particles. In the case of nitrided steels, the finest crack widths were reported to be ~10 µm (Ref 51, 278). The effect of crack width on visibility is discussed by Betz (Ref 51). For a large defect, about 1 cm in dimension, the minimum detectable width is considered to be ~5 µm.

Crack detection and visibility depend ultimately on the strength of the magnetic leakage field at the surface of the work piece. Extensive reviews have been presented in particular in Ref 222; data are presented using slots of known length, depth, and width, cut into a specimen.

Table 5.5 Estimates of surface crack visibility

Depth of surface crack		Visibility in arbitrary units from 10 (clearly visible) to 0 (invisible) Magnetization currents	
in.	mm	200 A/in.	800 A/in.
0.31	7.9	8	9
0.20	5.2	5	8
0.11	2.8	4	5
55 × 10⁻³	1.4	2	5
32 × 10⁻³	0.81	2	3
14 × 10⁻³	0.35	1	2
12 × 10⁻³	0.30	<1	1
8.7 × 10⁻³	0.22	<1	1
5 × 10⁻³	1.12	<1	1

Source: Ref 269

The magnetic leakage field and crack visibility increases with crack depth and with the applied magnetic field. Crack visibility increases with crack lengths up to about 0.5 mm, above which it is essentially constant.

Subsurface Cracks. Detection of subsurface cracks depends on very many factors apart from the depth of the crack. These factors cover:

- Defect shape
- Defect tightness; whether the crack is under severe compression
- Granular nature of the crack
- Surface condition of the part

In general, the most appropriate conditions to observe subsurface defects require the use of a circular field, half-wave rectified dc, dry powder method, and of course a trained inspector. In such a case, a weld ($\frac{1}{2}$ in. thick plate) can be usefully examined for defects $\frac{1}{4}$ in. below the surface.

Using the wet method and ac, an inclusion should be visible at about 5×10^{-3} inch (125 to 250 μm) below the surface. Full-wave rectified dc and the wet powder method should enable defects to be seen at 50×10^{-3} in. (~1 mm).

In Figure 5.8(a), some typical dimensions of detectable subsurface defects are illustrated (Ref 230). The size of the defect relative to the thickness of the testpiece is important. In general, the depth of the defect below the surface should be less than 10% of the thickness of the article, and the defect length should be nearly comparable to the defect depth as shown.

In Fig. 5.8(b), the visibility of subsurface cracks in shell cases is shown. This was part of a detailed examination of such shell cases, using direct current magnetization via a central conductor through the shell case axis, with wet fluorescent magnetic particles. Some typical results are presented.

Defect Indicator Standards

Values of magnetic leakage fields, necessary to obtain the highest sensitivity and resolution for MPI are not available. Instead testpieces are used based on artificial discontinuities as mag-

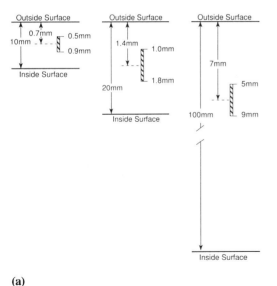

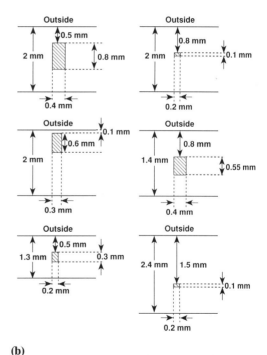

(a) **(b)**

Fig. 5.8 Subsurface defects. (a) The dimensions illustrate defects that have been detected below the surface, where the size of the defect relative to the thickness of the testpiece is important. However, there is a complexity of factors that govern the detection of subsurface cracks. (b) The visibility of subsurface cracks in shell casings has been reported in Ref 269. The tests employed direct current magnetization with the current passing along a central conductor through the shell casing, using wet fluorescent magnetic particles. Six typical results are shown where the shape of the defects is idealized.

netic field indicators. The various types of test-pieces are:

- Ketos ring (Fig. 5.9)
- Pie gage (Fig. 5.10)
- Shims (Fig. 5.11a-c)
- Artificial flaw indicators (AFI); quantitative quality indicators QQI (Fig. 5.11d)

The Ketos ring consists of a ring containing a series of holes (1.75 mm diam) drilled at different depths from the ring edge, as shown in Fig. 5.9(a), where a cross section and plan view are presented. The holes are numbered at fixed depths as listed in the table accompanying Fig. 5.9(b). The magnetizing current passes through the center of the ring via a central

conductor and the magnetic particles are spread over the outer surface of the ring. The recommended number of holes that should be visible for various magnetic powders and magnetizing currents are given in Fig. 5.9(c). The Ketos ring leaves much to be desired as a control of defect detection sensitivity; paste-on-defects, such as QQI (see Fig. 5.11d), provide useful additional tests. The QQI is especially useful for developing test procedures when using multidirectional magnetization and the wet continuous method on machined parts. A standard reference material (SRM 1853) is available from NIST (Ref 4, 274, 279), and this is in the form of a test ring that can be used to produce leakage fields of known magnitude.

The pie gage indicator, shown in Fig. 5.10 consists of a disk made from a high permeability ferromagnetic material that is separated into six (or eight) segments by gaps containing nonmagnetic material. The orientation of the magnetic

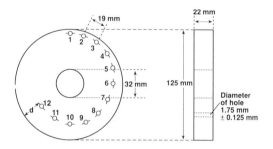

(a)

Distances of holes from ring edge of Ketos ring; holes are 1.75 mm diam

Hole No.	Distance from edge (*d*), mm
1	1.8
2	3.6
3	5.3
4	7.1
5	8.9
6	10.7
7	12.4
8	14.2
9	16.0
10	17.8
11	19.6
12	21.3

(b)

The required magnetizing currents for the Ketos ring test, with the minimum number of holes that should be visible when using full-wave rectified current along a central conductor passing through the hole in the Ketos ring

Magnetizing current, A	Minimum No. holes indicated
Black suspension (wet)	
1400	3
2500	5
3400	6
Dry powder	
1400	4
2500	6
3400	7
Fluorescent suspension (wet)	
1400	3
2500	5
3400	6
Source: Ref 64	

(c)

Fig. 5.9 Ketos ring (MPI). The Ketos ring is made of AISI 01 tool steel from annealed round stock, hardness 90 to 95 HRB. Two views of the ring are given in (a) and the distances (*d*) of the holes from the ring edge are listed in the accompanying table (b). The magnetizing current passes along a conductor through the center of the ring (see the accompanying table) (c). This ring is used extensively (Ref 274). It is a magnetic field indicator, but in reality determines the efficiency of the complete MPI testing procedure, including the capabilities and characteristics of the observer/inspector. The Ketos ring is likely to be surplanted by paste-on-defects such as QQI or other devices. A similar test ring is available from NIST (Ref 279).

field is perpendicular to the gap displayed most clearly when the magnetic particles are applied in the presence of a magnetic field. This indicator is most useful when using the dry method to inspect steel plate and welds.

Shims are shown in Fig. 5.11. The shim is placed on the test surface with the slits perpendicular to the magnetic field direction. When the magnetic particles are applied to the surface of the shim, the slits should become visible. There are various types of shims, also known as paste-on discontinuities, artificial discontinuity standards, magnetic discontinuity standards, shim discontinuity standards, and block discontinuity standards.

The Burmah-Castrol shim flux indicator is shown in Fig. 5.11(a). An iron core, 0.1 mm thick, is encased in a brass coating, 0.05 mm thick, giving a rugged, reusable device. The slits in the iron, which can be of various widths, are filled with nonmetallic material.

The Raised Cross in Fig. 5.11(b) consists of a circular disk of iron, covered with a brass protective plate, with four radiating spacer gaps forming a cross in the disk. Some of the spacer gaps are always suitably oriented with respect to the magnetic field, which permits determination of the magnetic field direction.

The Japanese Block discontinuity standards are linear or circular slots, as shown in Fig.

5.11(c), where the slots can be of a variety of thicknesses.

The artificial flaw indicators (AFI) also known as the quantitative quality indicators (QQI) (Fig. 5.11d) are suitable for multidirectional magnetic fields. Their efficiency and effectiveness are enhanced when supplementary measurements are made using a Hall effect gaussmeter. A videotape description of the use of QQI is available from Magnaflux Corporation

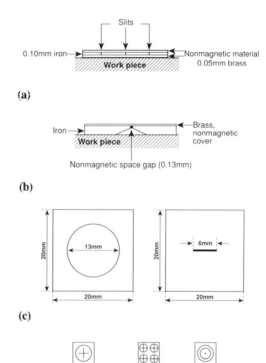

(a)

(b)

(c)

(d)

Fig. 5.11 Discontinuity shims as magnetic flux indicators. Source: Ref 69, 269. (a) Burmah-Castrol shim flux indicator. The slits are of different widths usually containing a nonmagnetic material. (b) Raised Cross. This circular shim contains four spacer gaps that form a cross so that the spacer avoids the case of being parallel to the direction of the magnetic field. (c) Japanese Block discontinuity standards. These are provided in a variety of thicknesses and slot depths, with circular or linear shapes. (d) Quantitative quality indicators (QQI). This shim should be glued or held with the flaw side in intimate contact with the cleaned part surface. Circular and crossed bar flaw configurations are cut into the shims at depths of 0.0006 in. (15 μm), which is 30% the shim thickness. Three types are shown, suitable for longitudinal, circular, and multidirectional magnetic fields (Ref 230). These shims are also known as artificial flaw indicators (AFI).

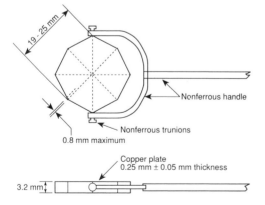

Fig. 5.10 Pie gage magnetic field indicator. The gage is a disk of high magnetic permeability, divided into six or eight segments separated by gaps containing a nonmagnetic material of different thicknesses. The pie gage is placed in contact with the test surface, and the magnetic particles spread over the upper surface of the pie gage.

157

(Ref 230, 275). The shims are able to indicate discontinuities as small as 0.25×0.05 mm, when a magnetic field of ~10 gauss (1 mT) is applied to the test object; see Fig. 5.11(d).

Part II Technical Discussion

The Magnetization Hysteresis Loop

The application of a magnetic field of H A/m can be carried out by passing an electric current through a solenoid (see Fig. 5.3b). The specimen becomes magnetized with a magnetic induction of B (gauss); for ferromagnetic solids, B can be replaced by J the ferromagnetic magnetization (see the section "Magnetic Leakage Field" in this chapter).

Magnetization is shown in Fig. 5.12. Starting with the externally applied magnetic field $H = 0$, the first application of a magnetic field causes all of the domains of magnetic poles to align until near-saturation is reached at point P_I, with all the magnetic dipoles nearly parallel to the applied magnetic field. On reversing H, the magnetic dipoles in the material begin to reverse, and point P_{II} where $H = 0$ is known as the retentivity of the material. On continuing to apply the field H in the reverse direction, the flux density becomes zero at point P_{III}, and this value of the applied field is known as the coercive force. Further increases in H in the reverse direction causes the magnetization of the material to increase to near-saturation at P_{IV}. Cycling the applied field leads to the hysteresis loop shown in the figure.

Demagnetization. To remove the magnetization of a part, the externally applied magnetic field H must be reversed and reduced successively to arrive at a minimum value of B or J at $H = 0$. This is shown in Fig. 5.13. In practice, about 25 to 30 reversals are used in the demagnetization process.

Ferromagnetic Magnetization

The materials examined by MPI are always ferromagnetic so that the induced magnetization M is related to the magnetic induction by:

$$B = \mu_0 (H + M) \cong \mu_0 M = J$$

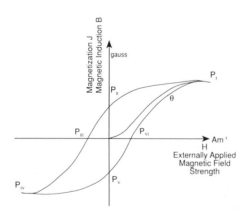

Fig. 5.12 The magnetization hysteresis loop. The required level of magnetization for MPI is recommended to be beyond the level of maximum permeability (the point θ). The magnetization level P_I for most steels and cast iron is on the order of $B \sim 10,000$ G (= 1 T) at $H \sim 1000$ A/m. For ferromagnetic materials, the magnetic induction B can be considered nearly equal to the magnetization J of the part. Magnetic terms are defined in Appendix 5.1. The point P_{II} is known as the retentivity, and P_{III} as the coercive force.

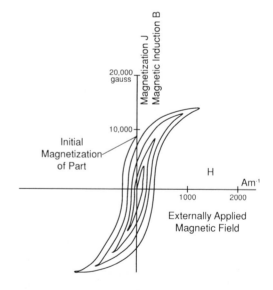

Fig. 5.13 Demagnetization. The externally applied magnetic field H (Am^{-1}) must be reversed and reduced successively to arrive at a minimum value of magnetization at $H = 0$. About 25 to 30 reversals are required.

(See the section "Magnetic Leakage Field" and Fig. 5.14.)

The SI unit of magnetic field H is amperes/meter (A/m). The cgs unit for H is oersted (Oe). In air, oersted and gauss are equivalent. To convert from amperes/meter to gauss, multiply the number of A/m by $4\pi/10^3$.

The SI unit for ferromagnetic magnetization J and magnetic induction B is Tesla. The cgs unit is

Gauss [where 10^4 gauss = 1 Tesla]

$$\left[= \frac{1\ \text{magnetic line}}{\text{cm}^2} \cong 6.47\ \text{lines/square inch} \right]$$

The magnetic field produced by closely wound coils consisting of n turns of coil radius r (cm) and coil length b (cm) carrying a current i (amperes) is given by (Ref 280):

- For $r \gg b$, $H = (2\pi\ i\ n)/10r$ gauss (at the center of the coil)
- Long solenoid of length $b \gg r$, in center: $H = (0.4\pi\ i\ n)/b$ gauss; at ends: $H = (0.2\pi\ i\ n)/b$ gauss
- For an infinitely long solenoid, $b \to \infty$, $H = 0.4\ \pi i\ n'$ gauss, where $n' = n/b$ = number of turns per unit length

These values of H are those obtained without a sample. Depending on sample geometry, the effective value of H can vary considerably from these values.

Required Level of Magnetization

The required level of magnetic induction B for MPI is discussed in Ref 274 and is often recommended to be at or greater than the levels at the point of maximum permeability, such as θ in Fig. 5.12. The applied magnetic field H should, in any case, be greater than the coercive force (points P_{III} or P_{IV}). However, the strength of the magnetic leakage fields, and hence the sensitivity of the method, will continue to increase for H well beyond the point P_I. This is due to demagnetization effects that occur in the vicinity of the defect resulting in a smaller effective value of H.

The actual level of current needed for the greatest sensitivity must usually be determined experimentally, especially for parts with complex shapes, and will be at the largest value of H that does not lead to a general surface layer background of magnetic particles that hinders the observation of indications; this value of H is likely to be beyond the "knee" of the hysteresis curve, such as point θ; see Fig. 5.12.

Magnetic Particle Inspection Specifications (Ref 51, 271)

There are a large number of independent variable factors that can affect the results of magnetic particle testing, and precise specifications have been prepared to cover all the conditions of testing to ensure reliable, uniform, and reproducible results.

A list of MPI specifications is given in Appendix 5.2, and these have been prepared by many organizations, including ASTM, ASME, SAE, American Welding Society, and various government organizations.

The specification used widely for many purposes is MIL-STD-1949 "Military Standard Inspection, Magnetic Particle" (U.S. Army Materials Laboratory, Watertown, MD). This specification is presented in a modified, abbreviated form; it represents an excellent synopsis of the procedures in general use in MPI. Recently, this standard has been superceded by ASTM E 1444, which is similar in almost all respects.

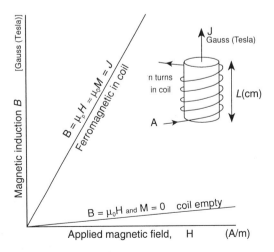

Fig. 5.14 For MPI, the material is always ferromagnetic. Consequently and as a simplification, the magnetic induction B can be replaced by magnetization J.

Lighting Intensities (Based on Section 4.8 of MIL-STD-1949)

Visible Light Intensities. The intensity of the visible light at the surface of the parts undergoing inspection should be maintained at a minimum of 1000 lux (see the section "Surface Inspection" in this chapter). Fluorescent magnetic particle inspection should be performed in a darkened area with a maximum ambient visible light level of 20 lux (for example, a 10 W lamp at a distance of 6 ft).

Black Light. This is the popular term used for a UV light of wavelengths ranging from about 320 to 380 nm. The black light intensity at the examination surface shall be 1000 $\mu W/cm^2$. This can be provided by a 100 W UV lamp at 15 in. distance.

Special Internal Black Light Equipment. The inspection of internal surfaces at depths greater than twice the internal diameter requires sufficient blacklight intensity in the 320 to 380 nm wavelength range to register at least 1000 $\mu W/cm^2$ on the inspection surface.

Magnetic Particle Materials (Based on Section 4.9 of MIL-STD-1949B)

The particles used in magnetic particle inspection are finely divided ferromagnetic materials that have been treated so as to be visible against the background of the surfaces under inspection. They can be either colored for use with visible light or coated with a fluorescent material for use with black light. The particles can be used as a free-flowing dry powder (dry method), for suspension at a given concentration in a suitable liquid (wet method), for suspension in a polymerizable material (magnetic rubber method), or in a slurry (magnetic painting). The particles should have a high magnetic permeability and a low retentivity. Careful control of particle size, shape, and material is required to obtain consistent results.

Dry and Wet Particle Requirements. Dry and wet particles need to meet the requirements of AMS 3040 (see Appendix 5.2). The particles should show indications as listed in Fig. 5.9(c) using the following procedure:

1. A conductor of a diameter between 25 and 31 mm and length greater than 40 cm is placed through the center of the ring, which is at the midpoint of the length of the conductor.

2. The ring is circularly magnetized by passing the current specified in Fig. 5.9(c) through the conductor.

3. The magnetic particles are applied to the ring surface while the current is flowing.

4. The ring is examined within 1 min after current application under a visible light of not less than 1000 lux. The minimum number of hole indications are specified in Fig. 5.9(c).

Magnetization Methods (Based on Section 5.2 of MIL-STD-1949B)

The magnetizing current for magnetic particle inspection can be full-wave rectified, half-wave rectified, or alternating current. Alternating currents are used for the detection of defects open to the surface. Full-wave rectified currents have the deepest possible penetration and should be used when inspecting for defects below the surface by the wet magnetic particle method. Half-wave rectified current is advantageous for dry powders, because this gives increased mobility and sensitivity to the particles by its pulsating unidirectional field.

Magnetic Field Directions. Discontinuities are difficult to detect by the magnetic particle method when they make an angle less than 45° with the direction of magnetization. To ensure detection of discontinuities in any direction, each part must be magnetized in at least two directions perpendicular to each other, by circular magnetization in two or more directions, by longitudinal magnetization in two or more directions, or by multidirectional magnetization. Artificial discontinuities can be used to determine magnetic field direction.

Magnetization can be accomplished by passing the electric current directly through the testpiece using head- and tailstock, prods, clamps, or magnetic leeches. The electrical current must not be on when contacts are being made or removed and excessive heating must not occur at the contact area. Prods should not be used for inspection of aerospace components or on finished surfaces. Indirect part magnetization can be carried out by coils, cable wraps, yokes, or a central conductor. Induced current magnetization (toroidal or circumferential field) is accomplished by inductively coupling the testpiece to a coil, and this method can be advantageous on ring-shaped parts with a diameter to thickness ratio greater than 5.

Magnetic Field Strength (Based on Section 5.3 of MIL-STD-1949B)

Factors that determine the required field strength include the size, shape, and magnetic permeability of the part, the technique of magnetization, the method of particle application, and the type and location of defects sought. The appropriate magnetic field strength can be determined by:

- Testing parts having known or artificial defects of the type, size, and location specified
- Using a Hall-effect gaussmeter capable of measuring the peak values of the tangential field

Tangential applied field strengths in the range of 2.4 to 4.8 kA/m are adequate field strengths for magnetic particle inspection.

Prod Current Levels. When using prods on material 19 mm in thickness or less, 3.5 to 4.5 A/mm (half-wave rectified) of prod spacing should be used. For material greater than 19 mm thickness, 4.0 to 5.0 A/mm (half-wave rectified) of prod spacing should be used. Prod spacing should not be less than 50 mm or greater than 200 mm. The effective width of the magnetizing field when using prods is $^1/_4$ of the prod spacing on each side of a line drawn through the prod centers: see Fig. 5.19. *Note*: The use of prods on crack-sensitive materials can lead to stress-corrosion cracking.

Direct Magnetization. When magnetizing by passing current directly through the part, for example using "head shots," the current should be from 12 to 32 A/mm (300 to 800 A/in.) of part

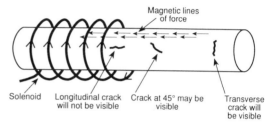

(a)

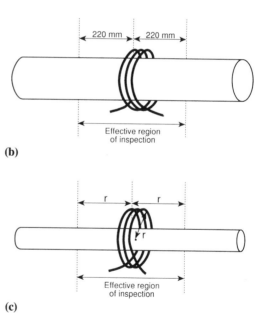

(b)

(c)

Fig. 5.16   A noncontact method of magnetization is provided by the use of a solenoid or a flexible coil that provides longitudinal magnetization of the workpiece, suitable for the examination of transverse cracks, as shown in (a). The solenoid can be used with a high fill factor, as in (b), or a low fill factor, as in (c). The effective regions of inspection are indicated for (b) and (c). Different magnetization currents apply as discussed in specification MIL-STD-1949B (Ref 271).

Fig. 5.15   The estimated region of effective magnetization for a small internal conductor of diameter Δ is as shown, according to specifications MIL-STD-1949B. This is a conservative estimate. Often the entire circumference is effectively magnetized. This can be checked using shims or a gaussmeter.

diameter; see Fig. 5.18. The diameter of the part should be taken as the largest distance between any two points on the outside circumference of the part. Normally, currents will be 30 A/mm or lower with the higher currents used to inspect for inclusions or to inspect low permeability alloys such as precipitation-hardening steels. When testing to locate inclusions in precipitation hardened steels, even higher currents, up to 40 A/mm, may be used.

Central Conductor Magnetization. Magnetization can be carried out by passing a current through a conductor that passes through the inside of the part. In this case, alternating current can be used when the sole purpose of the test is to inspect for surface discontinuities on the inside surface of the part.

When a small internal conductor is located near the part, the distance along the circumference (interior or exterior) that is effectively magnetized can be taken to be four times the diameter of the central conductor (Fig. 5.15). In many situations, a much larger area than that shown in Fig. 5.15 can be examined effectively. This can be established using shims or a gaussmeter.

Magnetization using Coils. Longitudinal magnetization is often accomplished by passing a current through a coil encircling the part to be tested, that is, a "coil shot." This produces a magnetic field parallel to the axis of the coil (Fig. 5.16). The part can fill the magnetic field interior of the coils completely or partially. A "fill factor" (the area fraction of the coil interior filled by the part under examination) needs to be considered. For low or intermediate fill factor coils, the effective field extends a distance on either side of the coil center approximately equal to the radius of the coil (Fig. 5.16c). For cable wrap or high fill factor coils, the effective distance of magnetization is 220 mm on either side of the coil center (see Fig. 5.16b).

Longitudinal Magnetization with Low Fill Factor Coils. When the cross-sectional area of the coil is ten or more times the cross-sectional area of the part being inspected, then the current in amperes through the coil, i, needs to be

For parts positioned to one side of the coil:

$$i = 45,000 \frac{\Delta}{nb} \qquad \text{(Eq 5.1)}$$

where b is the length of the part (the units of b and Δ are the same), Δ is the diameter of the part and n is the number of coil turns.

For parts positioned in the center of the coil:

$$i = \frac{1690r}{n[6b/\Delta - 5]} \qquad \text{(Eq 5.2)}$$

where r is the radius of the coil in mm. If the part has hollow portions, replace Δ with Δ_{eff} as given in Eq 5.5 and 5.6. These formulas hold only if b/Δ is greater than 2 and less than 15. If b/Δ is less than 2, pole pieces of the same diameter and material as the part being tested are added to each end of the part to increase b/Δ to at least 2. The value of 15 for b/Δ should be used if b/Δ is greater than 15.

Longitudinal Magnetization with Cable Wrap or High Fill Factor. When the cross-sectional area of the coil is less than twice the cross-sectional area including hollow portions of the part under test, then the current in amperes through the coil, is to be:

$$i = \frac{35,000}{n[b/\Delta + 2]} \text{ only for } 2 < b/\Delta < 15 \qquad \text{(Eq 5.3)}$$

Longitudinal-Magnetization for Intermediate Fill Factor Coils. When the cross-sectional area of the coil is between two and ten times the cross-sectional area of the part being inspected, the current through the coil, i, is to be:

$$i = i_{\text{H}} \frac{10 - \chi}{8} + i_{\text{E}} \frac{\chi - 2}{8} \qquad \text{(Eq 5.4)}$$

where i_{E} is the value of i calculated for low fill factor coils using Eq 5.1 or 5.2, i_{H} is the value of i calculated for high fill factor coils using Eq 5.3, and χ is the ratio of cross-sectional area of the coil to the cross-sectional area of the testpiece.

Calculating the b/Δ Ratio for Hollow or Cylindrical Parts

When calculating the b/Δ ratio for a hollow or cylindrical part, Δ is to be replaced with an effective diameter, Δ_{eff}, calculated using:

$$\Delta_{\text{eff}} = 2[(A_{\text{t}} - A_{\text{h}})/\pi]^{1/2} \qquad \text{(Eq 5.5)}$$

where A_{t} is the total cross-sectional area of the part, and A_{h} is the cross-sectional area of the hollow portions of the part. For cylindrical parts this is equivalent to:

$$\Delta_{\text{eff}} = [\Delta_{\text{o}}^2 - \Delta_{\text{i}}^2]^{1/2} \qquad \text{(Eq 5.6)}$$

where Δ_{o} = the outside diameter of the cylinder, and Δ_{i} = the inside diameter of the cylinder.

Particle Application (Based on Section 5.4 of MIL-STD-1949B)

Continuous Method. In the dry continuous method, magnetic particles are applied to the part while the magnetizing force is applied. In the wet continuous method, the magnetizing current should be applied simultaneously or immediately after applying the suspension of magnetic particles.

Residual Magnetization Method. In the residual magnetization method, the magnetic particles are applied to the test part after the magnetizing force has been discontinued. The residual method is not as sensitive as the continuous method, but it can be useful, for example, in detecting fatigue cracks on the surface of material with a high retentivity.

Prolonged Magnetization. When using polymers, slurries, or paints, prolonged or repeated periods of magnetization may be necessary because of lower magnetic particle mobility due to higher levels of viscosity.

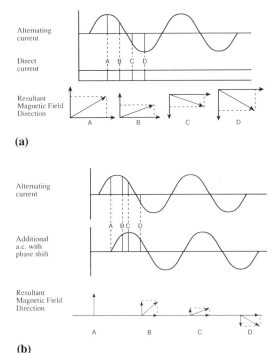

(a)

(b)

Fig. 5.17 Multidirectional magnetization results from the use of an ac supply with a dc component (a), or out-of-phase multi-alternating current components (b). Multidirectional magnetization enables defects in all directions to be examined (Ref 69, 269).

Dry Magnetic Particle Application. When using dry particles, the magnetizing current needs to flow prior to the application of the magnetic particles to the surface under test and continued until powder application has been completed and any excess blown off. The duration of the magnetizing current needs to be at least $\frac{1}{2}$ s and short enough to prevent any damage to the part. The dry powder needs to be applied so that a light, uniform dustlike coating settles on the surface of the test part while the part is being magnetized. Specially designed powder blowers or shakers using compressed air or hand power need to be used. The applicators need to introduce the particles into the air so that they reach the part surface in a uniform cloud with a minimum of force. After the powder is applied and before the magnetizing force is switched off, excess powder needs to be removed by a dry air current with sufficient force to remove the excess particles, but not too strong to disturb particles held by a magnetic leakage field. In order to recognize the broad, fuzzy, lightly held powder patterns formed by near surface discontinuities, the formation of indications must be carefully observed during powder application and during removal of the excess powder.

Wet Magnetic Particle Application. Fluorescent or nonfluorescent particles suspended in a liquid vehicle at the required concentration need to be applied either by gently spraying or flowing the suspension over the area to be inspected. Proper sequencing and timing of part magnetization and application of particle suspension are required to obtain proper formation and retention of indications. This requires that the stream of suspension be diverted from the part just before energizing the magnetic circuit. The magnetizing current needs to be applied for a duration of at least $\frac{1}{2}$ s for each application with a minimum of two shots being used. The second shot should follow the first in rapid succession, after the flow of suspension has been interrupted and should occur before the part is examined for indications. Weakly held indications on highly finished parts are readily washed away, and thus care must be exercised to prevent high-velocity flow over critical surfaces.

Magnetic Slurry/Paint Applications. Magnetic paints or slurries are applied to the part with a brush, squeeze bottle, or aerosol container before or during magnetization. The slurry is a very thick viscous mixture of magnetic particles in oil. This method is used for special examinations, such as overhead, outside in the rain, or underwater examination.

Magnetic Polymer Applications. While the magnetic particles are still mobile, the part should be magnetized to the specified level. Polymerizable materials containing magnetic particles need to be in contact with the test part during curing. This may require repeated periods of magnetization. This method is used for special examination such as the interior of holes that cannot be readily tested by the wet or dry method.

Magnetization Methods

The electric current supplied to the electromagnet can be either on during the application of the magnetic particles to the workpiece or off, in which case the magnetic powder inspection will depend on the residual magnetism in the workpiece.

The magnetic field level for MPI must generally be suficient to produce a magnetization M of the workpiece of 7000 G (0.7 T), and effective magnetic field H of 2400 A/m is usually sufficient to attain this level of magnetization in most ferromagnetic materials. However, a value of $H \sim 700$ A/m will often provide adequate magnetization for many low-alloy steels. Other specifications, such as for the aerospace industry, require effective fields $H \sim 8960$ A/m (Ref 69).

The magnetizing current may be alternating or direct current; full-wave or half-wave rectified. Alternating currents produce a skin effect with the electric current concentrated at the surface and are therefore useful for detecting surface defects. Full-wave rectified current has a deeper penetration into the part and is used to detect subsurface defects. Half-wave rectified alternating currents are advantageous for dry magnetic powder examination due to the pulsating nature of the electric and magnetic fields which result in higher mobility of the dry powder.

Multidirectional magnetic fields are obtained when using electric currents that have alternating and direct current components, or which have multi-alternating current components when the magnetic field direction varies with time over 360°. This is very important because defects in the workpiece are very difficult to detect if the direction of magnetization is parallel or nearly parallel to the direction of the flaw. Multidirectional magnetization or magnetization in two perpendicular directions are advantageous and can overcome the limitations imposed by single directional magnetization. Multidirectional magnetizations obtained by electric currents with multiple components are shown in Fig. 5.17.

Configurations for Magnetic Field Production. Different configurations are employed for the production of magnetic fields where each configuration has a different potential in crack determination. The contact method shown in Fig. 5.18 and 5.19 involves direct contact between the electrodes carrying the electric current and the test specimen. The noncontact method does not require contact of the test specimen and the electric current carrying conductor as illustrated in Fig. 5.16 and 5.20.

Direct contact does involve the risk of electric arcing at the contact points. The configuration for the examination by direct contact of a conducting bar for contact current magnetization or "head shot" for longitudinal cracks is illustrated in Fig. 5.18. The magnetic field at the surface of the conducting bar of diameter Δ_A using an electric current i is given by $H_A = i/\pi\Delta_A$. The contacting heads can be fixed or movable.

The examination by direct contact of a plate test specimen is shown in Fig. 5.19 using prods. The hand-held electrodes pass current directly through the plate resulting in circular magnetic lines of force as shown in (a). The magnetic field H_B from an electric current i through prods at a separation distance of d_B can be estimated by $H_B = i/\pi d_B$ (see Ref 69 and 271). The effective width

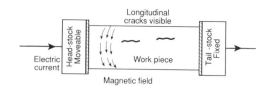

Fig. 5.18 Contact current magnetization using electric current in a straight conductor workpiece. The direction of the magnetic field and possible cracks are shown. Longitudinal cracks are perpendicular to the magnetic field direction and thus will result in a magnetic leakage field detectable by MPI. Transverse cracks will not be visible. The contact heads can be fixed or movable. Normally the tailstock is fixed and the headstock is moved to adjust to the length of the part under examination. An electric current varying from 12 to 32 A/mm (300 to 800 A/in.) of the part diameter is generally used.

of the magnetizing field from prods is estimated as $d_B/4$ on each side of the straight line drawn between the prods. Recommended values of electric current are given in specification MIL-STD-1949B for test specimens of varying thicknesses using different prod distances. Test materials of thicknesses less than 19 mm require 35 to 45 A/cm (90 to 120 A/in.) of prod spacing. Above 19 mm test specimen thickness, 40 to 50 A/cm (100 to 130 A/in.) of spacing should be used. Prod spacings should be between 5 and 20 cm. The use of prods on crack-sensitive materials can lead to stress-corrosion cracking due to arcing at the contact points.

Examination by a noncontact method can be carried out using a solenoid, or coils, as shown in Fig. 5.16. A longitudinal magnetization arises which is suitable to display transverse defects. This method is discussed in specification MIL-STD-1949B. High or low fill factors are determined by the ratio of the cross-sectional area of the testpiece and the internal diameter of the coil used, as shown in Fig. 5.16(b) and (c). Different magnetizing currents are recommended for these two situations. For high fill-factors, or for a cable-wrap, the effective distance of magnetization is considered to be 22 cm on either side of the coil center, as in Fig. 5.16(b). Low fill-factor coils provide an effective magnetic field over a distance from the coil equal to the coil radius, as shown in Fig. 5.16(c). Other arrangements of magnetizing solenoids and coils are discussed in Ref 222.

Another method of magnetizing the workpiece is by the use of a yoke, which is a U-shaped electro- or permanent magnet (Fig. 5.20). The magnetic lines of force are arranged as in Fig. 5.20(a) along the surface of the workpiece, and within the workpiece as in Fig. 5.20(b). How-

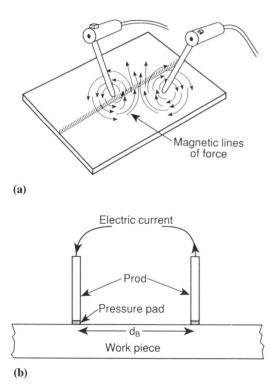

(a)

(b)

Fig. 5.19 Prods are a contact method of hand-held electrodes through which the magnetizing current flows into the test specimen.

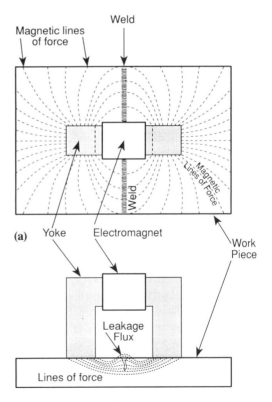

(a)

(b)

Fig. 5.20 A yoke is a U-shaped electromagnet, or a permanent magnet, that is used to induce a magnetic field in the workpiece. The magnetic lines of force are shown across the surface perpendicular to a weld direction (a), and below the surface perpendicular to a surface crack (b). Field strengths of ~5 G are required for magnetic particle examination.

ever, this method has a relatively limited magnetizing ability.

Some special applications of magnetizing yokes and electric cable magnetization methods are discussed in Ref 69 and 222 for the examination of very long workpieces such as wire ropes, or for the examination of circular workpieces.

Magnetic Field Gaussmeters

A magnetic field can be measured using a Hall-effect gaussmeter. The principle of the measurement is shown in Fig. 5.21(a); a voltage V_z is generated in a conductor when an electric current J_Y flows in a conductor in a magnetic field H_x:

$$H_x = \frac{V_z}{J_y R_H}$$

where R_H is the Hall coefficient of the conductor.

Hall-effect gaussmeters are quite accurate, very reliable, and practical. A typical commercial gaussmeter is shown in Fig. 5.21(b). Descriptions of the use of gaussmeters are given in Ref 222, 265, 273, and 274 and some practical notes taken from those references are listed below:

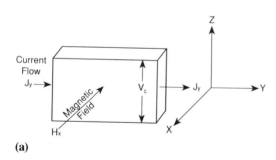

(a)

(b)

Fig. 5.21 Hall-effect gaussmeter. (a) Magnetic field generation. (b) Commercial gaussmeter.

- The magnetizing force H_r tangent to the surface is the same just inside and just outside the part. The magnetic induction B tangent to the surface of the part is not.
- The plane of the Hall-gaussmeter probe needs to be perpendicular to the surface part.
- The size of the Hall-gaussmeter probe needs to be relatively small compared to the part.
- When the surface is highly curved, the magnetic field varies rapidly with distance from the part.
- Magnetic field measurements are inaccurate where there are sharp changes of geometry of the part, such as at sharp corners or at the part end.

Applications of Magnetic Particle Inspection

MPI can be used in virtually all phases of the iron and steel industry. Reviews of observed defects and their probable cause are tabulated in Tables 5.6 to 5.11; further details are available from Ref 70 and 275. There are defects produced during the solidification of the molten metal (Table 5.6), and some likely to arise during the processing of rolling or forging (Table 5.7), casting (Table 5.8) or welding (Table 5.9). The finishing processes can give rise to other defects during machining, heat treatment, grinding, straightening, plating, or pickling (Table 5.10). The use of the metal part will give rise to other defects due to fatigue, corrosion, or stressing with plastic deformation (Table 5.11).

The range of magnetization methods and their applications are listed in Table 5.12. Reference to the different applications of MPI are given in Table 5.13.

ASTM Reference Standards

Reference photographs and standards are listed in Table 5.14 prepared by ASTM. Magnetic particle inspection of a gear wheel is compared with liquid penetrant inspection in Chapter 1, Fig. 1.1. Indication terms and magnetic property definitions are given in Appendix 5.1. Magnetic particle inspection specifications are listed in Appendix 5.2. A glossary of some terms used in MPI is given in Appendix 5.3.

Appendix 5.1: Indication Terms and Magnetic Property Definitions

Indication Terms (Ref 275)

defect. Any discontinuity that interferes with the usefulness or service of a part.

discontinuity. Any interruption in the physical configuration or composition of a part. It may or may not be a defect.

evaluation. Determination of whether an indication will be detrimental to the service of a part.

false indications. Particle patterns are held by gravity or surface roughness, not by magnetic attraction.

indication. An accumulation of magnetic particles that forms on the surface of the part during the inspection.

interpretation. Determination of the probable cause of an indication.

nonrelevant indications. Caused by flux leakage and has no relation to a discontinuity that is considered to be a defect. Examples are magnetic writing, changes in cross section of the part, or a heat-affected zone in welding.

relevant indications. Particle accumulations that form at discontinuities that would not be present in a perfect part. They may or may not be considered defects.

Magnetic Terms (Ref 280)

coercive force. The magnetic force H required to bring the magnetic flux density B

Table 5.6 Discontinuities produced during the solidification of molten metal (surface or subsurface)

	Probable cause
Porosity blow holes	Entrapped gases during solidification. These roll out into seams and laminations.
Inclusions	Nonmetallic contaminants introduced during castings. These roll out into stringers.
Hot tears	Restraint from core or mold during cooling. These can appear as a ragged line with many branches, sometimes not noted until after machining.
Pipe	The absence of molten metal during the final solidification, results usually in an elongated void at the center line of the finished product. MPI will not detect this defect.
Cold shuts	The meeting of two streams of liquid metal that do not fuse together can give rise to visible, sometimes sharp but mostly fuzzy MPI indications.
Segregation	Localized differences in material composition. MPI indications are mostly sharp, sometimes fuzzy.

Source: Ref 275

Table 5.7 Discontinuities produced by rolling or forging (surface or subsurface)

	Probable cause
Seam (surface)	Elongation of unfused surface discontinuity—sharp MPI indication.
Laminations and stringers (surface or subsurface)	Elongation and compression of a solidification discontinuity—sometimes at the surface after machining, stringers are fuzzy, laminations sharp MPI indications.
Cupping (subsurface)	Internal stresses during cold drawing, sometimes due to uneven temperature between the surface and center of stock during rolling. MPI will not detect.
Cooling cracks (surface)	Uneven cooling of drawn products can give rise to an apparent seam. Sharp MPI indications.
Lap (forged or rolled) (surface)	Material folded over and compressed. Crescent shape from poorly aligned forging dyes—straight from folding of excess metal during rolling. Usually sharp MPI indication.
Flash line tear (surface)	Improper trimming of flash. Sharp MPI indication.
Forging burst (surface or subsurface)	Forming processes at too high or too low temperatures. Sharp or fuzzy MPI indications.
Forging cracks (surface)	Forging using excessive pressure. Sharp MPI indications.
Flakes (subsurface)	Internal ruptures due to rapid cooling caused by dissolved hydrogen gas. Sometimes seen after machining.

Source: Ref 275

Table 5.8   Discontinuities produced during casting (surface or subsurface)

	Probable cause
Porosity, blow holes (surface or subsurface)	Trapped gases. MPI will not detect.
Shrink cracks (surface or subsurface)	Uneven cooling, at section changes. Mostly sharp MPI indication, seldom subsurface.
Inclusion (surface or subsurface)	Sand from mold, trapped by hot metal. MPI will not detect.
Cold shut (surface)	Meeting of two streams of metal that do not fuse together. Usually visible. Can be sharp but usually fuzzy MPI indication.
Misrun (surface or subsurface)	Incomplete filling of mold. Usually MPI visible.
Hot tears (surface)	Uneven rate of cooling between inside and outside of casting. Sharp MPI indications.

Source: Ref 275

Table 5.9   Typical discontinuities produced during welding

	Probable cause
Porosity (surface or subsurface)	Gas is entrapped during solidification. MPI does not detect.
Lack of penetration (subsurface)	Inadequate penetration of weld joint root by the weld metal. Wide fuzzy MPI indication.
Lack of fusion (subsurface)	Filler metal fails to coalesce with base metal. Usually fuzzy MPI indication.
Inclusions (subsurface)	Nonfiller metal contaminants (slag tungsten, oxide) is entrapped in weld pool during solidification. MPI does not detect.
Undercut/overlap (surface)	At toe of weld caused by over- or undersized weld pool (amperage, travel speed, electrode size) usually visible sometimes as sharp MPI indication at toe of weld.
Shrink cracks (surface or subsurface)	Cold or hot cracking, can give rise to long, straight, continuous cracks, usually at the surface. Sharp or fuzzy MPI indication, depends on depth.
Crater cracks (surface)	Usually star-shaped at point of electrode withdrawal. Sharp MPI indication.

Source: Ref 275

Table 5.10   Discontinuities produced during finishing processes on surface

	Probable cause
Machining tears.	Improper machining, dull tools, localized heat. Sharp MPI indications.
Heat-treating cracks	Uneven heating or cooling causes stresses above tensile strength sharp MPI indication.
Grinding cracks, checks	Localized overheating during grinding network of shallow cracks that normally stop at part edges. Sharp MPI indications.
Quench cracks	Sudden cooling, network of deep cracks that break over the edges of parts. Sharp MPI indications.
Plating/pickling	Residual stresses being relieved. Sharp MPI indication.
Straightening cracks	Stress due to bending—sharp MPI indications.
Embrittlement cracks	Absorption of hydrogen. Associated with plating. Sharp or fuzzy MPI indications.

Source: Ref 275

Table 5.11   Discontinuities produced by fatigue, corrosion, stresses with plastic deformation

	Probable cause
Fatigue cracks (surface)	Cyclically applied stress below the ultimate tensile strength.
Stress-corrrosion cracks (surface)	Static tensile load plus corrosive environment.
Hydrogen cracking (surface or subsurface)	Tensile or residual stress plus hydrogen-enriched environment.
Creep cracks (surface or subsurface)	Stresses below yield plus elevated temperature.
Copper penetration (surface)	Heat plus stresses below yield combined with low melting metal causes grain boundary penetration.
Overstress cracks (surface)	Accidents, one-time excessive stress.

Source: Ref 275

Table 5.12   Magnetization methods and magnetic particle inspection

Application	Advantages	Limitations
Coils		
Medium-size parts whose length predominates, such as a crankshaft or camshaft	Longitudinally magnetized to locate transverse discontinuities	Part should be centered in coil to maximize length effectively magnetized during a given shot
Large castings, forgings, or shafts	Longitudinal field easily attained by wrapping with a flexible cable	Multiple processing may be required
Miscellaneous small parts	Easy and fast, especially where residual methods are applicable. Noncontact with part. Relatively complex parts can usually be processed with same ease as simple cross section	Length-to-diameter (L/D) ratio is important in determining adequacy of ampere-turns; effective ratio can be altered by utilizing pieces of similar cross-sectional area. Sensitivity diminishes at ends of part because of general leakage field pattern. Quick break of current is desirable to minimize end effect on short parts with low L/D ratios
Yokes		
Large surface areas for surface discontinuities	No electrical contact. Highly portable. Can locate discontinuities in any direction	Time consuming
Miscellaneous parts requiring inspection of localized areas	No electrical contact. Good sensitivity to surface discontinuities. Highly portable. Wet or dry method can be used.	Yoke must be properly positioned relative to orientation of discontinuity. Relatively good contact must be established between part and poles of yoke
Central conductors		
Miscellaneous short parts having holes through which a conductor can be threaded, such as bearing rings, hollow cylinders, gears, large nuts, large clevises, and pipe couplings	No electrical contact, so that possibility of burning is eliminated. Circumferentially directed magnetic field is generated in all surfaces surrounding the conductor	Size of conductor must be ample to carry required current
Long tubular parts such as pipe, tubing, hollow shafts. Large valve bodies and similar parts	No electrical contact. Both inside and outside surfaces can be inspected. Entire length of part is circularly magnetized	Sensitivity of outer surface to indications may be somewhat diminished relative to inner surface for large-diameter and thick-wall parts
Direct contact, head shot		
Solid, relatively small parts (cast, forged, or machined) that can be inspected on a horizontal wet-method unit	Fast, easy process. Complete circular field surrounds entire current path. Good sensitivity to surface and near-surface discontinuities	Possibility of burning part exits if proper contact conditions are not met
Direct contact, clamps, and cables		
Large castings and forgings	Large surface areas can be inspected in a relatively short time	Large current requirements (8000-20,000 A)
Long tubular parts such as tubing, pipe, and hollow shafts	Entire length can be circularly magnetized by contacting end-to-end	Effective field is limited to outer surface so process cannot be used to inspect inner surface
Prod contacts		
Welds, for cracks, inclusions, open roots, or inadequate joint penetration	Circular field can be selectively directed to weld area by prod placement. In conjunction with half-wave current and dry powder, provides excellent sensitivity to subsurface and surface discontinuities	Only small area can be inspected at one time. Arc burn can result from poor contact
Large castings or forgings	Entire surface area can be inspected in small increments using nominal current values	Coverage of large surface areas requires a multiplicity of shots, which can be very time consuming. Arc burn can result from poor contact
Induced current		
Ring-shaped parts, for circumferential discontinuities	No electrical contact. All surfaces of part are subjected to toroidal magnetic field. 100% coverage is obtained in a single magnetization *(continued)*	Laminated core is required through ring to enhance magnetic path

Nondestructive Testing

Table 5.12 Magnetization methods and magnetic particle inspection *(continued)*

Application	Advantages	Limitations
Induced current		
Balls	No electrical contact. Permits 100% coverage for indications of discontinuities in any direction by reorientation of ball	For small-diameter balls, use is limited to residual method of magnetization
Disks and gears	No electrical contact. Good sensitivity at or near periphery or rim. Sensitivity in various areas can be varied by selection of core or pole piece	Type of magnetizing current must be compatible with magnetic hardness or softness of metal inspected

Note: This table is an abbreviated, modified version from Ref 129.

of a material to zero; see P$_{III}$ in Fig. 5.12 The magnetic field required to demagnetize a solid.

diamagnetic materials. Magnetic permeability is less than 1. Metals that repel magnetic fields; e.g., copper, brass, bismuth.

ferromagnetic magnetization. $J = \mu_oM \cong B$, when $M \gg H$.

ferromagnetic materials. Magnetic permeability μ is very much greater than 1. Metals that have a high attraction to permanent magnets; e.g., iron, steel, cobalt, nickel. MPI is carried out only on ferromagnetic materials with large values of μ.

magnetic induction or magnetic flux density (B). In units of Tesla (SI) or gauss (cgs emu). The strength of the magnetic field in the part. The ratio $B/H = \mu$ is the magnetic permeability of the material; for ferromagnetic materials such as iron μ is a function of H, and is a maximum where B increases most rapidly with H, that is at the knee of the magnetization hysteresis curve.

magnetic lines of force. This is used to describe a magnetic field. A magnetic line of force is such that its direction at every point is the same as the direction of the force that would act on a small magnetic pole placed at the point. A magnetic line of force is defined as starting from a north pole and ending on a south pole. Magnetic lines of force are always perpendicular to the direction of electric current flow.

magnetic permeability (μ_p). The ease with which a material can be magnetized, that is, the ability of a material to conduct magnetic lines of force. The permeability of free space is μ_o where $\mu_p = \mu_R\mu_o$ and μ_R is the relative permeability of the material.

magnetic susceptibility (χ_m). The ratio of *magnetization M* in a solid produced by a magnetizing force H.

magnetization (M). Magnetic moment per unit volume of solid. $B = \mu_o(H + M)$ in SI units or $B = H + 4\pi M$ in (cgs) units.

magnetizing force (H). The strength of the magnetic field in air (or vacuum): in units of amps/meter (SI) or oersted (cgs emu).

paramagnetic materials. Magnetic permeability μ is slightly greater than 1. Metals that have only a very slight attraction to magnetism, usually insufficient to attract a permanent magnet; e.g., aluminum, titanium.

permeability of free space μ_o. see *magnetic permeability.*

reluctance. The magnetic resistance or opposition of a material to conduct magnetic lines of force. This is the inverse of permeability.

retentivity (residual magnetism). The ability of a material to retain magnetism following the application of a magnetic field; see P$_{II}$ on Fig. 5.12.

Appendix 5.2: Magnetic Particle Inspection Specifications

American Society for Testing and Materials (ASTM) 1916 Race Street, Philadelphia, PA 19103, USA

A 275 Magnetic Particle Examination of Steel Forgings

A 456 Magnetic Particle Inspection of Large Crankshaft Forgings

E 125 Standard Reference Photographs for Magnetic Particle Indications on Ferrous Castings

E 1316G Magnetic Particle Inspection

E 1444 Magnetic Particle Examination

E709 Magnetic Particle Examination

American Society of Mechanical Engineers (Boiler and Pressure Vessel Code) ASME United Engineering Center, 345 East 47th Street, New York, NY 10017, USA

Sec IPower Boilers

Sec II Material Specifications

Sec III Nuclear Power Plant Components

Sec V Nondestructive Examination

Sec VIII (Div 1) Unfired Pressure Vessels

Sec VIII (Div 2) Alternative Rules for Pressure Vessels

Table 5.13   General references to magnetic particle inspection

	Ref
Weldments	51, 129, 230
Crater cracks	
Lack of fusion	
Lack of penetration	
Hard heat affected zone	
Castings	51, 129, 230
Cold shuts	
Hot tears	
Shrinkage cracks	
Forgings	51, 129, 230
Flakes	
Forging bursts	
Forging laps	
Flash line tears	
Formation discontinuities	129, 230
Ingot mold pipe, nonmetallic inclusions	
Blow holes	
Billets	129, 230
Hollow cylinders	
Chain links	
Primary processing—seams	129, 269
Laminations, cuppings, cooling cracks	
Automobile equipment	129
Manufacturing and fabrication	230, 269
Machining tears	
Heat-treatment cracks	
Straightening and grinding cracks	
Plating, pickling, and etching cracks	
Service discontinuities	269
Overstress cracking	
Fatigue cracking, corrosion	

Definitions of these terms will be found in Ref 70.

Sec XI Rules for In-Service Inspection of Nuclear Power Plant Components

American Welding Society 550 NW Lejeune Road, Miami, FL 33126, USA

D1.1 Structural Welding Code

D14.6 Welding of Rotating Elements of Equipment

Society of Automotive Engineers (SAE) 400 Commonwealth Drive, Warrendale, PA 15096, USA

AMS 2300 Premium Aircraft-Quality Steel Cleanliness, Magnetic Particle Inspection Procedure (also MAM 2300)

AMS 2301 Aircraft Quality Steel Cleanliness, Magnetic Particle Inspection Procedure

AMS 2303 Aircraft Quality Steel Cleanliness, Martensitic Corrosion Resistant Steels, Magnetic Particle Inspection Procedure

AMS 2641 Magnetic Particle Inspection

AMS 3040 Magnetic Particle Inspection, Material Dry Method

AMS 3041 Magnetic Particles, Wet Method, Oil Vehicle

AMS 3042 Magnetic Particles, Wet Method, Dry Powder

AMS 3043 Magnetic Particles, Wet Method, Oil Vehicle, Aerosol Canned

AMS 3044 Magnetic Particles, Fluorescent Wet Method, Dry Powder

AMS 3045 Magnetic Particles, Fluorescent Wet Method, Oil Vehicle, Ready to Use

AMS 3046 Magnetic Particles, Fluorescent Wet Method, Oil Vehicle, Aerosol Packaged

United States Department of Defense DOD Standardization Documents Order Desk Bldg. 40, 700 Robbins Avenue, Philadelphia, PA 19111, USA

MIL-STD-271 Nondestructive Testing Requirements for Metals (ACN-1)

MIL-STD-1949B Magnetic Particle Inspection

MIL-I-6867 Magnetic Inspection Units

MIL-M-23527 Magnetic Particle Inspection Unit, Lightweight

Table 5.14   ASTM standards for magnetic particle inspection

Standard No.	Description
E 125	Magnetic Particle Indications on Ferrous Castings (1985)
E 709	Magnetic Particle Examination (1991)
E 1444	Magnetic Particle Examination (1991)

MIL-I-83387 Magnetic Rubber, Inspection Process

DoD-F-87935 Fluid, Magnetic Particle Inspection, Suspension (Metric)

MIL-M-47230 Magnetic Particle Inspection Soundness Requirements for Materials, Parts and Weldments

British Standards Institution Linford Wood, Milton Keyes, U.K.

BS 6072 Method for Magnetic Particle Flaw Detection

Deutsche Gesellschaft für Zerstorungsfreie Prufung Unter den Eichen 87, 1000, W. Berlin 45, Germany

Guidelines for Magnetic Particle Flaw Detection

Japanese Institute of Standards University of Tokyo, Komaba, Neguro-Ku 461, Tokyo 153, Japan

JIS G 0565 Methods of Magnetic Particle Testing of Ferromagnetic Materials and Classification of Magnetic Particle Indications

Appendix 5.3: Glossary of Some Terms Used in Magnetic Particle Inspection (Ref 61, 190) (see also Appendix 5.1)

black light. Invisible electromagnetic radiation in the near ultraviolet range of wavelength (330 to 390 nm).

central conductor magnetization. The inside surface of a tubular specimen can be examined by a magnetic field produced by a conductor threaded through the specimen.

circular magnetization. The electric current is passed through the specimen to produce a circular magnetic field.

coercive force. The magnetizing field, H, at which the magnetic flux density, B, is equal to zero; the magnetic field required to demagnetize a material (Appendix 5.1).

coil magnetization. A specimen placed in a current-carrying coil becomes magnetized along the coil axis.

ferromagnetic magnetization. see Appendix 5.1.

fill factor. The ratio of the cross-sectional area of the specimen to the cross-sectional area of the encircling coil.

fluorescent magnetic particle inspection. Employing finely divided fluorescent particles that fluoresce in *black light*.

gaussmeter. An instrument for measuring magnetic flux density B.

Hall effect. A potential difference is established between opposite surfaces of a conductor under the influence of a transverse magnetic field (Fig. 5.21).

hysteresis. The lag of the magnetization of a ferromagnetic behind the applied magnetic field.

inductance. The property of a circuit where an electromotive force develops to oppose changes of current flow.

lines of force. A conceptual representation of magnetic flux; the line pattern produced when iron filings are sprinkled on paper in a magnetic field (see *magnetic lines of force*, Appendix 5.1).

magnetic field strength, H. See Appendix 5.1.

magnetic flux density, B. See Appendix 5.1.

magnetic leakage field (flux leakage field). The magnetic field that leaves or enters the surface of a part at a discontinuity.

magnetic permeability. The ease with which a magnetic field can be induced in a solid; see Appendix 5.1.

prodmagnetization. The magnetization current can be passed through a small region of the specimen using two prods (Fig. 5.19).

residual field inspection. Application of magnetic particles to the part after the magnetization current has been removed.

residual magnetic field (retentivity, remanence). The field that remains in a magnetized material after the magnetizing force has been removed (Fig. 5.12).

skin effect. See Appendix 6.1; the tendency of alternating currents to flow near to the surface of a material.

solenoid. A long cylindrical coil of wire through which an electric current is passed to produce a magnetic field.

yoke magnetization. A magnetizing field is produced by applying the ends of a U-shaped soft-iron armature to the specimen (Fig. 5.20).

Chapter VI

Eddy Current Testing*

Part I General Discussion

Introduction and Historical Background

A varying electric current flowing in a coil gives rise to a varying magnetic field. A nearby conductor resists the effect of the varying magnetic field, and this manifests itself by an eddy current flowing in a closed loop in the surface layer of the conductor so as to oppose the change causing a back electromotive force (emf) in the coil (Fig. 6.1a and b). Cracks and other surface conditions modify the eddy currents generated in the conductor so that the back emf is altered (see Fig. 6.1c).

Eddy current testing (ET) is only possible because of the availability of very sensitive electronic devices that are able to detect the very small changes of the magnetic fields involved. The original discoveries, observations, and explanations of eddy currents are listed in Table 6.1. Jean-Bernard Léon Foucault is generally credited with the first clear demonstration of eddy currents, by showing that electrical currents

are set up in a copper disk moving in a nonuniform magnetic field. Hughes is considered to have been the pioneer in using eddy currents to inspect metals. Farrow was involved in extensive testing of tubes on an industrial scale. More recently, impedance plane diagrams have been introduced, known as Förster diagrams, and are very useful in analyzing the eddy current effects.

Eddy currents are also known as Foucault currents or induced currents and can only exist in conducting materials.

Outline of Eddy Current Testing (ET)

Inspection System. An eddy current inspection system consists of:

- Source of varying magnetic field, for example, a coil carrying an alternating current of frequencies ranging from well below 1 kHz to above 10 MHz (a pulsed source may also be used)
- Sensor to detect minute changes in the magnetic field (~0.01%), for example, inspection coil or Hall gaussmeter
- Electronic circuitry to aid the interpretation of the magnetic field change

The inspection system can be manual or fully automated. A basic test system is shown in Fig.

*This chapter has been reviewed and revised by Robert Levy, Intercontrôle (13, rue du Capricorne, 94583 Rungis Cedex, France) and chairman of the commission for NDT methods and equipment of Cofrend (1, rue Gaston Boissier, 75724 Paris Cedex 15, France).

Nondestructive Testing

6.1(c) for a flat specimen. A pancake-type inspection coil is used in this case; see Fig. 6.2(a). Tubes and cylinders are examined using solenoid coils as shown in Fig. 6.2(b) and (c). The solenoid can be either on the outside or the inside of the tube.

An ac source is applied to an inspection excitation coil so that the magnetic lines of force penetrate into the specimen conducting surface, providing a good magnetic coupling. Eddy currents circulate in the specimen surface and are modified by the presence of discontinuities (e.g., surface cracks). Eddy currents flow parallel to the plane of the windings of the coil. The detector instrument will note changes in the time variation of voltage and current in the inspection coil (Fig. 6.1c). However, many other factors, such as shape and temperature, cause changes so that the measurement must be carried out under strictly controlled conditions.

Depth of Penetration. Alternating currents and eddy currents travel along the surface of conductors and penetrate very little into the specimen (Ref 270). The decrease with depth of the electric current is illustrated in Fig. 6.3(a), where the drop-off with depth is exponential. The standard depth of penetration δ is where the electric current has decreased by the inverse of the exponential factor, that is a factor of $(1/e) = 36.8\%$. The variation of δ with ac frequency, ν (in Hz) for a plane wave on a planar surface, is given by (Ref 278):

$$\delta = \frac{1}{\sqrt{\pi\nu\,\mu_o\,\mu_r\,\sigma}} \qquad \text{(Eq 6.1)}$$

where σ (in Siemens/m) is the conductivity, μ_o (in Henry/m) is the permeability of free space, and μ_r is the relative magnetic permeability. For nonferromagnetic materials, μ_r is equal to 1. Values of δ versus frequency are plotted in Fig. 6.3(b), and some values are given in Table 6.2 for several conductors.

Oscilloscope Displays. The measurements of the inspection coil are essentially of the alterna-

Table 6.1 Historical outline of eddy current testing

Individual	Year	Description	Ref
H.C. Oersted	1819	Change of electric current affected a magnet	281
W. Sturgeon	1823	Copper wire around a horseshoe produced an electromagnet	
Gamby	1824	Oscillations of suspended bar magnet damped by presence of metal plate	
J.B. Foucault	1830	Demonstrated existence of eddy currents	211
M. Faraday	1832	Law of electromagnetics induction	282
D.E. Hughes	1879	Electric pulses from a microphone coil to induce eddy currents in metals for NDT	283
F. Krantz	1920	Wall thickness measurements	
C. Farrow	1925	Eddy current inspection testing of steel tubes on an industrial scale	
Reutlingen Institute, Germany	1948	Development of eddy current instrumentation	
H.G. Doll	1949	Eddy current in geology	
F. Förster	1954	Impedance plane diagram. Used model of mercury conductor with plastic strips as discontinuities	69, 284

Table 6.2 Depth of penetration values for several conductors

Metal	Electrical conductivity room temperature (σ), 10^6 S/m	Standard depth of penetration $\delta \times 10^{-3}$ in.		
		1 kHz	0.1 MHz	10 MHz
Copper (International Annealed Copper Standard IACS)	58	82	8.2	0.82
Aluminum	35.4	105	10.5	1.05
Zirconium	2.5	396	39.6	3.96
Titanium	1.8	467	46.7	4.67
Stainless steel (304)	1.4	531	53.1	5.31
Graphite	0.1	1980	198	19.8

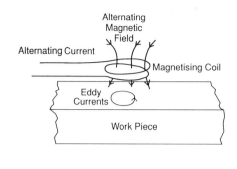

(a)

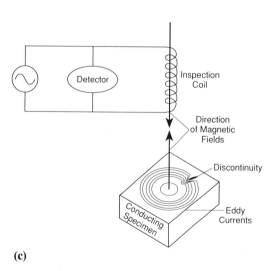

(b)

tive current i, which lags behind the applied alternating voltage V by a phase angle ϕ. This measures the impedance Z of the circuit, which is the opposition that the circuit presents to the flow of an alternating current. The impedance Z consists of two components, R, the resistive component, and X_L reactance, the reactive component. This is discussed further in the section "Physical Principles" in this chapter. These two components are measured by the inspection coil circuit and can be fed separately to the X-plates and Y-plates of an oscilloscope, providing a two-dimensional representation of the inspection coil measurements of R and X_L. This is shown in Fig. 6.4. The ordinate on the oscilloscope corresponds to the reactive component X_L, and the abscissa to

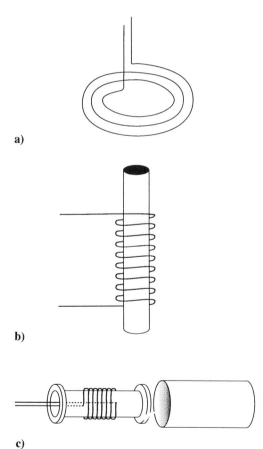

(c)

Fig. 6.1 Schematic of eddy current testing of a specimen. (a) and (b) Instantaneous view of an alternating current in the testing coil producing a magnetic field as shown. The specimen can be flat or cylindrical. (c) The magnetic field from the inspection coil is opposed by an induced magnetic field from the eddy current. A surface crack modifies the eddy current and hence also the induced magnetic field.

Fig. 6.2 The inspection coils have different configurations depending on the specimen shape. (a) A flat surface is normally examined by a flat pancake-type coil. (b) A cylindrical specimen is examined using an encircling coil. (c) The interior of a tube can be examined by an inside, inserted, or bobbin coil.

the resistive component R. Z is given by the vector, known as "phasor," shown in Fig. 6.5(a).

In the absence of a conductor, that is to say in air, the impedance will be given by the phasor Z_0; see Fig. 6.5(b). In the presence of a conducting material, the eddy currents change the impedance of the inspection coil circuit from Z_0 to Z_1 (Fig. 6.5b) the magnitude of $|Z|$ and the phase angle ϕ are now given by the phasor Z_1 on the oscilloscope.

Selection of Inspection Frequencies. The frequencies of ac used in eddy current testing range from a few kHz to more than 5 MHz. At the lower frequencies, the penetration δ is relatively high (see Eq 6.1), but the sensitivity to the detection of discontinuities is relatively low; the reverse is true at higher frequencies. In the case of ferromagnetic materials where δ is reduced, the use of lower frequencies is inevitable.

Selecting a frequency to inspect a specimen depends on the electrical conductivity, the magnetic permeability and the dimensions of the specimen. In the case of a cylindrical shape of radius r, frequency selection depends on the characteristic frequency v_c (also known as limit frequency) which is derived by considering the strength of the magnetic field penetrating into the specimen. It has been shown by Förster (Ref 284) that

$$v_c = \frac{1}{2\pi \, r^2 \, \mu_r \, \mu_0 \, \sigma} \qquad \text{(Eq 6.1a)}$$

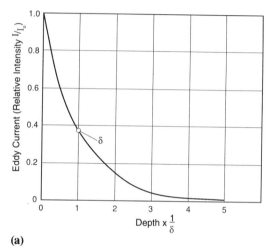

(a)

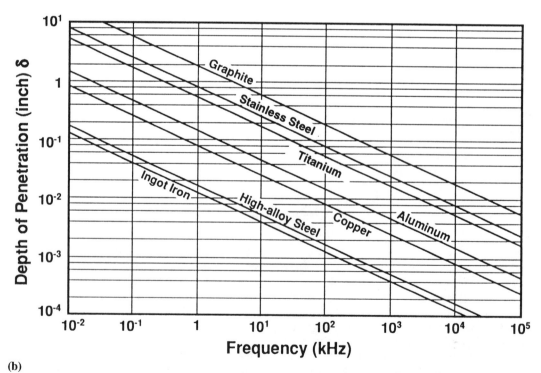

(b)

Fig. 6.3 Skin effect; depth of penetration. (a) The significance of the standard depth of penetration δ of an alternating current of frequency v (in Hz) is shown where δ corresponds to the depth at which the current density decreases by $1/e$ (37%) (see Eq 6.1). (b) Values of δ_{inches} versus frequency v (in Hz) for several conductors (Ref 66). See also Table 6.2.

Useful charts have been prepared that aid in the selection of v/v_c for particular types of examination in terms of specimen conductivity, magnetic permeability, and size (e.g., Ref 129, p 174).

An estimate of frequency for the examination of thin-walled tubes can be obtained by considering specimen thickness Δ to be equal to the penetration δ. Then Eq 6.1 can be written

$$v = \frac{1}{\Delta^2 \pi \, \mu_o \, \mu_r \, \sigma}$$

As an example, consider stainless steel tubing of wall thickness $\Delta = 4 \times 10^{-3}$ m, $\sigma = 1.5 \times 10^6$ S/m, $\mu_o \, \mu_r = 1.3 \times 10^{-6}$ H/m. Then

$$v = \frac{1}{(10^{-3})^2 \, \pi \, 1.5 \times 10^6 \times 1.3 \times 10^{-6}} = 0.16 \text{ MHz}$$

Interpretation of Oscilloscope Displays

The oscilloscope display of the impedance Z is affected by factors depending on the specimen itself and the experimental condition. Specimen-dependent factors include:

- Electrical conductivity of the specimen.
- Magnetic permeability: Unmagnetized ferromagnetic materials become magnetized resulting in a large change in impedance,

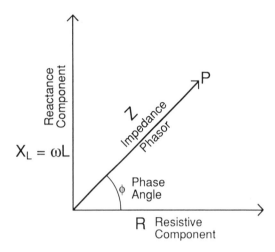

(a)

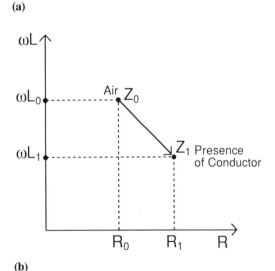

(b)

Fig. 6.5 Impedance-plane diagram (phasor diagram). The signal of the resistance R component of the impedance Z is applied to the horizontal X plates, and the reactance ωL component of Z to the vertical Y plates; see part (a). The eddy current changes the impedance Z and the phase angle ϕ of the inspection coil circuit so that in the presence of a conductor, the point on the oscilloscope is displaced from Z_0 (for air only) to Z_1; see part (b). The vector Z is known as the phasor. See also Eq 6.10 and 6.11 in the section "Physical Principles" in this chapter.

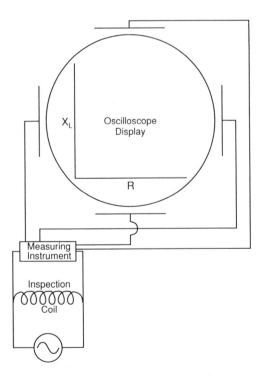

Fig. 6.4 The inspection (induction) coil circuit measures the resistive component R, which is fed to the X-plates and the reactance component X_L, which is fed to the Y-plates of the oscilloscope.

which can mask other effects. This problem can be overcome by magnetizing the specimen as close as possible to saturation using a dc electromagnet.

- Specimen thickness: A thick specimen may not be fully traversed by the magnetic field, which limits inspection by ET to specimens ~1 cm thick. A thin specimen may be traversed by the magnetic field, and specimen thickness should be less than ~3 δ.

Factors dependent on experimental conditions are:

- ac frequency (ν) in Hz
- Electromagnetic coupling between the coil and the specimen. Lift-off is the separation between a flat specimen and the inspection coil. Even a small change in lift-off has a pronounced effect. In the case of a cylindrical specimen inside an encircling coil, "lift-off" is replaced by the fill-factor, which is the fraction of the inspection coil area filled by the specimen. Alternatively, the coil can be inside the cylindrical specimen, an ID or bobbin probe (see Fig. 6.2c), when the internal wall of the cylinder is examined. The fill-factor for ID coils depends on the extent to which the inside area of the part is filled. In all cases, it is important to keep the fill-factor constant and as high as possible. The inspection coil must be kept centered to avoid favoring one wall of the tube.
- Inspection coil size and number of turns. The magnetic field from the inspection coil is approximately the same size as the coil. Narrow coils are more responsive to the small changes due to discontinuities. To detect discontinuities or changes in properties of the specimen, it is important to provide a strong magnetic field strength to induce adequate eddy currents. This can be done by increasing the number of turns in the coil and by increasing the electric current.

The different effects on the oscilloscope display are shown in Fig. 6.6 to 6.8. It is assumed that all other factors are held constant; that is, using the same inspection circuit, coil, frequency, lift-off, and a thick specimen, not traversed by the magnetic field. One factor only is allowed to vary.

The effect of conductivity increase in the specimen is shown in Fig. 6.6. A very good conductor such as copper (IACS, International An-

nealed Copper Standard) will have an impedance as shown at C; a poorer conductor (bronze) will have an impedance at B, as shown. The absence of a conductor (air) gives rise to an impedance, as shown at A. The curve ABC is the plot of impedance with increasing specimen conductivity.

The curve CDA in Fig. 6.6 shows the effect on copper (100% IACS) of increasing the lift-off separation between a flat specimen surface and inspection coil (all other factors constant). Other specimens of lower electrical conductivity will give rise to the family of curves as shown; for example, BEA is the lift-off curve for bronze. For cylindrical specimens in solenoids, the lift-off factor is replaced by a fill-factor, which is given by the fraction of the area of the solenoid cross section filled by the cylinder. If r_1 and r_2 are the radii of cylindrical specimen and solenoid, respectively, the fill-factor is given by $(r_1/r_2)^2$. Different fill-factors change the impedance in a characteristic way, similar to different lift-off; curves analogous to Fig. 6.6 are available in references such as Ref 69 and 211.

The effect of frequency is shown in Fig. 6.7. The resistive and inductive components of Z are changed, as well as the penetration of the magnetic field into the specimen. As the frequency is increased, the impedance position moves as shown for bronze at 20 kHz, 100 kHz, and 1 MHz (all other factors constant). It can be seen that bronze is closer to copper at 1 MHz. As a consequence, it is easier to distinguish between good conductors such as aluminum and copper at frequencies lower than 20 kHz. On the other

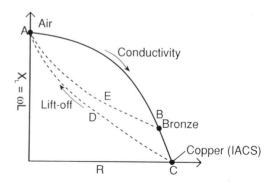

Fig. 6.6 Conductivity curve and the family of lift-off curves on the impedance-plane diagram (IACS, International Annealed Copper Standard). The effects of lift-off on copper (IACS) and on bronze specimens are shown.

hand, at 20 kHz, graphite and air are closer together and thus less easily distinguished.

The thickness of the specimen also has an effect on the impedance vector Z, as illustrated in Fig. 6.8. The outer curve for brass was determined by measuring Z for increasing numbers of sheets of brass, until increasing the brass thickness had no further effect on the impedance Z. It can be seen that the effect of thickness becomes smaller and smaller on increasing the numbers of brass sheets. Ultimately, the whole of the coil's magnetic field is trapped within the specimen; this corresponds to a "thick" specimen of thickness greater than approximately 3δ.

Figure 6.8 is a general impedance plane diagram showing the effects of conductivity, thickness, and lift-off on the impedance vector.

Testing Circuits

Impedance Testing. A simple method, impedance testing, measures the magnitude of the impedance with no information about the phase change. This can be carried out using a simple circuit as shown in Fig. 6.9. The balanced bridge method is set up so that there is no signal through the meter when the inspection coil is against the surface of a specimen of good condition. When the inspection coil is in the presence of a discon-

tinuity, the bridge is now unbalanced resulting in a potential difference across the meter.

Impedance Plane (Phasor) Method. The signal from the inspection (excitation) coil can be resolved into resistive and reactance components and these are displayed separately on the oscilloscope. This is known as the impedance plane (phasor method). The oscilloscope screen becomes the impedance plane.

A point is seen on the oscilloscope screen corresponding to the phasor position Z_0 when no conductor is present, as shown in Fig. 6.5(b). In the presence of a conductor, the point moves to Z_1. This displacement from Z_0 to Z_1 can only be interpreted if one factor is allowed to vary, and all other factors are held constant. Furthermore, the change from Z_0 to Z_1 must be calibrated using specimens of known behavior. Thus, in Fig. 6.10(a), Z_0 is observed with inspection coil in air, then placed on different metals to obtain the new positions of the spot on the oscilloscope screen,

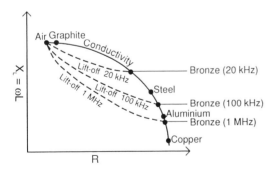

Fig. 6.7 The impedance vectors for several conductors are shown at 20 kHz. The variations of impedance with frequency is shown for bronze with values given at 100 kHz and 1 MHz. At higher frequencies, high conductivity metals have impedances very close together. At lower frequencies, the low conductivity materials have impedances very close to one another. The impedance vector for (IACS) copper does not vary appreciably with frequency. Lift-off curves are shown for bronze at different frequencies. Note: At higher frequencies, the lift-off curves are approximately perpendicular to the conductivity curve.

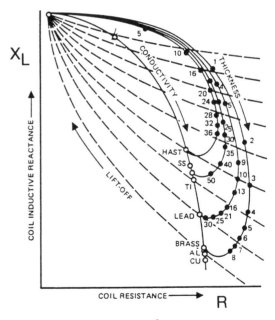

Fig. 6.8 Sheets of metal (10^{-3} in. thick) are stacked until there is no further change in impedance. The specimen is now thick enough to entrap the whole of the applied magnetic field (all other factors held constant). The conductivity curve is for thick specimens. Data shown were taken at 120 kHz. (Ref 66). HAST, superalloy; SS, stainless steel; Ti, titanium. Reprinted with permission of The American Society for Nondestructive Testing.

179

corresponding to the conductivities of the metals (all other factors constant).

Again, as in Fig. 6.10(b), the impedance vector Z_0 is observed in air, and moves to Z_1 when placed on a good metal surface. When discontinuities are present, the spot for the impedance vector Z will be displaced. Examples of the effects of different defects are given in Fig. 6.10(b), based on calibration with known defects.

A differential method, as shown in Fig. 6.11(b), is often used where the coils I and II are coupled closely together, resulting in a signal whenever one of the two coils moves over a discontinuity. The signals observed on the oscilloscope screen (in impedance plane) are typically as shown in Fig. 6.11. Other variations of this technique are described in Ref 66 and 211.

Practical Tips

1. Induced eddy currents in a conductor produce a magnetic field that opposes the magnetic field produced by the test coil, resulting in a change of impedance. It is this impedance change that is to be detected with a high degree of accuracy by the measuring circuitry.
2. During eddy current testing, attention must be paid to other conducting surfaces. This applies to other parts of the specimen under test, as well as conducting table tops, metal supports, and so forth. Multifrequency ET is suited to cases in which other conducting surfaces are present.
3. The presence of a discontinuity gives rise to a local, brief change in the impedance as the test coil is moved across the surface.
4. It is recommended that equipment be checked frequently against a known standard to allow for "drift" to be detected.
5. Changes in test coil impedance can be caused by change in the specimen of

Properties of the Conductor

- Alloy composition
- Heat treatment, hardness, grain size
- Magnetic permeability

Dimensions

- Specimen thickness, diameter
- Specimen eccentricity
- Distance from other conductors

Surface Condition

- Surface coating
- Corrosion
- Specimen temperature

Discontinuities

- Cracks
- Inclusions
- Dents
- Holes
- Scratches, and so forth

6. Eddy currents always flow parallel to the plane of the windings of the test coil producing them; a discontinuity or crack parallel to the eddy current flow has less effect than one perpendicular to the eddy current flow for nonferrous materials.
7. In planar surfaces, deep-lying discontinuities have less effect than discontinuities near the surface.
8. The eddy current flows in a plane at right angles to the magnetic field of the coil.
9. An alternating current in the inspection coil produces an alternating magnetic field of approximately twice the diameter as the coil.

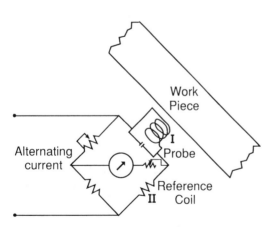

Fig. 6.9 Impedance testing can be carried out using a balanced bridge circuit with a simple meter when the probe inspection coil I is balanced against a reference coil II.

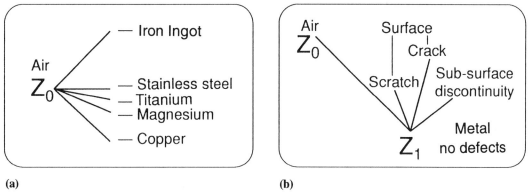

(a)   **(b)**

Fig. 6.10 Vector point signal on oscilloscope screen. The changes in the impedance plane can be recognized if one factor only is varying. (a) Electrical conductivity changes in specimen (all other factors constant). The impedance vector point will change as shown. The inspection coil and system needs to be calibrated using metals of known conductivity. (b) Various defects (all other factors constant). The various changes are recognized by experience and by calibration with known defects.

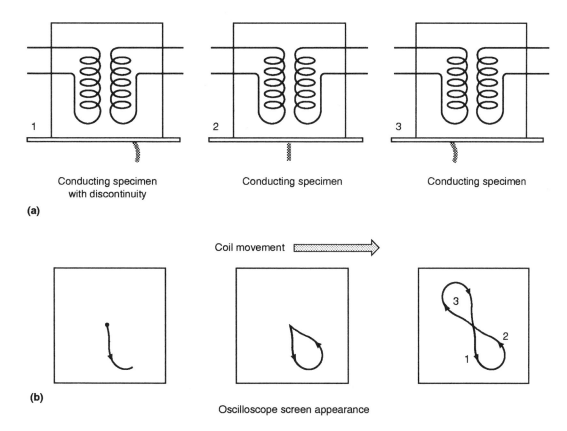

Fig. 6.11 Differential testing (impedance plane). The two inspection coils are linked electrically in opposition. The circuit becomes unbalanced if one coil moves opposite a discontinuity. (a) Three different positions, 1, 2, and 3, of the coils as they are moved alongside the specimen with discontinuity; in this case, from left to right. (b) Corresponding view of the oscilloscope screen indicating the direction and manner of the formation of the Lissajous figure (Ref 286).

10. Penetration (skin effect). The higher the ac frequency, electrical conductivity, and magnetic permeability, the shallower the penetration. This is called the skin effect. The lower the ac frequency, electrical conductivity, and magnetic permeability, the deeper the penetration.
11. In cylinders (and rods), using an encircling coil, eddy currents flow around the circumference of the cylinder.
12. Ferromagnetic materials; saturation magnetization using a dc solenoid is recommended.
13. Advantages of eddy current testing

- Instantaneous results
- Sensitive to a range of physical properties
- Contact between inspection coil and specimen not required
- Equipment small and self-contained
- Can be miniaturized to observe discontinuities as small as 1 mm^3

14. Eddy current test results depend on specimen factors and testing conditions. Specimen factors are:

- Electrical conductivity
- Discontinuities in specimen
- Magnetic permeability (ferromagnetic materials)
- Dimensions and shape of specimen
- Metal condition (alloy, hardness, homogeneity, grain size)

Pertinent testing conditions are:

- Distance between coil and specimen lift-off
- Alternating current frequency, coil size, number of turns

15. Limitations of eddy current testing

- Electrical conductors only
- Depth of penetration restricted, probably only 1 cm from surface
- Complication with ferromagnetic materials
- Interpretation depends on skill of operator
- Sensitive to many parameters

16. Eddy current testing is normally used for examination of (nonmagnetic):

- Tubing cracks for tube diameters up to 1 in., and wall thickness below 0.15 in. (3 mm)

- Inclusions in thin wall tubing and rods of small diameters
- Lack of penetration in nonferrous welded pipe and tubing
- Gas porosity in thin-walled welded pipe and tube
- Mandrel drag of thin-walled pipe and tubing
- Coatings and metal sheet thicknesses

17. Eddy current testing can be used, but is not the normal technique for the examination of:

- Bursts in nonferrous wrought materials less than 0.25 in. thick
- Heat treatment cracks in nonferrous surfaces
- Intergranular corrosion in tubes and pipes depending on the metal involved

18. Eddy current testing is approximately as sensitive as liquid penetrant inspection and magnetic particle inspection to surface cracks and can be carried out almost instantaneously.
19. The impedance change in the inspection coil due to a crack can be as small as 0.01%.
20. Eddy current testing is most convenient when all factors are fixed and controlled with only one factor at a time being allowed to vary when using the impedance method.
21. End effects. Eddy currents are distorted at the ends of specimens. Inspections should not be within the coil diameter from an edge depending on factors such as test coil size and frequency.
22. Similar defects in a wide range of materials will give similar indications on a phasor diagram (Förster diagram) providing the inspection (excitation) frequencies are the same ratio of the characteristic frequency v_c.

Part II Technical Discussion

Physical Principles

An eddy current testing circuit is equivalent to a resistance R and an inductance L; see Fig. 6.12(a).

For dc: $V = iR$ (Ohm's Law)   (Eq 6.2)

For ac: $V = iZ$ (Eq 6.3)

This is derived as follows:

In the case of ac, $I = I_0 \sin \omega t$ (Eq 6.4)

where ν is the frequency in Hz, ω is angular frequency in radians/s so that $\omega = 2\pi\nu$.

Then:

$$V - L\frac{di}{dt} = iR \qquad \text{(Eq 6.5)}$$

because the inductance opposes the applied voltage (see for example Ref 270). The induced voltage V_L

$$V_L = -L\frac{di}{dt} = -L\omega i_0 \cos \omega t = -X_L i_0 \cos \omega t \qquad \text{(Eq 6.6)}$$

where $X_L = \omega L$. *Note*: The current i and the induced voltage V_L are always 90° out of phase; see Fig. 6.12(b). From Eq 6.5 and 6.6:

$$V - X_L i_0 \cos \omega t = i_0 R \sin \omega t \qquad \text{(Eq 6.7)}$$

Therefore:

$$V = i_0 (R \sin \omega t + X_L \cos \omega t) \qquad \text{(Eq 6.8)}$$

If we write $i = V/Z$, where Z is the impedance, then

$$Z = R \sin \omega t + X_L \cos \omega t \qquad \text{(Eq 6.9)}$$

The magnitude of

$$|Z| = \sqrt{R^2 + X_L^2} = \sqrt{R^2 + \omega^2 L^2} \qquad \text{(Eq 6.10)}$$

and the phase angle ϕ between Z and R is

$$\phi = \tan^{-1}\frac{X_L}{R} = \tan^{-1}\frac{\omega L}{R} \qquad \text{(Eq 6.11)}$$

Mutual Inductance. Eddy currents are induced in a conductor by the mutual inductance of a neighboring electric circuit, as shown in Fig. 6.13(a). The inspection electric circuit can be represented by χ_1 and the "object under test" becomes the second equivalent circuit χ_2. There is no source emf in circuit χ_2 except that induced by mutual inductance.

The changes in the resistance ΔR and in the self-inductance ΔL of the test circuit can be measured by a balanced bridge circuit method. Using alternating current theory, ΔR and $\omega\Delta L$ are shown to be related and depend on the mutual

inductance M_1 and resistance of the specimen R_1. It can be shown that (Ref 211, 270):

$$\Delta R = \frac{M_1^2 \omega^2 R_1}{R_1^2 + \omega^2 L_1^2} \qquad \text{(Eq 6.12)}$$

$$\omega\Delta L = \frac{M_1^2 \omega^3 L_1}{R_1^2 + \omega^2 L_1^2} \qquad \text{(Eq 6.13)}$$

$$\Delta R = -\frac{R_1}{\omega L_1} \omega\Delta L \qquad \text{(Eq 6.14)}$$

$$(\Delta R)^2 + \omega\Delta L + \frac{1}{2}\frac{M_1^2\omega^2}{L_1} = \frac{1}{4}\frac{M_1^4\omega^4}{L_1^2} \qquad \text{(Eq 6.15)}$$

The following should be noted:

- The slope of ΔR versus $\omega\Delta L$ depends on R_1. At high resistance, ΔR and $\omega\Delta L$ tend to produce comparable effects. However at low resistance, ΔR and $\omega\Delta L$ effects do not produce changes in the same direction.
- Changing values of M produce a family of circles. It is easier to differentiate changes of ΔR at lower resistance values of R_1, because then the effects of $\omega\Delta L$ and ΔR do not occur in the same direction (Fig. 6.13b). This is discussed further in Ref 211.

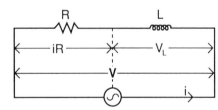

(a)

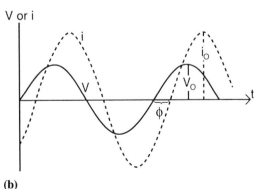

(b)

Fig. 6.12 (a) ac circuit with resistance R and inductance L. (b) Phase relations between V and i in an R-L circuit. V and i are not in phase because $V = i_0 (R \sin \omega t + X_L \cos \omega t)$.

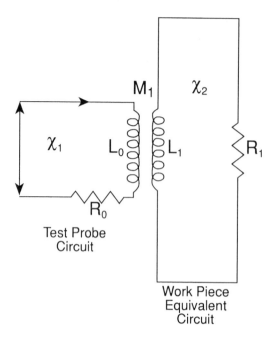

(a)

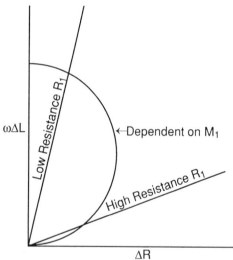

(b)

Fig. 6.13 Mutual inductance and eddy currents. (a) Circuit χ_2 is the equivalent circuit of the object under examination, represented by a resistance R_1 and a self-inductance L_1. This is coupled by a mutual inductance M_1 to the testing probe circuit χ_1, which consists of a self-inductance L_0, resistance R_0, alternating current i and voltage V. There is a phase difference ϕ between the alternating current and voltage and this is modified by the mutual inductance M_1. (b) Two-dimensional plot of ΔR versus $\omega \Delta L$; see Eq 6.12 to 6.15.

Eddy Current Inspection Coils

There are three methods of undertaking eddy current inspection using a

- *Single Coil as a Combined Induction-Receiver.* The change of impedance of the coil (or coils) is determined in the same coil (or coils) used to generate the magnetic field; see Fig. 6.14(a).
- *Separate Induction-Receiver Coils.* The induced magnetic field is measured by a separate coil; see Fig. 6.14(b) and (c). Decreasing the size of the inspection coils is an advantage, and also the coil can be enclosed in a magnetic shield using mu-metal, when the coil is considered focused. Several coil systems can be used; see Ref 65.
- Hall effect device is used to sense the eddy current magnetic field; see Fig. 5.21. This is known as the flux leakage method (Ref 278).

Applications

Most eddy current instruments are dedicated to a particular application, such as the detection of cracks, inspection of tubes, metal sorting, or the determination of coating thickness or conductivity. There are multipurpose instruments where impedance amplitude and phase changes are observed. The specimens are very often pipes, bars, tubes, and wires. Generally, ferromagnetic materials need to be fully saturated and magnetized using a dc solenoid to avoid any change in the magnetization state during eddy current testing. Subsequently, it is usually necessary to demagnetize the specimen.

Crack (Discontinuity) Testing (Flat Specimens). Some steps to undertake flaw detection are as follows:

1. Knowledge of probable type, position, and frequency of flaws should be considered.
2. Frequency choice. For surface flaws, the frequency should be as high as possible (MHz). For subsurface flaws, lower frequencies are chosen, although this means lower sensitivity. For ferromagnetic materials, a lower frequency is chosen to overcome lower penetration.
3. Selection of coil. Many shapes are available. A coil should be chosen to fit the geometry

of the specimen. The coil should be as small as possible.

4. The inspection coil probe is placed on the specimen and the impedance noted. A small-tipped coil in a ball-nosed probe can be used.
5. As the inspection coil is moved over the surface, a local change in impedance will occur as the probe moves over a discontinuity.
6. Cracks within 1 cm of the surface can be detected depending on the frequency used, the specimen conductivity and magnetic permeability.
7. Skill is required to maintain the same surface contact, pressure, and probe angle. This is accomplished using automated scanners, or with fixtures to help maintain orientation and pressure.
8. The instrument needs to be checked and calibrated using standard specimens with narrow machined slots.

When testing tubes, bars or wires, encircling coils are used, and very rapid, automatic systems are available; see ASTM Standards E 309, E 1033, and E 690 (Ref 61) and Table 6.3.

Electrical conductivity can be measured in a specimen if the effects of thickness and lift-off are suppressed; see ASTM Standard E 1004 (Ref 61) and Table 6.3:

- Use higher frequency where the effects of lift-off are approximately perpendicular to the conductivity curves (Fig. 6.7).
- A frequency is chosen where the depth of penetration is less than the specimen thickness (Fig. 6.3b and Table 6.2).

- Nonconducting shims can be added to calibrate the lift-off when measuring thickness of nonconducting coatings.
- This can be repeated at another frequency to observe minimum change in impedance.

When lift-off and thickness effects have been reduced to a minimum, the impedance is compared to the calibrated conductivity-impedance curve (Ref 66).

Metal Sorting. The impedance value is established for the specimen using a reference standard. Lift-off and thickness effects are reduced to a minimum. Several measurements should be carried out on the specimen to avoid interference by discontinuities; see ASTM Standards E 566 and E 703 (Ref 61) and Table 6.3.

Thickness of Coatings. The impedance value is established using a reference standard of known thickness. Lift-off and thickness of substrate effects need to be reduced to a minimum; see ASTM Standard E 376 (Ref 61) and Table 6.3.

Nonconducting coatings tend to have gradual changes in thickness leading to gradual changes in the impedance values so that absolute measurements are necessary. This is different from the abrupt changes given by discontinuities, for which differential methods can be used.

Surface coatings can be observed if there is a substantial difference in electrical conductivity compared to the substrate conductor.

Surface Condition (Corrosion, Heat Damage, Hardness). Standard calibrated specimens should be available for comparison. The electrical conductivity of some alloys (Ti, Al) change with thermal treatment.

Table 6.3 ASTM standards for eddy current testing

Standard	Description
E 243	Eddy Current Examination of Copper and Copper-Alloy Tubes (1990)
E 309	Eddy Current Examination of Steel Tubular Products using Magnetic Saturation (1987)
E 376	Coating Thickness by Magnetic-Field or Eddy Current Test Methods (1989)
E 426	Eddy Current Examination of Tubular Products (1988)
E 566	Eddy Current Sorting of Ferrous Metals (1988)
E 571	Eddy Current Examination of Nickel and Nickel Alloy Tubular Products (1988)
E 690	Eddy Current Examination of Nonmagnetic Heat Exchanger Tubes (1991)
E 703	Eddy Current Sorting of Nonferrous Metals (1992)
E 1004	Eddy Current Measurements of Electrical Conductivity (1991)
E 1033	Eddy Current Examination of Continuously Welded Ferromagnetic Pipe and Tubing above the Curie Temperature (1991)
E 1312	Eddy Current Examination of Ferromagnetic Bar above the Curie Temperature (1989)

Note: These standards are published annually (Ref 61).

Other Eddy Current Testing Procedures

Multifrequency (see Ref 278, section 20). One method of overcoming interfering signals from other conductors in the vicinity of the specimen under test is by the use of multi-excitation frequencies. Several frequencies are applied simultaneously to the transducer producing extensive eddy current data of the region of interest in the specimen. The interfering signal needs to be identified in advance, and analysis permits this signal to be suppressed. In practice, up to four frequencies have been used; in most cases, two excitation frequencies are mixed with the suppression of one interfering signal applied.

This method has been developed by several organizations, such as Intercontrôle (Ref 286) for example, for the examination of steam generator tubes in the presence of support plates.

Pulsed. Deeper penetration is obtained in ferromagnetic materials by the use of pulsed excitation. The pulse duration is important, using large capacitors at relatively low voltages of ~50 V (Ref 278).

Remote (Far) Field Low Frequency. This technique permits the detection of discontinuities in thick ferromagnetic tubes. The inspection excitation coil is kept distant from the detector coil to avoid any electromagnetic coupling between them; this is in contrast to the close proximity of the coils required by the other methods, as shown in Fig. 6.11 and 6.14. The phase of the detector

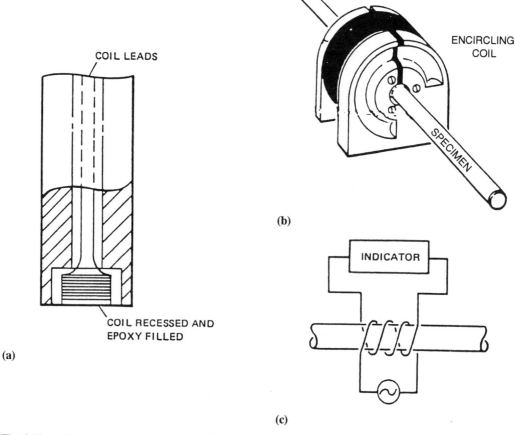

Fig. 6.14 Eddy current inspection probes (Ref 66). (a) Typical single coil probe, combined inspection (excitation) and receiver for flat plate examination. Encircling coil for pipe, tube, bar, or wires are shown in (b) typical form and (c) arrangement of outer heavy coil for excitation, with finer inner coil as receiver. Reprinted with permission of The American Society for Nondestructive Testing.

eddy current signal relative to that of the excitation frequency provides a measure of the pipe wall characteristics (Ref 278, section 20).

ASTM Standards

Detailed test procedures, codes, and calibration standards are described in the *Annual Book of ASTM Standards* (Ref 61). Nondestructive testing is reviewed on an annual basis in Volume 03.03. The codes are very useful reviews of all aspects of the testing methods. The references to the eddy current standards are listed in Table 6.3.

Useful glossaries of terms used in eddy current testing are also available in the *Annual Book of ASTM Standards* and in Ref 278, as well as in Appendix 6.1 to this chapter. A film describing eddy current testing has been prepared by Intercontrôle (Ref 286).

Appendix 6.1: Glossary of Some Terms Used in Eddy Current Testing (Ref 61, 190, 290)

characteristic frequency. Conventional quantity used as a frequency unit. It is derived from a mathematical model for the eddy current distribution in cylindrical conductors. The value is dependent on the characteristics of the specimen such as electrical conductivity, magnetic permeability, and diameter.

coil fill factor. For an encircling coil, it is the ratio of the external cross-sectional area of the specimen under test to the internal cross-sectional area of the coil. For an internal coaxial coil, it is the ratio of the external cross-sectional area of the coil to the internal cross-sectional area of the specimen under test.

combined transmit receive probe. Probe in which the functions of excitation (inspection, induction, or transmission) and reception are fulfilled by the same coil(s).

depth of penetration. (skin effect, standard depth of penetration). The depth at which the magnetic field strength or intensity of induced eddy currents has decreased by $\frac{1}{e}$ of its surface value, where $e \cong 2.718$ is the base of natural logarithms.

eddy current. Electric current induced in a conducting material by a changing magnetic field.

effective depth of penetration. The limit in depth for detecting discontinuities; this is approximately three times the standard depth of penetration.

effective permeability. The permeability of a complex component, assuming the field strength in the magnetizing coil to be constant throughout the coil.

electromotive force (emf). The force, measured in volts, supplied by a source of electric current that drives electric charges around a circuit.

electromagnetic coupling. Electromagnetic interaction between two or more circuits. In eddy current examination the specimen to be tested becomes a circuit. See also *magnetic coupling; mutual inductance.*

encircling coils. Coils that surround the part to be tested; also known as annular, circumferential, or feed-through coils.

end effect (edge effect). The disturbance of the magnetic field and eddy currents due to an abrupt change in specimen geometry.

fill factor. (1) For internal testing, the ratio of the effective cross-sectional area of the internal probe coil to the cross-sectional area of the tube interior. (2) For external testing, the ratio of the cross-sectional area of the specimen to the effective cross-sectional area of the primary encircling coil.

IACS. The International Annealed Copper Standard.

ID coil. A coil assembly for internal testing by insertion into the test piece; also known as inside coils, inserted coils, or bobbin coils.

impedance. The opposition of a circuit to the flow of an alternating current; the complex quotient of voltage divided by current.

impedance plane diagram. A graphical presentation of the variations in the impedance of a test coil as a function of changing frequency, conductivity, coil size, etc.

lift-off. The spacing between the test coil and the specimen.

lift-off effect. The effect of a change in magnetic coupling between the specimen and the test coil as the distance between them is varied.

Lissajous figure. The locus of the displacement of a point that undergoes the effect of two (or more) periodic motions. Generally, the two periodic motions are at right angles to each other, and of the same frequency, giving rise to elliptical formations.

magnetic coupling. See *electromagnetic coupling*; *mutual inductance*.

magnetic field strength H **(magnetic intensity).** A measure of the magnetic field from the flow of electric currents; measured in A/m; see Appendix 5.1.

magnetic flux density B **(magnetic induction).** The magnetization of a ferromagnetic material; see Appendix 5.1.

magnetic leakage flux. The magnetic field from a discontinuity at the surface of a specimen; see the section "Magnetic Leakage Field" in the chapter "Magnetic Particle Inspection."

magnetic permeability. The ease with which a magnetic field can be induced in a solid; see the section "Magnetic Leakage Field" in the chapter "Magnetic Particle Inspection."

magnetic saturation. The limit of magnetization of a specimen where an increase in the magnetizing force produces no increase in the magnetic flux density in the specimen; see Fig. 5.12.

mutual inductance. A current flowing in one circuit induces an electromotive force in a neighboring circuit; see Fig. 6.13(b).

phasor. A vector representing the impedance of a circuit, giving both the amplitude and the phase relationship of the impedance.

probe coil. A coil that is placed with its end near a test specimen; see Fig. 6.11.

reactance (X). The part of the impedance of an electric circuit given by the self-inductance L of the circuit and the angular frequency ω of the alternating current where $X = \omega L$; the reactance is measured in ohms.

relative permeability (μ_r). A magnetic property of a medium, given by the ratio of the self-inductance of a circuit surrounded by the medium to the self-inductance of the circuit in free space; see Appendix 5.1.

resistance (R). The ability of a solid or circuit to resist the flow of an electric current; measured in ohms.

resistivity. The resistance per unit length of the material of unit cross section.

self inductance (L). The property of a circuit where a changing electric current induces an electromotive force opposing the charge in the circuit.

skin effect. The tendency of alternating currents to flow near to the surface of a material; see *depth of penetration*.

solenoid. A cylindrical coil used to produce a magnetic field.

Symbols and Abbreviations

A_h	cross-sectional area of hollow portion of part	$\overline{D}$	average particle size
A_t	total cross-sectional area	D_R	danger range of an unshielded radioactive source
A	amplitude		
A_w	atomic weight	DAC	distance amplitude correction
AA	activation analysis	d	distance
AFI	artificial flaw indicators	dc	direct current
A_s	activity of radioactive source in Curies	E	Young's modulus
ac	alternating current	E	energy
B	magnetic induction	E_g	total gamma radiation energy in MeV
B_D	beam diameter	E_M	energy of γ-ray
B_o	light beam incident intensity	EF	exposure factor
B_T	light beam transmitted intensity	EPR	electron paramagnetic resonance
Bq	Becquerel	ET	eddy current testing
BIT	backscatter image tomography	e	electronic charge, 1.602×10^{-19} C
b	solenoid length	e	exponential factor
C_S	radiographic contrast	emf	electromotive force
C	capacitance	esu	electrostatic unit of charge
CAT	computer assisted tomography	F	electric field
CS	cross section	F	effective focal spot width
Ci	Curie	FS	film speed
C_T	computed tomography	F_z	depth of field
CW	continuous wave	f	focal length
c	velocity of light, 2.998×10^8 ms^{-1}	ftc	footcandle
D	photographic density	G	Gauss

G	shear modulus	m	mass per unit area
G	film gradient; gamma	mA-s	milliamp second
Gy	gray	$m_o^+ c^2, m_o^- c^2$	rest masses
g	acceleration due to gravity	min	minimum; minutes
H	magnetizing force A/m	N_o	Avogadro's Number, 6.022×10^{23} mol^{-1}
H	accumulated dose in rems	N	near-field distance
HAZ	heat affected zone	NAA	neutron activation analysis
HVL	half-value layer	N_E	number of electrons in one gram
HVT	half-value thickness	NDE	nondestructive evaluation
Hz	Hertz	NDT	nondestructive testing
h	hours	NMR	nuclear magnetic resonance
h	height	n	neutrons
h	Planck's constant, 6.626×10^{-34} J	n_t	number of disintegrations at time t
I	intensity	O	object, specimen
I_B	background intensity	P	excess local pressure
I_o	intensity incident beam	PE	pulse echo
I_s	scattered intensity	ph	photoelectric effect
I_z	intensity beam at distance z	P_S	sound path
IQI	image quality indicator	p	penumbra
IACS	International Annealed Copper Standard	p_o	reduced penumbra
ICRP	International Commission Radiation Protection	Q	particle displacement velocity
i	electric current in amperes	QQI	quantitative quality indicators
J	ferromagnetic magnetization (Tesla, Gauss)	R	receiver; Roentgen
j	Compton incoherent scattering	R	resistive component of impedance
K	constant	RA	radioactivity
KE	kinetic energy	R/K	crack width; resistance
k	wave number	R_E	energy fraction reflected
kV	kilovolt	R_F	Roentgen per hour at one foot
L	self inductance	Rhm	Roentgen per hour at one meter
L/D	length to diameter	R_H	Hall coefficient
L_o	distance: source to object	RT	reconstructive tomography
L_s	distance: source to film	r	radius
LPI	liquid penetrant inspection	rbe	relative biological effectiveness
l	distance: object to screen	rem	Roentgen equivalent man (mammal)
lm	lumen	S	umbra
M	magnetization	S_A	specific activity
M_d	depth focal point (immersion testing)	S_K	skip distance
M_g	magnification	S_P	signal amplification
M_I	mutual inductance	S_R	radioactive source strength in Curies
MPI	magnetic particle inspection	s	seconds

T	Tesla; transmitter	δ	standard depth of penetration of alternating electric current
T	temperature		
TBq	teraBecquerel	ε	exposure
T_E	energy fraction transmitted	ε'	strain
T/R	transmitter and receiver	ζ	sensitivity
TVL	tenth-value layers	η	photodisintegration
Th	Thomson scattering	θ	contact angle
t	time	θ'	mean free path
U	unsharpness	Λ	decay constant radioactive isotope
U_f	film unsharpness	λ	wavelength
U_g	geometric unsharpness	μ	linear absorption coefficient
U_m	motion unsharpness	μ_{dB}	absorption coefficient in decibels
U_s	screen unsharpness	μ_m	mass absorption coefficient
U_T	total unsharpness	μ_{Np}	absorption coefficient in neper
u	age of individual in years	μ_o	magnetic permeability of free space
V	applied voltage; sound velocity	μ_p	magnetic permeability
V_{BL}	bulk wave velocity	μ_R	relative magnetic permeability
V_L	sound velocity: longitudinal waves	μ_E	electronic absorption coefficient
V_S	sound velocity: shear waves	μ_a	atomic absorption coefficient
W	object size	μ_g	gram-atomic absorption coefficient
W_P	distance of specimen surface to transducer active element	μ_s	scattering component of absorption
		μ_τ	coefficient of true absorption
W'	size limit of detection	ν	frequency, Hertz (Hz)
w	weight fraction	ν_L	frequency (lines/mm)
$X_L = \omega L$	reactance	ν_c	characteristic (or limit) frequency for eddy current inspection of coil
Y	wave amplitude		
Z	specific acoustic impedance; impedance electric circuit	ξ	pair and triplet formation
		ρ	density
Z_0	impedance in the absence of a conductor	σ	Poisson's Ratio
Z_o	atomic number	σ	electrical conductivity, Siemens m^{-1}
z	distance	σ'	stress
α	piezoelectric constant, mV^{-1}	τ	half-life radioactive isotope
α	scattering angle	$\tau = 1/\nu$	period
β	piezoelectric constant, Vm^{-1} Pa^{-1}	ϕ	phase angle
γ	surface tension	ϕ_o	work function
Δ	diameter	χ	efficiency
Δ_o	specimen thickness	χ_m	magnetic susceptibility
		ω	rotation frequency, radians/s

Data Tables

Properties of the elements

Symbol	Name	Atomic No.	Atomic weight	Density, g/cm³ 20 °C 1 atmosphere	Melting point, K	Boiling point, K	Symbol
Al	Aluminum	13	26.98	2.7	933.2	2740	Al
Sb	Antimony	51	121.75	6.69	903.7	1650	Sb
Ar	Argon	18	39.95	0.138	83.7	87.4	Ar
As	Arsenic	33	74.92	5.73	...	...	As
Ba	Barium	56	137.34	3.6	1123	1413	Ba
Be	Beryllium	4	9.01	1.8	1550	...	Be
Bi	Bismuth	83	208.98	9.95	544.4	1830	Bi
B	Boron	5	10.81	2.5	2600	...	B
Br	Bromine	35	79.90	(300 K) 3.1	265.9	331.9	Br
C	Carbon	5	12.01	...	...	...	C
	Diamond	...	...	3.51	...	...	
	Graphite	...	...	2.25	...	...	
Ca	Calcium	20	40.08	1.54	1120	1513	Ca
Cd	Cadmium	48	112.40	8.65	594.2	1038	Cd
Ce	Cerium	58	140.12	3.92	1077	1673	Ce
Cl	Chlorine	17	35.45	...	...	...	Cl
Cr	Chromium	24	52.00	7.2	2160	2753	Cr
Co	Cobalt	27	58.93	8.9	1765	3170	Co
Cs	Cesium	55	132.90	1.873	301.6	960	Cs
Cu	Copper	29	63.55	8.93	1356	2610	Cu
F	Fluorine	9	19.00	1.7×10^{-3}	53.5	85.01	F
Gd	Gadolinium	64	157.25	7.9	1585	3000	Gd
Ga	Gallium	31	69.72	5.95	302.9	...	Ga
Ge	Germanium	32	72.59	5.36	1210.5	...	Ge
Au	Gold	79	196.97	19.32	1336.1	3239	Au
Hf	Hafnium	72	178.49	13.300	2423	5700	Hf
He	Helium	2	4.003	0.166×10^{-3}	0.95	4.21	He
H	Hydrogen	1	1.00797	0.08987×10^{-3}	14.01	20.4	H
In	Indium	49	114.82	7.31	429.8	2300	In
I	Iodine	53	126.9	4.94	386.6	457.4	I
Ir	Iridium	77	192.2	22.42	2716	4800	Ir
Fe	Iron	26	55.85	7.87	1808	3300	Fe
Kr	Krypton	36	83.80	3.49×10^{-3}	116.5	120.8	Kr

(continued)

Properties of the elements *(continued)*

Symbol	Name	Atomic No.	Atomic weight	Density, g/cm³ 20 °C 1 atmosphere	Melting point, K	Boiling point, K	Symbol
La	Lanthanum	57	138.91	6.15	1190	3742	La
Pb	Lead	82	207.19	11.34	600.4	2017	Pb
Li	Lithium	3	6.94	0.534	452	1590	Li
Mg	Magnesium	12	24.31	1.741	924	1380	Mg
Mn	Manganese	25	54.94	7.44	1517	2370	Mn
Hg	Mercury	80	200.59	13.59	234.3	629.7	Hg
Mo	Molybdenum	42	95.94	10.2	2880	5380	Mo
Nd	Neodymium	60	114.24	6.96	1297	3300	Nd
Ne	Neon	10	20.18	0.839×10^{-3}	24.5	27.2	Ne
Ni	Nickel	28	58.71	8.9	1726	3005	Ni
Nb	Niobium	41	92.91	8.57	2741	5200	Nb
N	Nitrogen	7	14.01	1.165×10^{-3}	63.3	77.3	N
Os	Osmium	76	190.2	22.48	3300	4900	Os
O	Oxygen	8	16.00	1.33×10^{-3}	54.7	90.2	O
Pd	Palladium	46	106.4	12.0	1825	3200	Pd
P	Phosphorus	15	30.97	2.2	317.2	552	P
Pt	Platinum	78	195.09	21.45	2042	4100	Pt
K	Potassium	19	39.10	0.86	336.8	1047	K
Pr	Praseodymium	59	140.91	6.8	1208	3400	Pr
Pm	Promethium	61	145	...	1308	3000	Pm
Pa	Protactinium	91	231	15.4	1500	4300	Pa
Ra	Radium	88	226	5.0	970	1410	Ra
Rn	Radon	86	222	9.73×10^{-3}	202	211.3	Rn
Re	Rhenium	75	186.2	20.5	3450	5900	Re
Rh	Rhodium	45	102.91	12.44	2230	4000	Rh
Rb	Rubidium	37	85.47	1.53	312.0	961	Rb
Ru	Ruthenium	44	101.07	12.4	2520	4200	Ru
Sm	Samarium	62	150.35	7.5	1345	2200	Sm
Sc	Scandium	21	44.96	3.0	1812	3000	Sc
Se	Selenium	34	78.96	4.81	490	958	Se
Si	Silicon	14	28.09	2.3	1680	2628	Si
Ag	Silver	47	107.87	10.5	1234	2485	Ag
Na	Sodium	11	22.99	0.97	371	1165	Na
Sr	Strontium	38	87.62	2.6	1042	1657	Sr
S	Sulfur	16	32.06	2.07	386	717.7	S
Ta	Tantalum	73	180.95	16.6	3269	5698	Ta
Tc	Technetium	43	98.91	11.4	2500	4900	Tc
Te	Tellurium	52	127.60	6.24	722.6	1260	Te
Tb	Terbium	65	158.92	8.3	1629	3100	Tb
Tl	Thallium	81	204.37	11.86	576.6	1730	Tl
Th	Thorium	90	232.04	11.5	2000	4500	Th
Tm	Thulium	69	168.93	9.3	1818	2000	Tm
Sn	Tin	50	118.69	7.3	505.1	2540	Sn
Ti	Titanium	22	47.90	4.54	1948	3530	Ti
W	Tungsten	74	183.85	19.32	3650	6200	W
U	Uranium	92	238.03	19.05	1405.4	4091	U
V	Vanadium	23	50.94	6.1	2160	3300	V
Xe	Xenon	54	131.30	0.055	161.2	166.0	Xe
Yb	Ytterbium	70	173.04	7.0	1097	1700	Yb
Y	Yttrium	39	88.91	4.6	1768	3200	Y
Zn	Zinc	30	65.37	7.14	692.6	1180	Zn
Zr	Zirconium	40	91.22	6.5	2125	3851	Zr

Metals

Name	Density, g/cm3 20 °C	Melting point, K	Specific heat, J/kg · K	Thermal expansion, 10^{-6}/°C	Thermal conductivity, W/m · K	Electrical resistivity, $\Omega \cdot m$	Tensile strength, MPa	Yield strength, MPa	Young's modulus, GPa	Poisson's ratio
Aluminum	2.71	932	913	23	201	2.65	80	50	71	0.34
Antimony	6.68	904	205	10	18	40	…	…	78	…
Bismuth	9.80	544	126	13	8	115	…	…	32	0.33
Brass (70Cu-30Zn)	8.50	1300	370	18	110	8	550	450	100	0.35
Bronze (90Cu-10Sn)	8.80	1300	360	17	180	30	260	140	…	…
Cobalt	8.90	1765	420	12	69	6	~500	…	…	…
Copper	8.93	1356	385	17	385	1.7	150	75	117	0.35
Gold	19.30	1340	132	14	296	2.4	120	…	71	0.44
Invar (64Fe-36Ni)	8.00	1800	503	0.9	16	81	480	280	145	0.26
Iron, pure	7.87	1810	106	12	80	10	300	165	206	0.29
Iron, cast gray	7.15	1500	500	11	75	10	100	…	110	0.27
Iron, cast white	7.70	1420	…	11	75	10	230	…	…	…
Iron, wrought	7.85	1810	480	12	60	14	~370	150	197	0.28
Lead	11.34	600	126	29	35	21	15	12	18	0.44
Magnesium	1.74	924	246	25	150	4	190	95	44	0.29
Manganin	8.50	…	400	18	22	45	…	…	120	0.33
Monel (70Ni-30Cu)	8.80	1600	…	14	210	42	520	240	…	…
Nickel	8.90	1726	460	13	59	59	300	60	207	0.36
Platinum	21.45	2042	136	9	69	11	350	…	150	0.38
Silver	10.50	1230	235	19	419	1.6	150	180	70	0.37
Stainless steel (18Cr-8Ni)	7.93	1800	510	16	15	96	600	230	210	…
Steel (1020)	7.86	1700	420	11.5	63	15	460	300	210	0.29
Steel (1040)	7.80	1700	…	…	50	17	3000	…	210	0.29
Tantalum	16.6	3.629	…	6.7	50	14	950	…	150	…
Tin	7.30	505	226	23	65	11	30	…	40	0.36
Titanium	4.54	1950	523	9	23	53	620	480	…	0.36
Zinc	7.14	693	385	31	111	5.9	150	…	110	0.25

Nondestructive Testing

Nonmetals

Name	Density, g/cm3 20 °C	Specific Melting point, K	Thermal heat, J/kg · K	expansion, 10^{-6}/°C	Tensile strength, MPa	Young's modulus, GPa
Rubber IR polyisoprene	0.91	300	1600	220	0.15	0.02
Silicon carbide	3.17	...	...	4.5	...	...
Sulfur	2.07	386	730	64	0.26	...
Titanium carbide	4.50	...	...	7	28	345
Alumina ceramic	3.80	2300	800	9	29	345
Bone	1.85	...	...	...	...	28
Brick						
Building	2.30	...	...	9	0.6	...
Fireclay	2.10	...	...	4.5	0.8	...
Paving	2.50	...	...	4.0	...	...
Silica	1.75	...	...	...	0.8	...
Carbon						
Graphite	2.30	...	710	7.9	5.0	207
Diamond	3.30	...	525	~0	900	1200
Concrete	2.40	...	3350	12	0.1	14
Cork	0.24	...	2050	...	0.05	...
Glass						
Crown	2.60	1400	670	9	1.0	71
Flint	4.20	1500	500	8	0.8	80
Ice	0.92	273	2100	31	2.0	...
Magnesium oxide	3.60	3200	960	12	...	207
Marble	2.60	...	880	10	2.9	...
Epoxy resin (EP)	1.12	...	1400	39	...	4.5
Melamine formaldehyde (MF)	1.50	...	1700	40	0.3	9
Naphthalene	1.15	350	1310	107	0.4	...
Nylon (PA)	1.15	470	1700	100	0.25	...
Paraffin wax	0.90	330	2900	110	0.25	...
Polyvinylchloride (PVC)	...	...	...	...	...	...
Nonrigid	1.25	485	1800	150	...	0.01
Rigid	1.70	485	1000	55	...	2.8
Polyvinylidene chloride (PVDC)	...	470	...	190	...	...
Phenol formaldehyde (PF)	1.30	...	1700	40	0.2	6.9
Polyethylene	...	...	...	...	...	...
Low density (LDPE)	0.92	410	2300	250	...	0.18
High density (HDPE)	0.96	410	2300	250	...	0.43
Polypropylene (PP)	0.90	450	2100	62	...	1.2
Polystyrene (PS)	1.05	510	1300	70	0.08	3.1
Urea formaldehyde (UF)	1.50	...	...	27	...	10.3

S.I. prefixes

Factor	Prefix	Symbol
10^{12}	tera	T
10^{9}	giga	G
10^{6}	mega	M
10^{3}	kilo	k
10^{1}	deka	da
10^{-1}	deci	d
10^{-2}	centi	c
10^{-3}	milli	m
10^{-6}	micro	μ
10^{-9}	nano	n
10^{-12}	pico	p
10^{-15}	femto	f
10^{-18}	atto	a

CGS units with special names

erg	erg	10^{-7} J
dyne	dyn	10^{-5} N
poise	P	1 dyn s/cm^2 = 0.1 Pa · s
stokes	St	1 cm^2/s
gauss	Gs, G	10^{-4} T
oersted	Oe	$1000/4\pi$ A/m
maxwell	Mx	10^{-8} Wb
stilb	sb	cd/cm^2 = d
phot	ph	10^4 lx

Constants

Quantity	Value	Units
Velocity of light, c	2.997 925(3)	10^8 m · s^{-1}
Mass hydrogen atom	1.673 43(8)	10^{-27} kg
Charge of electron, e	1.602 10(7)	10^{-19} C
Wavelength associated with 1 eV	1.239 81	10^{-6} m
Planck constant, h	6.625 6(5)	10^{-34} J · s
Avogadro constant, N_0	6.022 52(28)	10^{23} mol^{-1}
Molar volume of ideal gas at S.T.P.	2.241 36(30)	10^{-2} m^3 mol^{-1}
Base of natural logarithm, e	2.718 281 828 4	
$\log_e 10 = 2.30259$	$\log_{10}$ e = 0.434294	
Gravitational acceleration, g	9.806 65	m s^{-2}
π	3.1 41 592 653 59	
Roentgen, R	2.58×10^{-4}	C· kg^{-1}
Rad	10^{-2}	J · kg^{-1}
Curie, Ci	3.7×10^{10}	disintegrations/s
Energy equivalent of electron mass, mc^2	0.51 MeV	
Gas constant, R	1.987 Cal · mol^{-1} · K^{-1}	
	8.314×10^7 ergs · mol^{-1} · K^{-1} ≡ 8.314 J · mol^{-1} · K^{-1}	
	82.06 cm^3 atmospheres mol^{-1} · K^{-1}	
	0.082 litre atm mol^{-1} · K^{-1}	
Boltzmann constant, k	1.38054×10^{-23} J · K^{-1}	
	0.33×10^{-23} Cal · K^{-1}	
	8.65×10^{-5} eV · K^{-1}	

Conversions

Unit	Conversion factor	Units
Electrical		
ampere	1	coulomb/s
coulomb	1.036×10^{-5}	faraday
faraday/s	96516	coulomb
watt	1	joule/s
Energy		
Btu	1.0548×10^3	joule
eV	1.6021×10^{-19}	joule
erg	7.375×10^{-8}	ft · lbf
joules	0.7376	ft · lbf
Length		
angstroms (Å)	10^{-8}	cm
cm	0.3937	inch
m	1.0926	yard
mil	10^{-3}	inch
rod, perch, pole	5.5	yard
furlong	40	rod
mile	5280	foot
mile	8	furlong
Mass		
g	3.527×10^{-2}	oz
g	2.205×10^{-3}	lb
kg	0.0685	slugs
amu (physical ^{16}O scale)	1.66024×10^{-27}	kg
Power		
horsepower	42.42	Btu/min
horsepower	33,000	ft · lbf/min
horsepower	4.655×10^{15}	MeV/s
watt	10^7	ergs/s
watt	3.414	Btu/h
Pressure		
atmosphere	14.696	lb/in.^{2} (psi)
atmosphere	1.0133	bar
atmosphere	1.01325×10^5	Pa
bar	14.504	lb/in.^{2} (psi)
bar	10^5	Pa
lb/in.^{2} (psi)	6.89373×10^3	Pa = N · m^{-2}
Volume		
cm^3	6.1023×10^{-2}	in.^{3}
ft^3	2.832×10^{-2}	m^3
gal (U.S.)	231.0	in.^{3}

Energy equivalencies

Energy	J	eV	Calorie	kilowatt-hour
1 Joule (J)	1	6.242×10^{-18}	0.2389	2.778×10^{-7}
1 eV	1.602×10^{-19}	1	3.828×10^{-20}	4.45×10^{-26}
1 calorie	4.186	2.613×10^{17}	1	1.63×10^{-6}
1 kilowatt-hour	3.6×10^6	2.247×10^{24}	8.6×10^5	1

Conversion factors

To convert from	To	Multiply by
Atmosphere	Pascal (Pa)	$1.013\,25 \times 10^5$
Btu	Joule (J)	$1.055\,056 \times 10^3$
Btu/h	Watt (W)	$0.2\,930\,711$
Calorie	Joule (J)	$4.186\,8$
Centipoise	Pascal second (Pa $\cdot$ s)	1.0×10^{-3}
ft/s^2	Meter per second^2 (m/s^2)	3.048×10^{-1}
Inch	Meter (m)	2.540×10^{-2}
Ounce (avoirdupois)	Kilogram (kg)	$2.834\,952 \times 10^{-2}$
Pound (lb avoirdupois)	Kilogram (kg)	$4.535\,924 \times 10^{-1}$
lb/in.^{3}	Kilogram per meter^3 (kg/m^3)	$2.767\,99 \times 10^4$
Torr (mm Hg 0 °C)	Pascal (Pa)	$1.333\,22 \times 10^2$
W h	Joule (J)	3.6×10^3
Yard	Meter (m)	9.144×10^{-1}

REFERENCES

1. L. Mullins, Evolution of NDT, *Progr. Appl. Mater. Res.*, Vol 5, E.G. Stanford, J.H. Fearon, and W.J. McGonnagle, Ed., 1964, p 205-212
2. L. Mullins, *Electron. Eng.*, 1945, p 17
3. *Ultrasonic Nondestructive Testing*, Monograph No. 9, The Institute of Metals, UK
4. Office of Nondestructive Testing, National Institute of Standards and Technology (NIST), Gaithersburg, MD, USA
5. D. Graham and T. Eddie, *X-ray Techniques in Art Galleries and Museums*, A. Hilger, 1985
6. S.T. Fleming, *Authenticity in Art*, Institute of Physics, United Kingdom, 1975
7. H.H. Anderson and S.T. Picraux, Ed., Ion Beam Analysis in the Arts and Archaeology, *Nucl. Instrum. Meth. Phys. Res. B*, Vol 14 (No. 1), Jan 1986
8. A. Gilardoni, *X-rays in Art*, Como, Italy, 1977
9. E.V. Sayre, Ed., Materials Issues in Art and Archeology, *Materials Research Soc. Symp. Proc.*, Vol 123, 1988
10. E. Bielecki, et al., MQS Inspection, Inc., 5512 West State Street, Milwaukee, WI
11. C.F. Bridgman and S. Keck, The Radiography of Paintings, *Med. Radiogr. Photogr.*, Vol 37, 1961, p 62-70
12. A.E. James, et al., Radiographic Analysis of Paintings, *Med. Radiogr. Photogr.*, Vol 63, 1987, p 1-24
13. R.V. Ely, Ed., *Microfocal Radiography*, Academic Press, 1980
14. V.P. Guinn, JFK Assassination Bullet Analysis, *Ann. Chem.*, Vol 51, 1979, p 484A-492A
15. D.E. Gray, Ed., *American Institute of Physics Handbook*, 2nd ed., McGraw-Hill, 1963
16. B. Keisch, The Atomic Fingerprint: Neutron Activation Analysis, *Science*, Vol 160, 1968, p 413-415
17. I. Perlman, F. Asaro, and H.V. Michel, Nuclear Applications in Art and Archeology, *Ann. Rev. Nucl. Sci.*, 1972, p 383
18. P. Meyers, *Archaeometry*, Vol 11, 1969, p 67
19. L. Ryback and A.H. Youmans, New Nuclear Logging Methods, *Bull. Ver. Schweiz. Petrol. Geol. u-Ing.*, Vol 35 (No. 34), 1968
20. D.A. Bromley, Neutrons in Science and Technology, *Phys. Today*, 1983, p 30-39
21. C.M. Bernston, Nondestructive Testing Sheds Light on Preserving America's Past, *Mater. Eval.*, Vol 43, 1985, p 1180-1186
22. The Liberty Bell, *Med. Radiogr. Photogr.*, Vol 52, 1976, p 30-33
23. P.H.K. Gray, Radiography of Ancient Egyptian Mummies, *Med. Radiogr. Photogr.*, Vol 43, 1967, p 34-44
24. W.M.F. Petrie, "Deshaheh 1897: Fifteenth Memoir of the Egypt Exploration Fund," Published by Egypt Exploration Fund, London, U.K., 1898
25. NDT Leads Archeological Search for Clean Air, *Mater. Eval.*, Vol 51 (No. 3), 1993, p 388-389
26. D. Koralewski, Radiographs Reveal Iron Skeleton in Freedom's Feather Headdress, *Mater. Eval.*, Vol 51, 1993, p 1088-1089
27. H.H. Murray and W.D. Johns, Measurement of Coating Thickness and Weight, *Tappi*, Vol 44, 1961, p 217
28. J.F. Cameron and J.R. Rhodes, X-ray Spectrometry with Radioactive Sources, *Nucleonics*, Vol 19 (No. 6), 1961, p 53
29. F.A. Hasenkamp, Radiographic Laminography, *Mater. Eval.*, Aug 1974, p 170-180
30. R.J. Urich, *Principles of Underwater Sound*, McGraw-Hill, 1975

31. Model SPATE 8000, Ometron Ltd, Park Road, Chislehurst, Kent BR7 5AY, England

32. H.C. Wright, *Infrared Techniques*, Clarendon Press, Oxford, 1973

33. D.J. McDougall, Ed., *Thermoluminescence of Geological Materials*, Academic Press, 1968

34. C. Martin, P. Fauchais, and A. Borie, Detection of Adhesion Defects by Infrared Thermography, *8th World Conference NDT* (Cannes, France), 1978

35. *Polarized Light Selected Reprints*, American Association of Physics Teachers, American Institute of Physics, 1979

36. L. Cartz and G. Trischan, "Separation of Plastics Using Polaroid Light," US Patent 5,141,110, 1990

37. J.P. Duncan, Non-Coherent Optical Techniques for Surface Survey, Chapter 8 in Vol 2 of *Research Techniques in Nondestructive Testing*, R.S. Sharpe, Ed., Academic Press, 1st ed., 1970, 7th ed., 1984

38. R.P. Feynman, R.B. Leighton, and M. Sands, Harmonics, Section 50; Waves, Section 51, *The Feynman Lectures on Physics*, Vol 1, Addison-Wesley, 1963

39. A.F. Brown, *Seeing with Sound*, Vol 35, Endeavor, 1976, p 123-128

40. B. Dibner, *Röntgen and the Discovery of X-rays*, F. Watts, 1968

41. N. Morton, Thomas Young and the Theory of Diffraction, *Physics Education*, Vol 14, 1979, p 450-453

42. P.P. Ewald, Ed., "Fifty Years of X-ray Diffraction," International Union Crystallography, 1962

43. J.H. Elliwell, "Laue, Bragg and All That," *Phys. World*, Vol 2, 1989, p 29-32

44. V.E. Pullin, *Engineering Radiography*, Bell, 1934

45. O. Glasser, *Wilhelm Konrad Roentgen*, Bale, Sous A. Danielson, 1933

46. E. Rutherford, *Radioactivity*, Cambridge University Press, 1905

47. E.P. Thompson, *Roentgen Rays and Phenomena of the Anode and Cathode*, Van Nostrand, 1896

48. *NMR and EPR; Selected Reprints*, American Association of Physics Teachers, American Institute of Physics, 1965

49. C.E. Betz, "Principles of Penetrants," Magnaflux Corp., Chicago, 1963

50. Lord Rayleigh, *The Theory of Sound*, Vol 1, Macmillan, 1877

51. C.E. Betz, "Principles of Magnetic Particle Testing," Magnaflux Corporation, Chicago, 1967

52. S.M. Saxby, Magnetic Testing of Iron, *Engineering*, Vol 5, 1868, p 297

53. E.G. Stanford, J.H. Fearon, and W.J. McGonnagle, Ed., *Progr. Appl. Mater. Res.*, Vol 8, 1968; Vol 7, 1967; Vol 5, 1964

54. H.A. Colwell, *History of Electrotherapy*, Heinemann, 1922

55. *Radiological Health Handbook*, United States Department of Health, Education and Welfare, 1970

56. *Physical Aspects of Irradiation*, No. 85, *NBS Handbook*, Superintendent of Documents, US Government Printing Office, March 1964

57. *Radiation Quantities and Units*, No. 84, *NBS Handbook*, Superintendent of Documents, U.S. Government Printing Office, 1962

58. *National Bureau of Standards Handbooks*, Washington, DC
No. 42, *Safe Handling of Radioactive Isotopes*, 1949
No. 50, *X-ray Protection Design*, 1952
No. 51, *Radiological Monitoring Methods and Instruments*, 1952
No. 54, *Protection Against Radiations from Radium, Cobolt 60 and Cesium 137*, 1954
No. 55, *Protection Against Betatron-Synchrotron Radiations Up to 100 Million Electron Volts*, 1954
No. 59, *Permissible Dose from External Sources of Ionizing Radiation*, 1954
No. 60, *X-ray Protection*, 1955
No. 72, *Measurement of Neutron Flux and Spectra for Physical and Biological Applications*, 1960
No. 75, *Measurement of Absorbed Dose of Neutrons and Mixtures of Neutrons and Gamma Rays*, 1960

59. H. Berger, Ed., *Nondestructive Testing Standards: A Review*, STP 624, ASTM, 1976

60. *Materials Evaluation*, Journal of American Society for Nondestructive Testing, PO Box 2958, Clinton, IA, USA

61. Nondestructive Testing, Vol 03.03, Section 3, *Annual Book of ASTM Standards*, American Society for Testing and Materials (revised annually), 1916 Race St., Philadelphia, PA 19103, USA

62. *NDT Training Program Continuing Education in NDT Classroom Training Handbooks*, American Society for Nondestructive Testing

63. "Liquid Penetrant NDT," CT-6-2, American Society for Nondestructive Testing

64. "Magnetic Particle Testing," CT-6-3, American Society for Nondestructive Testing

65. "Ultrasonic Testing," CT-6-4, American Society for Nondestructive Testing

66. "Eddy Current Testing," CT-6-5, American Society for Nondestructive Testing

67. "Radiographic Testing," CT-6-6, American Society for Nondestructive Testing

68. "Document for NDT Personnel Training," SNT-TC-1A, American Society for Nondestructive Testing

69. R. Halmshaw, *Nondestructive Testing*, Edward Arnold, 1987

70. R.S. Sharppe, et al., Ed., *Quality Technology Handbook*, 4th ed., Butterworth, 1984

71. J.F. Bussiére, P. Monchalin, C.O. Ruud, and R.E. Green, Ed., *Nondestructive Characterization of Materials II*, Plenum, 1986

72. E. Downham, Vibration Monitoring, Chapter 9, in Vol 2 of *Research Techniques in Nondestructive Testing*, R.S. Sharpe, Ed., Academic Press, 1984

E.M. Uygur, Nondestructive Dynamic Testing, Chapter 6, in Vol 4 of *Research Techniques in Nondestructive Testing*, R.S. Sharpe, Ed., Academic Press, 1984

73. P. Höller, Ed., *New Procedures in NDT*, Springer-Verlag, 1983

74. W.J. Baxter, Exoelectron Emission from Metals, Chapter 12, in Vol 3 of *Research Techniques in Nondestructive Testing*, R.S. Sharpe, Ed., Academic Press, 1984

75. B. Culshaw, Fibre Optic Sensing Techniques, Chapter 6, in Vol 7 of *Research Techniques in Nondestructive Testing*, R.S. Sharpe, Ed., Academic Press, 1984

76. C.M. Vest, "Holographic NDE: Status and Future," NBS-GCR-81-318, Office of Nondestructive Evaluation, National Institute of Standards Technology (formerly National Bureau of Standards), Dept. of Commerce, Washington, DC 20234, USA

77. T.D. Beynon, Neutron Holography, *Physics Bull.*, Vol 37, 1986, p 129-131

78. R.K. Erf, Ed., *Holographic NDT*, Academic Press, 1974

79. J. Davies, Case on Three Dimensional TV, *Phys. World*, Vol 2, 1989, p 22-23

80. G. Birnbaum and G.S. White, Laser Techniques in NDE, Chapter 8, in Vol 7 of *Research Techniques in Nondestructive Testing*, R.S. Sharpe, Ed., Academic Press, 1984

81. H.E. Boyer, Ed., *Nondestructive Inspection and Quality Control*, 8th ed., Vol 11, *American Society for Metals*, 1976 (See Ref 129 for 9th ed.)
P.R. Locher, Proton NMR Tomography, *Philips Tech. Rev.*, Vol 41, 1984, p 73-88

82. P.E. Mix, *Introduction to Nondestructive Testing: A Training Guide*, Wiley, 1987

83. "NMR A Perspective on Imaging," Publication 5485, General Electric Company, 1984

84. E.R. Andrew and B.S. Worthington, NMR Imaging, *Radiology of the Skull Brain*, Vol 5, T.H. Newton and D.G. Potts, Ed., Mosby, 1981

85. C. Sutton, Magnetic Window into Bodily Functions, *New Sci.*, Sept 1986, p 32-37

86. Y.K. Park, J.T. Waber, and C.L. Snead, Positron Annihilation Methods Dislocation Densities in Fe, *Mater. Lett.*, Vol 3, 1985, p 181-186
A.E. Hughes, "Probing Materials with Positrons," *Materials in Engineering,* 2, Sept. 1980

87. "Fulmer Tape Abrasivity Metals," Fulmer Research Institute, Slough, Berks, England.; *Phys. Bull.*, Vol 34, 1983, p 466

88. R.W. Astheimer, *Handbook of Infrared Radiation Measurement*, Barnes Engineering Co., Stamford, CT, 1983

89. R. Pochaczevsky, Assessment of Back Pain, *Orthopaedic Review*, Vol 12, 1983, p 45-58

90. J. Cohen, "Elements of Thermography for Nondestructive Testing," NBS Technical Note 1177, National Bureau of Standards, 1983

91. G. Gaussorgues, *La Thermographie Infrarouge*, Librairies Lavoisier, Paris, 1980

92. L.J. Inglehart, et al., Thermal-Wave NDT of Composites, *J. Appl. Phys.*, Vol 59, 1986, p 234-240

93. P. Cielo, et al., Thermoelastic Inspection of Layered Material, *Mater. Eval.*, Vol 43, 1985, p 1111-1116

94. J.H. Richardson, *Optical Microscopy for the Materials Sciences*, Marcel Dekker, 1971

95. Tolansky, *Properties of Metallic Surfaces*, Monograph, Vol 13, Institute of Metals, London, 1953

96. R. Barer, *Lecture Notes on the Use of the Microscope*, Blackwell, 1971

97. R.H. Greaves and H. Wrighton, *Practical Microscopical Metallography*, Chapman and Hall, 1967

98. L.C. Martin and B.K. Johnson, *Practical Microscopy*, Blackie, 1949

99. A.F.H. Allimond, *Manual of the Polarizing Microscope*, Cooke, Troughton & Simms Ltd., 1953

100. A.S. Kuo and H.W. Liu, Moire Method NDT, Chapter 17, *Nondestructive Evaluation of Materials*, J.J. Burke and V. Weiss, Ed., Plenum, 1979

101. J.H. Lewis and S. Blake, Xonics Electron Radiography in Industrial NDT Application, Chapter 7, in Vol 3 of *Research Techniques in Nondestructive Testing*, R.S. Sharpe, Ed., Academic Press, 1984

102. W.C. Roentgen, *Nature*, Vol 53, 1896, p 274; translated from article in *Sitzber Wurzburger Physik*, Medic, Ces., 1895, On a New Kind of Ray

103. L.E. Bryant, Ed., *Nondestructive Testing Handbook*, 2nd ed., Vol 3, *Radiography and Radiation Testing*, American Society for Nondestructive Testing; American Society for Metals, 1985

104. G.W.C. Kaye, *Practical Applications of X-rays*, Dutton, 1923

105. M. Stennbeck, *Wissenschaftliche Veroffentlichungen aus den Siemens-Werken*, Vol 17, Chapter 4, 1938, p 363

106. K.H. Kingdon and H.E. Tanis, *Phys. Rev.*, Vol 53, 1938, p 128

107. W.J. Oosterkamp, *Philips Tech. Rev.*, Vol 5, 1940, p 22

108. C.M. Slack and L.F. Ehrke, Field Emission X-ray Tube, *J. Appl. Phys.*, Vol 12, 1941, p 165

109. D.W. Kerst, *Phys. Rev.*, Vol 60, 1941, p 47

110. A. St. John and H.R. Isenburger, *Industrial Radiography*, 2nd ed., Chapman and Hall, 1943

111. H. Becquerel, *Comptes Rendu French Academy Sciences*, Vol 122, p 501

112. H.R. Clauser, *Practical Radiography in Industry*, Reinhold, 1952

113. G.L. Clark, Ed., *Encyclopedia of X-rays and γ-rays*, Reinhold, 1963
D. Graham and J. Thomson, *Grenz Rays,* Pergamon, 1980

114. R.L. Sproull, *Modern Physics*, John Wiley, 1967

115. V.E. Cosslett and W.C. Nixon, *X-ray Microscopy*, Cambridge University Press, 1960

116. R. Jenkins and J.L. DeVries, "Practical X-ray Spec-

trometry," Phillips Technical Library, 1967

117. M.M. Ter-Pogossian, *The Physical Aspects of Diagnostic Radiology*, Hoeber-Medical Division, Harper and Row, 1969

118. "Photon Cross Sections, Attenuation Coefficients and Energy Absorption Coefficients from 10 keV to 100 GeV," NSRDS-NBS 29, National Bureau of Standards, 1969
K. Lonsdale, Ed., *International Table for X-ray Crystallography, 3,* International Union of Crystallography, 1968, p 157-200
B.L. Henke, et al., X-ray Absoprtion 2-200 Å Region, *Norelco Reporter,* Vol 14 (No. 3-4), 1967, p 117-118

119. J.G. Brown, *X-rays and Their Applications*, Plenum, 1966

120. R.A. Quinn and C.C. Sigl, Ed., *Radiography in Modern Industry*, 4th ed., Eastman Kodak Co., 1987

121. "Industrial Radiography on Radiographic Paper," Risφ Report 37, Risφ National Lab., Denmark, 1977

122. Glasser, Quimby, Taylor, Morgan and Weatherwax, *Physical Foundations of Radiology,* Pitman, 1962

123. R. Halmshaw, *Industrial Radiology*, American Society for Nondestructive Testing, 1982
R. Halmshaw, Industrial X-rays Sharpen Their Image, *New Scientist*, No. 1477, 1985, p 44-46

124. R.S. Sharpe, *Research Techniques in Nondestructive Testing*, Academic Press, 1st ed., 1970, 7th ed., 1984

125. W.R. Hampe, Modern Fluoroscope Practices, *Nondestructive Testing*, Vol 14, 1956, p 36-40

126. D.T. O'Connor, Industrial Fluoroscopy, *Nondestr. Test.*, Vol 11, 1952

127. R. Halmshaw, Direct-View Radiological System in Research Techniques, Chapter 8, *Nondestructive Testing*, Vol 1, R.S. Sharpe, Ed., Academic Press, 1970

128. R. Halmshaw, Direct-View Radiological System in Research Techniques, Chapter 8, *Nondestructive Testing,* Vol 1, R.S. Sharpe, Ed., Academic Press, 1970
R. Halmshaw, Fundamentals of Radiographic Imaging, Real-Time Radiologic Imaging; Medical and Industrial Applications, ASTM STP 716, D.A. Garrett and D.A. Bracher, Ed., 1980, p 5-21
R. Halmshaw, *Physics of Industrial Radiography,* Elsevier, New York, 1966

129. *Nondestructive Evaluation and Quality Control, Metals Handbook*, ASM International, 9th ed., Vol 17, 1989 (See Ref 81 for 8th ed.)

130. "Flash Radiography," Technical Bulletin B23, Hewlett-Packard, 1973

131. M. Held, "Flash-X-radiography in Ballistics," *Materials Evaluation*, Vol 43, 1985, p 1104

132. F. Jamet and G. Thomer, *Flash Radiography*, Elsevier, 1976

133. L.E. Bryand, Ed., *Flash Radiography*, Symposium

(Houston), American Society of Nondestructive Testing, 1976

134. E.A. Webster and A.M. Kennedy, Ed., 1984 Flash Radiography Symposium, *American Society for Nondestructive Testing*, 1984

135. J.J. Burke and V. Weiss, Ed., *Nondestructive Evaluation of Materials*, Plenum, 1979

136. H. Berger, *Neutron Radiography*, Elsevier, 1965
A.A. Harms and D.R. Wyman, *Mathematics and Physics of Neutron Radiography,* D. Reidel, 1986
P. von der Hardt and H. Röttger, *Neutron Radiography Handbook,* D. Reidel, 1981

137. A.M. Koehler and H. Berger, Proton Radiography, Chapter 1, in Vol 2 of *Research Techniques in Nondestructive Testing*, R.S. Sharpe, Ed., Academic Press, 1984

138. I. Kaplan, *Nuclear Physics*, Addison-Wesley, 1962

139. W.J. McGonnagle, *Nondestructive Testing*, 1st ed., McGraw-Hill, 1961

140. G.L. Clark, Ed., *Encyclopedia of Microscopy*, Reinhold, 1961

141. H. Kölker, P. Henze, K.A. Schwetz, and A. Lipp, X-ray Microfocus and Dye Penetrant Techniques for Crack Detection in Ceramics, *3rd Int. Symp. Ceramic Materials for Engine* (Las Vegas, NV), 1988

142. M. Howells, et al., Soft-X-ray Microscopes, *Physics Today*, Aug. 1985, p 22-32
J.H. Underwood and D.T. Attwood, The Renaissance of X-ray Optics, *Physics Today*, 1984, p 44-52

143. R.G. Rosemeier, Characterization of Laser Materials by Real-Time X-ray Topography, *Proc. Int. Conf. Laser 1984*, Nov 1984, p 112-114

144. R.N. Pangborn, X-ray Topography, *Materials Characterization*, 9th ed., Vol 10, *Metals Handbook*, American Society for Metals, 1986, p 365-379

145. P.S. Prevey, X-ray Diffraction Residual Stress Techniques, *Materials Characterization*, 9th ed., Vol 10, *Metals Handbook*, American Society for Metals, 1986, p 380-392

146. G.L. Clark, Porosity Measurements by Radiation Absorption, *Anal. Chem.*, Vol 29, 1957, p 1539

147. G.E. Martin and E.W. Grohse, *Proc. 9th Ann. Denver Conf. X-ray Analysis*, 1969, p 319-334

148. G.H. Morrison and J.F. Cosgrove, Determination U^{235} by Gamma Spectrometry, *Anal. Chem.*, Vol 29, 1957, p 1770

149. R.Y. Parry, Combined β and Dielectric Gauge, *J. Brit. Inst. Radio Eng.*, Vol 14, 1954, p 427-432

150. P. Lublin, *Norelco Reporter*, Vol 3, 1956, p 58-61

151. NDE of Density Measurement, *Ceram. Bull.*, Vol 67, 1988, p 1869

152. C.F. Coleman and A.E. Hughes, Positron Annihilation, Chapter 11, in Vol 3 of *Research Techniques in Nondestructive Testing*, R.S. Sharpe, Ed., Academic Press, 1984

153. M.M. Ter-Pogossian, et al., Positron-Emission Transaxial Tomography for Nuclear Imaging, *Nucl.*

Med., 1975, p 89-97

154. K. Kouris, N.M. Spyrou, and D.F. Jackson, *Imaging with Ionizing Radiations*, Surrey University Press, 1982

155. P. Riemers, W.B. Gilboy, and J. Goebbels, Industrial Applications of Computerized Tomography, *NDT Int.*, Vol 17, 1984, p 197-207

156. F.F. Hopkins, et al., Tomographic Image Analysis, *Mater. Eval.*, Vol 40, 1982, p 1226-1228

157. E. Segal, et al., Dimensional Information from Industrial Computerized Tomography, *Mater. Eval.*, Vol 40, 1982, p 1268-1279

158. D. Chapman and P.D. Magnus, *High Resolution NMR*, Academic Press, 1966

159. P. Mansfield and P.G. Morris, *NMR Imaging in Biomedicine*, Academic Press, 1982

160. E. Fukushima and S.B.W. Roeder, *Experimental Pulse NMR*, Addison-Wesley, 1981

161. P.N.T. Wells, *Biomedical Ultrasonics*, Academic Press, 1977

F.W. Kremkau, *Diagnostic Ultrasound,* 2nd ed., Grune and Stratton, 1984

F.W. Kremkau, Ultrapuzzles, *J. Clin. Ultrasound,* 6, 85, 1980

162. O.I. Babikov, *Ultrasonics and Its Industrial Applications*, Consultants Bureau, 1960

163. A.R. Williams, *Ultrasound: Biological Effects and Potential Hazards*, Academic Press, 1983

164. A. Macoviski, *Medical Imaging Systems*, Prentice Hall, 1983

165. Brochure, General Electric Quality Technology Center, Computed Tomography Services, Cincinnati, Ohio, USA

166. D.R. Craig and M.P. Sirkis, Simplified Apparatus for Producing Transaxial Tomograms, *Mater. Eval.*, Oct 1978, p 20-23

167. A. Notea, Film-Based Industrial Tomography, *NDT Int.*, Vol 18, 1985, p 179-184; Vol 16, 1983, p 263-270

168. R.S. Holt, Compton Imaging, Vol 9, *Endeavor*, 1985, p 97-105

G. Harding, H. Strecker, and R. Tischler, X-ray Imaging with Compton-Scatter Radiation, *Philips Tech. Rev.*, Vol 41, 1983/84, p 46-59

169. D. Gifford, *Handbook of Physics for Radiologist and Radiographers,* J. Wiley, 1984

D.N. Chesney and M.O. Chessney, *X-ray equipment for student radiographers,* 3rd ed., Blackwell, 1984

170. E. Kotz, R. Linde, U. Tiemens, and H. Weiss, Flashing Tomosynthesis, *Philips Tech. Rev.*, Vol 38, 1978/79, p 338-346

171. D.J. Hagemaier, Aerospace Radiography—The Last Three Decades, *Mater. Eval.*, Vol 43, 1985, p 1262-1283

172. R.A. Brooks and C.D. Chiro, Theory of Image Reconstruction in Computed Tomography, *Radiology*, Vol 117, 1975, p 561-572

173. C.L. Morgan, *Basic Principles of Computed Tomography*, University Park Press, 1983

174. S. Takahashi, Ed., *Illustrated Computer Tomography*, Springer Verlag, 1983

175. G.T. Herman, *Image Reconstruction from Projections*, Academic Press, 1980

176. P.R. Locher, Proton NMR Tomography, *Philips Tech. Rev.*, Vol 41, 1983/84, p 73-88

177. W.J. Richards, et al., Neutron Tomography, *Mater. Eval.*, Vol 40, 1982, p 1263-1267

178. A. DeVolpi, et al., Neutron and Gamma Ray Tomography, *Mater. Eval.*, Vol 40, 1982, p 1273-1279

179. P.D. Edmonds, *Ultrasonics*, Academic Press, 1981

180. A.H. Compton and S.K. Allison, *X-rays in Theory and Experiment*, 2nd ed., D. Van Nostrand Company, 1935

181. L. Cartz, *J. Appl. Phys.* Vol 35, 1964, p 2274-2275

L. Cartz, Non-destructive Testing, *Advances in Materials Technology, Monitor,* No. 15, Unido, 1989, p 1-55

182. L.V. Azaroff, *Elements of X-ray Crystallography*, McGraw-Hill, 1968

183. N.F.M. Henry, H. Lipson, and W.A. Wooster, *The Interpretation of X-ray Diffraction Photographs*, Macmillan, 1953

184. "Protections Against X-rays up to Energies of 3 MeV and β and Gamma Rays from Sealed Sources," *International Committee on Radiological Protection*, Pergamon, 1960 (see Ref 185.)

185. Dr. F. D. Sowby, International Commission on Radiological Protection (ICRP), Clifton Avenue, Sutton SM2 5PU, England

186. USA National Commission on Radiation Protection, P.O. Box 4867, Washington, DC 20008

187. General Safety Standard for Installation using Non-Medical X-ray and Sealed Gamma Ray Sources Energies up to 10 MeV, *NBS Handbook 114*, US Department of Commerce, National Bureau of Standards, 1975

188. "Standards for Protection Against Radiation," U.S. Government Printing Office, 1971

189. B.D. Cullity, *Elements of X-ray Diffraction*, 2nd ed., Addison-Wesley, 1978

190. W.R. Hibbard, *Dictionary of Mining, Mineral and Related Terms*, Bureau of Mines, U.S. Dept. of Interior, 1967

191. J. Szilard, *Ultrasonic Testing*, John Wiley, 1982

192. R.D. Shockley, et al., *J. Acoust. Soc. America*, Vol 71, 1982, p 51-60

193. W.H. Munk, et al., *J Phys. Oceanogr.,* Vol 18, 1988, p 1876-1898

194. G. Haines, *Sound Underwater*, David & Charles, New York, 1974

195. B. Carlin, *Ultrasonics*, McGraw-Hill, 1960

196. *The NDT Yearbook*, British Inst. of Nondestructive Testing, UK, 1987

197. P. Langevin, British Patent 145,691, 1921

198. P. Biquard, *Les Ultrasons*, Presses Univérsitaires de France, 1951

199. R.W. Wood and A.L. Loomis, *Phys. Rev.*, Abstract,

Vol 29, 1927, p 273

200. G.W. Pierce, *Proc. Am. Acad.*, Vol 63, 1928, p 1

201. A. Behun, German Patent 464,516, 1921

202. A.N. Nicholson, *Proceedings of American Institute of Electrical Engineers*, Vol 38, 1919, p 1315

203. S. Sokolov, Propagation of Ultra-acoustic Oscillations in Various Bodies, *Elekt. Nach Tech.*, Vol 6, 1929, p 450-461

204. S. Sokolov, Ultrasonic Methods for Determining Internal Flaws in Metal Objects, *Zavod. Lab.*, Vol 4, 1935, p 1468-1473

205. O. Muhlhauser, German Patent 569,598, 1933

206. S.J. Sokolov, US Patent 2,164,125, 1939

207. G.A. Homés and I.H. Otts, Nondestructive Testing of Materials with Ultrasound, *Bull. Scient. AIM Belg.*, Vol 62, 1949, p 23-54

208. J. Krautkramer and H. Krautkramer, *Ultrasonic Testing of Materials*, 3rd ed., Springer-Verlag, 1983 (See Ref 219.)

209. M.G. Silk, *Ultrasonic Transducers for NDT*, Adam Hilger, 1984

210. F.A. Firestone, Flaw Detecting Device and Measuring Instrument, US Patent 2,280,226, 1942; and 2,280,130, 1940

211. Anon., Nondestructive Testing, A Survey, SP-5113, NASA, 1973

212. C.H. Desch, D.O. Sproule, and W.J. Dawson, The Detection of Cracks in Steel by Means of Supersonic Waves, *J. Iron Steel Inst.*, Vol 153, 1946, p 319-352

213. F.A. Firestone, The Supersonic Reflectroscope for Interior Inspection, *Met. Prog.*, Vol 48, 1945, p 505-512

F.A. Firestone, The Supersonic Reflectoscope, *J. Acoustical Soc. America*, 17, 1946, p 287-299

214. R. Pohlman, Possibility of Acoustic Imagery Analogous to Optics, *Z. Phys.*, 1939, p 697-709

215. S. Sokolov, An Ultrasonic Microscope, *Dokl. Akad. Nauk. USSR*, Vol 64, 1949, p 333-355

216. J.S. Sandhu and R.E. Thomas, Acoustograph Nondestructive Evaluation, *Proc. IEEE Ultrasonics Symposium* (Chicago, IL), 1988

217. P.A. Doyle and C.M. Scala, Crack Depth Measurement by Ultrasonics: A Review, *Ultrasonics*, Vol 16, 1978, p 164-170

218. J.P. Charlesworth and J.A.G. Temple, *Engineering Applications of US Time of Flight Diffraction*, Research Studies Press Ltd., John Wiley, 1989 (This is Vol 1 of US Inspec Eng Series, M.J. Whittle, Ed.)

219. J.H. Krautkämer, *U.S. Testing of Materials*, 2nd ed., Springer-Verlag, 1977

J. Krautkrämer, Flaw Size Determination With Ultrasound, *Arch. Eisenhütten*, 30, 1959, p 693-703

220. J.G. Crowther and R. Whiddington, *Science at War*, H.M.S.O., 1948

221. G.S. Kino, *Acoustic Waves*, Prentice Hall, 1987

L.E. Kinsler and P. Frey, *Fundamentals of Acoustics*, Wiley, 1962

222. D.E. Bray and R.K. Stanley, *Nondestructive Evalu-*

ation: A Tool for Design, Manufacturing and Service, McGraw-Hill, 1989

223. E. Meyer and E.G. Neumann, *Physical and Applied Acoustics*, Academic Press, 1972

224. H.J. Pain, *The Physics of Vibrations and Waves*, John Wiley, 1968

225. G.W.C. Kaye and T.H. Laby, *Tables of Physical and Chemical Constants*, 15th ed., Wesley, 1966

226. A.R. Selfridge, *Properties of Isotropic Materials*, Transactions on Sonics and Ultrasonics, SU-32, 1985, p 381-394

227. V.M. Ristic, *Principles of Acoustic Devices*, Wiley, 1983

228. Sonotech Couplants, P.O. Box 2189, Bellingham, WA 98227, USA

229. Panametrics, Inc., 221, Crescent Street, Waltham, MA, USA (Various brochures)

230. Brochures: "Outline of the Theory of Magnetic Particle Inspection," "Quantitative Quality Indicators QQI" (Videotape Description), Application Reports: R. Grutzmacher: 1955; K. Schroeder: 1995, 1959; D. Lorenzi: 1983; Magnaflux Corporation, 7300 W. Lawrence Ave., Chicago, IL

231. T.F. Hueter and R.H. Bolt, *Sonics*, John Wiley, 1955

232. E.R. Dobbs, Electromagnetic Generator of Ultrasonics, *Research Techniques in Nondestructive Testing*, Vol 2, R.S. Sharpe, Ed., Academic Press, 1st ed., 1970, 7th ed., 1984

233. L. Filipczynski, Z. Pawlowski, and J. Wehr, *Ultrasonic Methods of Testing Materials*, Butterworth, 1965

234. H.D. Megaw, *Ferroelectricity in Crystals*, Methuen, 1957

235. W.A. Wooster, Physical Properties and Atomic Arrangements in Crystals, *Rep. Prog. Phys.*, Vol 16, 1953, p 62-82

236. B. Hull and V. John, *Nondestructive Testing*, Macmillan Education, 1980

237. W.N. McDicken, *Diagnostic Ultrasonics: Principles and Use of Instruments*, John Wiley, 1976

238. M. Hussey, *Basic Physics and Technology of Medical Diagnostic Ultrasound*, Elsevier, 1984

239. NDT System, Av. de Teleport, Futurescope, 86960, France

240. *Ultrasonic Testing of Welds*, British Standards Institute, London, England, 1986

241. *Handbook on the Ultrasonic Examination of Austenitic Welds*, American Welding Society for the International Institute of Welding, 1986, American Welding Society, 550 NW Lejeune Rd., Miami, FL 33126, USA, The International Institute of Welding, 54, Princess Gate, Exhibition Rd., London, UK

242. A.F. Brown, Acoustic Wave Optics, *Phys. Bull.*, Vol 34, 1983, p 473-476

243. A. Vary, Conference Chairman, Analytical U.S. in Materials Research and Testing, *NASA Conference*, Publication 2383, 1984

244. S.J. Klima, D.J. Lesco, and J.C. Freche, "Application of Ultrasonics to Detection of Fatigue Cracks,"

TM X-52109, National Aeronautics and Space Administration, 1965

245. S.J. Klima and J.C. Freche, "Ultrasonic Detection and Measurement of Fatigue Cracks in Notched Specimens," TN D-4782, National Aeronautics and Space Administration, 1968

246. F.J. Hope and N. Inman, "Development of Nondestructive, Automatic Technique for Monitoring and Recording of Fatigue Crack Growth," CR-66320, National Aeronautics and Space Administration, 1967

247. W.N. Clotfelter, B.F. Hankston, and E.E. Zachary, "The Nondestructive Evaluation of Stress-Corrosion Induced Property Changes in Aluminum," TM X-53772, National Aeronautics and Space Administration, 1968

248. A.D. Cordellos, R.O. Bell, and S.B. Brummer, Use of Rayleigh Waves for the Detection of Stress-Corrosion Cracking (SCC) in Aluminum Alloys, *Mater. Eval.*, Vol 27, April 1969, p 85-90

249. R.W. Benson, J.R. Chapman, H.F. Huffman, and S. Pearsall III, *Development of Nondestructive Methods for Determining Residual Stress and Fatigue Damage in Metals*, Robert W. Benson and Associates, Inc., Nashville, TN, 1969

250. G. Martin, J.F. Moore, and J.W. Rocke, *Development of Nondestructive Testing Techniques for Honeycomb Heat Shields*, Vol I and Vol II, North American Aviation, Inc., Los Angeles, CA, 1966

251. G. Martin and J.F. Moore, *Design, Development, and Fabrication of Portable Instrumentation for Nondestructive Testing of Composite Honeycomb Materials*, North American Rockwell Corp., Los Angeles, 1969

252. W.N. Clotfelter, "Acoustic Techniques for the Nondestructive Evaluation of Adhesively Bonded Composite Materials," TM X-53219, National Aeronautics and Space Administration, 1965

253. Nondestructive Testing: Trends and Techniques, *Proceedings of the Second Technology Status and Trends Symposium*, SP-5082, National Aeronautics and Space Administration, 1966

254. C.F. Quate, Acoustic Microscopy, *Phys. Today*, Aug 1985, p 34-41

255. A.B. Wood, *A Textbook of Sound*, G. Bell, 1957

256. J. Hansen, An Ear for Detail, *New Sci.*, Jan 1983, p 148-150

257. A.M. Murdoch, "Development of a Ring-Beam Wheel Surface Wave Search Unit," CR-76361, National Aeronautics and Space Administration, 1966

258. Materials Selector 1990, *Mater. Eng.*, Section II, Testing and Evaluation, Dec 1989

259. *Thomas Register*, Thomas Publishing Co., 1990

260. J. Duncan and S.G. Starling, *A Text Book of Physics*, Macmillan, 1939

261. R.C. McMaster, Ed., *Nondestructive Testing Handbook*, 2nd ed., Vol 2, *Liquid Penetrant Tests*, American Society for Nondestructive Testing; American Society for Metals, 1982

262. R.J. Malins, D. McCall, and G.W. Rhodes, *Mater. Eval.*, Vol 51, 1993, p 338-351

263. R.T. Fricker, *Mater. Eval.*, Vol 31, Sept 1972, p 200

264. R.J. Lord and J.A. Hollaway, *Mater. Eval.*, Vol 34, Oct 1975, p 249

265. Military Specification Inspection Materials, Penetrants, MIL-I-25135E, ASD/ENES, Wright Patterson AFB, OH, 1989

266. *Nondestructive Testing*, The Institute of Metals, USA, 1985

267. Boiler and Pressure Vessel Code, Section XI, American Society for Mechanical Engineers

268. "Classroom Training Handbook—Liquid Penetrant Testing," CR-61229, National Aeronautics and Space Administration, 1968

269. P. McIntire, *Nondestructive Testing Handbook*, Vol 6, *Magnetic Particle Testing*, 2nd ed., American Society for Nondestructive Testing, 1989

270. D.H. Tomborlian, *Electric and Magnetic Fields*, Harcourt, Brace & World, 1965

271. U. Sen, "Considerations of Ultraviolet and Visible Light in Magnetic Particle Inspection," American Society for Nondestructive Testing, March 1993 Military Standard, Inspection, Magnetic Particle, MIL-STD-1949, A or B, U.S. Army Materials Technology Laboratory, Watertown, MD, USA. This has been replaced by ASTM Inspection Specification E-1444-93.

272. R.C. Anderson, "Magnetic Particle Inspection of Steel Castings for Pressure Applications," Steel Foundry Facts, No. 167, 1956, p 9-13

273. L.J. Swartzendruber, "Interpretation, Use and Improvement of MPI Specifications," American Society for Nondestructive Testing Meeting, March 1993 V. Deutsch and M. Vogt, A comparison of AC and DC fields for magnetic particle methods, *British Journal of Nondestructive Testing*, No. 4, July 1982

274. L.J. Swartzendruber, "Establishment of Magnetization Levels in Magnetic Particle Inspection," American Society for Nondestructive Testing, Symposium Book, March 1993, p 129-135

275. "Demagnetization Techniques" (Videotape with descriptive literature), A. Lindgren, L & L Consultants, Inc., 1629 Eddy Lane, Lake Zurich, IL 80047

276. S.G. Starling, *Electricity and Magnetism*, 8th ed., Longmans and Green, 1956

277. W.D. Rummel, D.H. Todd, S.A. Frescka, and R.A. Rathke, "Detection of Fatigue Cracks by NDT," CR-2369, National Aeronautics and Space Administration, 1974

278. R.C. McMaster, Ed., *Nondestructive Testing Handbook*, 2nd ed., Vol 4, *Electromagnetic Testing*, American Society for Nondestructive Testing; American Society for Metals, 1986

279. NIST Standard reference material SRM 1853 (similar to Ketos ring); available from NIST Office of Standard reference materials (See also Ref 274.)

280. E.G. Cullwick, *The Fundamentals of Electro-Magnetism*, Cambridge University Press, 1939

281. I. Asimov, *Understanding Physics*, Dorset Press, 1966

282. H.L. Libby, Principles & Techniques of Eddy Current Testing, *Nondestr. Test.*, Vol 14 (No. 6), 1956, p 12-18, 27

283. E.G. Stanford and J.H. Fearon, *Progress in Nondestructive Testing,* Vol 1, Macmillan, 1959, p 59-109

284. F. Förster and H. Breitfeld, *Z Metallurgy*, Vol 45 (No. 4), 1954, p 188

285. R.C. Anderson, *Inspection of Metals,* Vol 1, *Visual Examination,* American Society for Metals, 1983; Vol 2, *Destructive Testing,* ASM International, 1988

286. Robert Lévy, Intercontrôle, 13 rue du Capricorne, 94583, Rungis Cedex, France; La Cofrend (French Society for NDT), 1, rue Gaston Boissier, 75724, Paris, Cedex 15, France

287. F.W. Smith and P.K. Stumpf, Ultrasonic Generators, *Electronics,* 19, 1946, p 116

288. R.W. Miller, A Method for Evaluating Materials Used in Penetrant Flaw Detection, *J. Welding*, Jan. 1958, p 30

289. U. Sen., "Considerations of Ultraviolet and Visible Light in Magnetic Particle Inspection," American Society for Nondestructive Testing, March 1993

290. European Committee for Standardization (CEN), Document EN-1330-5. *Nondestructive Examination Terminology, Part 5.* Terms used in eddy current testing: see Ref 286.

Index

E